T0342553

Theory of Linear Poroelasticity with Applications to Geomechanics and Hydrogeology

PRINCETON SERIES IN GEOPHYSICS

ROBERT A. PHINNEY, *EDITOR*

Stress Regimes in the Lithosphere, *by Terry Engelder*

Geophysical Inverse Theory, *by Robert L. Parker*

Theory of Linear Poroelasticity with Applications to Geomechanics and Hydrogeology, *by Herbert F. Wang*

Theory of Linear Poroelasticity

with Applications to Geomechanics and Hydrogeology

Herbert F. Wang

PRINCETON UNIVERSITY PRESS · PRINCETON AND OXFORD

Copyright © 2000 by Princeton University Press
Published by Princeton University Press, 41 William Street,
Princeton, New Jersey 08540
In the United Kingdom: Princeton University Press,
3 Market Place, Woodstock, Oxfordshire OX20 1SY

ALL RIGHTS RESERVED

Library of Congress Cataloging-in-Publication Data

Wang, Herbert.
Theory of linear poroelasticity with applications to geomechanics and hydrogeology /
Herbert F. Wang.
p. cm.—(Princeton series in geophysics)
Includes bibliographical references and indexes.
ISBN 0-691-03746-9
1. Rock mechanics. 2. Rocks—Permeability. 3. Elasticity. 4. Fluid dynamics. I. Title.
II. Series.
TA706.W34 2000
624.1′5182—dc21 00-032635

This book has been composed in Times Roman

The paper used in this publication meets the minimum requirements of
ANSI/NISO Z39.48-1992 (R1997) (*Permanance of Paper*)

www.pup.princeton.edu

Printed in the United States of America

1 2 3 4 5 6 7 8 9 10

To Rosemary

and Michelle, Melissa, Michael, and Matthew

Contents

Preface

The goal of this book is to present the theory of linear poroelasticity and its applications in geomechanics and hydrogeology. The presentation presumes a working knowledge of both subjects, as well as mathematical background in partial differential equations and integral transforms. The theory of poroelasticity addresses the time-dependent coupling between the deformation of rock and fluid flow within rock. The basic theory is described by Biot's (1941) linear constitutive relations coupled with Darcy's (1856) law for fluid flow in a porous medium. A major purpose of this text is to synthesize and integrate into one place, and with one notation, a number of classical solutions and applications of this simplest linear theory. Emphasis is given to physical interpretation of poroelastic variables, governing equations, and problem solutions.

The introductory chapter recounts parallel developments in geomechanics, hydrogeology, and reservoir engineering that are unified by the theory and concepts of poroelasticity. In Chapters 2, 3, and 4 the constitutive and governing equations and their associated poroelastic material parameters are described. In the subsequent chapters the equations are specialized for different simplifying geometries: unbounded problem domains (Chapter 5), uniaxial strain (Chapter 6), plane strain (Chapter 7), radial symmetry (Chapter 8), and axisymmetry (Chapter 9). Example problems from geomechanics, hydrogeology, and petroleum engineering are incorporated throughout to illustrate poroelastic behavior and solution methods. Finally, outlines are given for finite-element and boundary-element formulations of the governing equations (Chapter 10).

This book has been approximately six years and three different editors in the making. Sara Van Rheenen initiated the project by developing a contract for the book, now many years expired but always a security blanket. During the first stages she was always only a few keystrokes away from an answer to various startup questions. Jack Repcheck took over during the middle years, providing assurances and continuity until he had a manuscript to send out for review. Finally, Kristen Gager inherited the production stages and its many details. To each I extend my gratitude.

This book represents my own personal odyssey through parts of the kingdom of poroelasticity. The influence of recent writings of Alex Cheng, and

Emanuel Detournay, Evelyn Roeloffs, John Rudnicki, and Paul Segall is evident throughout the book. I have learned much from former and present students at the University of Wisconsin-Madison: Evelyn Roeloffs, Doug Green, Kyle Lewallen, Dave Hart, and Tim Masterlark. Their research, and hence my learning, included support from several agencies, including the Department of Energy, National Science Foundation, United States Geological Survey, and State of Wisconsin. The book was strengthened by the critiques of rock physics seminar students and of visiting researcher Tomochika Tokunaga. I thank my collaborators Jim Berryman, Pat Berge, Steve Blair, and Brian Bonner at Lawrence Livermore Laboratory for the scientific sojourns I have enjoyed with them. C.S. Clay provided advice about Biot theory and book writing over many years. Mary Anderson and Jean Bahr participated in several hydrogeology seminars in which I tested some of this material. Extensive written reviews by Steve Martel, Shemin Ge, Evelyn Roeloffs, and Tom Tharp were invaluable in revising the manuscript.

I have dedicated this book to my family. I am indebted to my wife, Rosemary, for her love and support over the many years of our life together. My daughter, Michelle, got me started with LaTeX. She answered my call for help with "Dad, you wouldn't have another file named 'tmp1,' would you?" Indeed, I did. My son, Matthew, edited most of the first draft of figures in the book and created new ones in a unique style that I think are easily recognizable. If only the material of this book really allowed his full creativity to be exploited. He also diligently helped me proofread the references. I thank my daughter Melissa and son Michael for enlivening my life with their medical and artistic interests, respectively. And to my sons-in-law Michael Allan and Omar Baldonado, thanks for playing golf with me.

I know this book is not bug free. I regret any errors I have introduced or old ones I am propagating. I am happy to receive notice of mistakes or gaffs and will try to maintain a Web page of errata (www.geology.wisc.edu/~wang).

Theory of Linear Poroelasticity with Applications to Geomechanics and Hydrogeology

1

Introduction

1.0 CHAPTER OVERVIEW

The purpose of this book is to quantify, in the simplest linear approximation, how fluid extraction or injection produces stress changes in isotropic, fluid-saturated rock formations; and conversely, how loads applied to an aquifer by engineered structures, atmospheric pressure, earth and ocean tides, tectonic activity, or reservoir lake level changes produce water-level changes in wells. This coupling between changes in stress and changes in fluid pressure forms the subject of poroelasticity. Literal clues to poroelastic phenomena appear in metaphors such as "stressing an aquifer" in reference to pumping a well or "strain meter" in reference to a water well on a fault.

The term *poroelasticity* was first coined by J. Geertsma as a footnote in his 1966 paper entitled "Problems of rock mechanics in petroleum production engineering." Geertsma's footnote refers specifically to "Biot's work on the theory of the elasticity and viscoelasticity of fluid-saturated porous solids" as "typical of the macroscopic stress-strain relations to be encountered." Geertsma explicitly pointed out that "the mathematical description of the macroscopic theory of poroelasticity is similar to that used in the theory of thermoelasticity." This chapter provides a historical perspective on how the disciplines of geomechanics, hydrogeology, and petroleum engineering contributed to our present understanding of coupled fluid and mechanical behavior.

This chapter previews the constitutive equations of poroelasticity for the case of an isotropic applied stress field. It examines the physical significance of material coefficients such as drained compressibility, poroelastic expansion coefficient, specific storage coefficient, and Skempton's coefficient. The chapter concludes with a description of the analogy between poroelasticity and thermoelasticity.

1.1 HISTORICAL EXAMPLES

Poroelastic behavior can explain an initially unexpected connection between a causal event and its subsequent effect. What follows is a variety of historical examples of poroelastic phenomena:

- *Water Levels in Wells Correlate with Ocean Tides.* Pliny the Elder stated in A.D. 77 that near the temple of Hercules in Cadiz "there is a closed source similar to a well which occasionally rises and falls with the ocean, but at other times does the opposite" (Melchior, 1983, p. 2).
- *Water Levels Change in Well as Trains Pass.* F. H. King (1892) of the University of Wisconsin reported that water levels in a well near the train station at Whitewater, Wisconsin, went up as a train approached and went down as a train left the station. The water level fluctuation was greater for a heavy freight train than for a lighter and faster passenger train.
- *Water Levels in Boardwalk Wells Fluctuate with Ocean Tides.* In 1902 the United States Geological Survey (USGS) reported that water-level oscillations in wells in Atlantic City, New Jersey, were synchronous with ocean depths, because the weight of sea water at high tide compressed the underlying rock, thereby forcing pore water up the wells (Meinzer, 1928).
- *Texas Claims Oil Fields That Sink into the Sea.* The state of Texas claimed title to the part of the Goose Creek oil field near Galveston, Texas, that had become covered with water from Galveston Bay following oil production. The state used the argument that the submerged land belonged to the state. The counterclaim by landowners was based on a geologic study by Pratt and Johnson (1926, p. 582) showing that the subsidence could be attributed to the extraction of 100 million barrels of water, oil, and sand from the reservoir. The courts ruled against the state claim because the submergence was due to human action and not natural causes.
- *Water Levels Rise in Wells Near a Pumping Well.* Observers in the small fishing village of Noordbergum in northern Friesland (Netherlands) witnessed a curious rise in water levels when large pumps were turned on in nearby wells. The behavior lasted a few hours before the water levels dropped. When the large pumps were shut off, the reverse situation occurred—namely, water levels dropped farther before they recovered. Verruijt (1969) concluded that the reverse well fluctuations occurred because pumping instantly compressed the aquifer to force water up the well.
- *Lake Mead Triggers Earthquakes.* The completion in 1935 of Hoover Dam along the Colorado River created 100-meter-deep Lake Mead behind it. Shortly after completion, small earthquakes beneath the lake occurred as a result of a combination of factors (Roeloffs, 1988). One factor was that the additional weight of the dam and water was stressing faults to the failure point; another was that water from

the lake seeped into the fault, thereby reducing their resistance to slipping.

1.2 BASIC CONCEPTS

Two basic phenomena underlie poroelastic behavior:

- *Solid-to-fluid coupling* occurs when a change in applied stress produces a change in fluid pressure or fluid mass.
- *Fluid-to-solid coupling* occurs when a change in fluid pressure or fluid mass produces a change in the volume of the porous material.

As used here, the word *solid* refers to the skeletal framework of bulk, porous material. The terms *fluid pressure* and *pore pressure* are also used interchangeably. The solid-to-fluid and fluid-to-solid couplings are assumed to occur instantaneously in the quasistatic approximation in which elastic wave propagation is ignored. The simplest mathematical description of the two basic forms of coupling between solid and fluid is a set of linear constitutive equations. The equations relate strain and fluid-mass changes to stress and fluid-pressure changes. The poroelastic constitutive equations are generalizations of linear elasticity whereby the fluid pressure field is incorporated in a fashion entirely analogous to the manner in which the temperature field is incorporated in thermoelasticity (cf. Section 1.9). An increase of fluid pressure causes the medium to expand just as an increase of temperature causes it to expand.

The magnitude of the solid-to-fluid coupling depends on the compressibility of the framework of the porous material, the compressibility of the pores, the compressibility of the solid grains, the compressibility of the pore fluid, and the porosity. Negligible solid-to-fluid coupling occurs for a highly compressible fluid such as air. An example of solid-to-fluid coupling is the response of water levels in a well to the passage of nearby trains (Fig. 1.1).

Changes of fluid mass or fluid pressure in a porous material produce strains in the bulk, porous solid. A uniform change in fluid pressure throughout a porous body subjected to no external stresses or constraints (*free strain case*) produces a uniform strain and no poroelastic stresses. In general, consideration of boundary constraints means that the strain field is different from the free strain case, and poroelastic stresses exist within the body. An example of fluid-to-solid coupling is subsidence due to large amounts of fluid extraction from an aquifer or hydrocarbon reservoir (Fig. 1.2).

A nonuniform pore pressure distribution leads to time-dependent fluid flow according to Darcy's (1856) law. The time dependence of pore pressure pro-

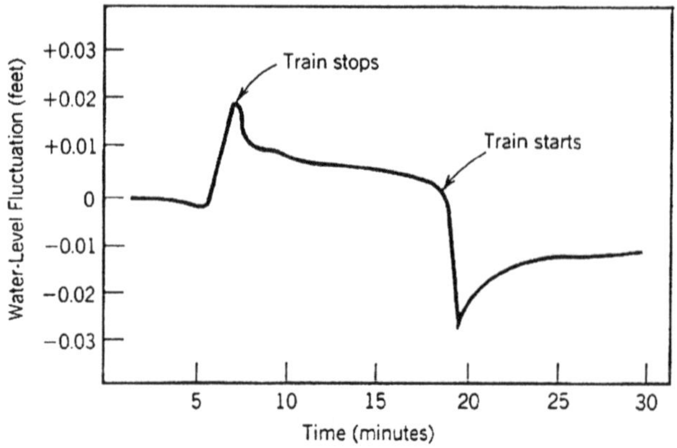

Figure 1.1: Water-level fluctuations due to a passing train. An approaching train compresses the aquifer, which quickly raises the pore pressure in the affected region. Fluid then flows away from the high-pressure region. As the train departs, the aquifer expands, thereby quickly reducing the pore pressure in the affected region. Fluid again flows in response to the pressure differences, but this time it builds up the pore pressure. The approximately equal and opposite behaviors demonstrate that the aquifer is elastic (Domenico and Schwartz, 1998, p. 65; Jacob, 1940).

duces time dependence of poroelastic stresses and strains, which in turn couple back to the pore pressure field. Quantifying these basic poroelastic concepts for application to time-dependent, coupled deformation and fluid-flow problems in hydrogeology, geomechanics, and petroleum engineering is the subject of this book. If only fluid-to-solid coupling were important, the problem would be mathematically simpler because the fluid-flow problem can be solved independently of the stress field. The stress field (and hence strain and displacement fields) could then be calculated as functions of position and time once the pore pressure field has been determined as a function of position and time. This one-way coupling is called the *uncoupled* problem. However, when the time-dependent changes in stress feed back significantly to the pore pressure, the two-way coupling is important, and the problem is called *coupled.* In the mathematically analogous subject of thermoelasticity, significant heating of material from stress changes does not occur for most materials, and hence most thermoelastic analyses ignore this direction of coupling, and are uncoupled. However, applied stress changes in fluid-saturated porous materials typically produce significant changes in pore pressure, and this direction of coupling is significant.

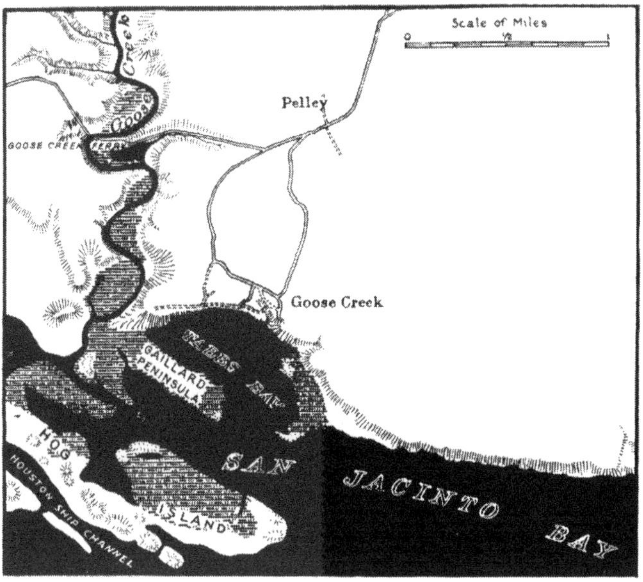

(a) Before Subsidence

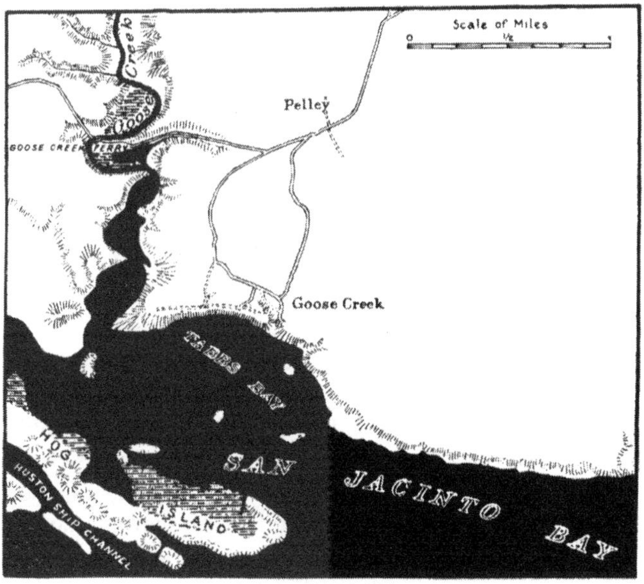

(b) After Subsidence

Figure 1.2: Vertical subsidence due to 100 million barrels of fluid (and sand) extracted from the Goose Creek oil field near Galveston, Texas (Pratt and Johnson, 1926, p. 582). Water-covered areas are shown in black. Subsidence is not purely elastic, as a significant amount is not recovered if fluids are reinjected.

1.3 BRIEF HISTORY

Important concepts of poroelasticity developed somewhat independently in geomechanics, petroleum engineering, and hydrogeology. One theme of this book is to highlight the unified description that poroelastic theory provides all three disciplines.

The scientific history of poroelastic concepts spans approximately one hundred fifty years. The early period, from the publication of Darcy's law[1] in 1856 to 1900, included observations of well behavior in response to various loading phenomena such as trains and tides. Increased exploitation of groundwater and hydrocarbon resources in the 1900–1930 period was the impetus for improved scientific and engineering understanding of the principles governing their occurrence and flow. Also, increased civil construction during this period became the impetus for improved understanding of the behavior of soil as a foundation material. The canonical geomechanics problem was soil consolidation, and the canonical hydrogeology problem was elastic storage in a confined aquifer. From 1930 to 1960, significant progress was made in each of the three disciplines in developing fundamental concepts, formulating or extending constitutive laws and governing equations, and obtaining analytical solutions. After 1960, more complex analytical solutions were obtained for problems in land subsidence and earthquake mechanics. Numerical solutions increased in importance as the digital computer enabled more realistic simulations of geological situations and of nonlinear and time-dependent constitutive properties.

1.4 GEOMECHANICS

Karl Terzaghi (1883–1963) sought to understand the behavior of soil as a foundation material by performing controlled laboratory experiments. The work, which led to his consolidation theory,[2] was conducted between 1916 and 1925, when he was assigned by the Austrian Department of Foreign Affairs in Vienna to lecture at Robert University in Istanbul. He published two influential textbooks, *Erdbaumechanik auf bodenphysikalischer Grundlage* in 1925 and *Theoretical Soil Mechanics* in 1943. The key experiment from which he developed the governing Eqn. 1.1 is shown schematically in Figure 1.3. A fully saturated soil sample is confined laterally in a cylinder of

[1] An entertaining and illuminating biographical sketch of Darcy is given by Freeze (1994).

[2] A bitter scientific dispute developed between Terzaghi and Fillunger initially over the theory of uplift in dams and subsequently over the theory of consolidation. Fillunger committed suicide in 1937 after learning that a committee of experts would support Terzaghi's theory (Skempton, 1960; de Boer and Ehlers, 1990; de Boer, 1996; de Boer et al., 1996).

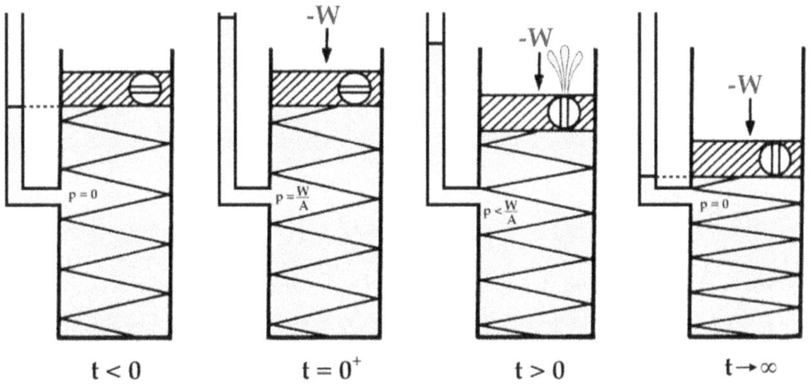

Figure 1.3: Schematic of uniaxially constrained soil consolidation (after Craig, 1997, p. 86). A compressive load $-W$ is applied suddenly at time $t = 0$ to a uniaxially confined sample of cross-sectional area A. The excess fluid pressure jumps to its undrained value W/A to support the load. Stress is transferred partially to the solid skeleton of the porous material (represented by the spring) until excess fluid pressure is again zero for long times and the load is carried entirely by the solid framework.

cross-sectional area A. An axial load $-W$ is applied suddenly at $t = 0$ and then held constant. (Tensile stresses are taken to be positive.) The water pressure throughout the sample jumps up by the amount $p = W/A$ at $t = 0^+$. A profile of excess pressure develops within the sample as water flows out the top drain, which is maintained at atmospheric pressure. Terzaghi derived the consolidation equation for this experiment to be the diffusion equation for excess (greater than hydrostatic) water pressure p,

$$\frac{\partial p}{\partial t} = c\frac{\partial^2 p}{\partial z^2} \tag{1.1}$$

where c is a diffusivity that is known as the consolidation coefficient, t is time, and z is distance along the soil column.

As will be demonstrated in Section 6.3, Eqn. 1.1 is independent of stress, because the theory of poroelasticity leads to the special result that the pore pressure field and applied stress field are uncoupled for the boundary conditions in Terzaghi's experiment. The time evolution of the pressure profile is exactly the same as the analogous thermal conduction problem of a sudden step change (Carslaw and Jaeger, 1959, pp. 96–97), which was noted by Terzaghi.

Terzaghi is generally recognized for elucidating the important concept of effective stress, which for soils is well approximated to be the difference between the applied stress and pore pressure, because the grains are incompressible relative to the bulk soil. Initially, the axial load is borne entirely by the fluid, but it is shifted to the skeletal frame as the excess pore pressure dissipates. A discussion of the effective stress concept is presented in Section 2.8.

1.5 HYDROGEOLOGY

Whereas Terzaghi sought general principles in the laboratory, developments regarding the poroelastic behavior of aquifers were based primarily on field observations. The first published record of water wells responding to passing trains was made by F. H. King (1892) in Wisconsin. He noted that a heavy freight train produced a greater rise in water level than did a lighter and faster passenger train, and that a locomotive alone did not produce a noticeable effect. O. E. Meinzer, in his 1928 paper "Compressibility and elasticity of artesian aquifers," sought to resolve the puzzle of the source of the large amounts of irrigation water pumped from the Dakota sandstone. He reasoned that recharge was insufficient to produce 3000 gallons per minute for 38 years within three townships, and that the compressibility of water alone was likewise insufficient. He concluded that elastic aquifer compression occurred as a result of the decline in fluid pressure, and that the reduction in pore volume was the principal source of water released from storage. He also recognized that aquifers were elastic because well levels recovered after they were shut down. In support of his hypothesis, Meinzer cited King's work on train-induced fluctuations in water levels, the in-phase response of water well levels to ocean tides in Atlantic City, New Jersey (Fig. 1.4), and the subsidence of the Goose Creek oil field. Meinzer also explicitly referenced Terzaghi's effective stress principle to equate pressure decline to an effective stress increase. Meinzer's insight was that he recognized that aquifers were compressible and that the laboratory-derived principle of effective stress could be applied to aquifers. No equations were presented in Meinzer's 1928 paper, although he incorporated calculations for the relative amounts of water released from aquifer compression versus water expansion.

The next development in hydrogeology was Theis's nonequilibrium or transient flow solution for drawdown of a well pumped at a constant rate. Theis first conceptualized the problem in terms of heat conduction and then sought advice from a former college classmate, Clarence Lubin, a mathematics professor at the University of Cincinnati:[3]

> The flow of ground water has many analogies to the flow of heat by conduction. We have exact analogies in ground water theory for thermal

[3] This passage is in a letter from Theis to Clarence Lubin dated December 19, 1934 (White and Clebsch, 1994, p. 51).

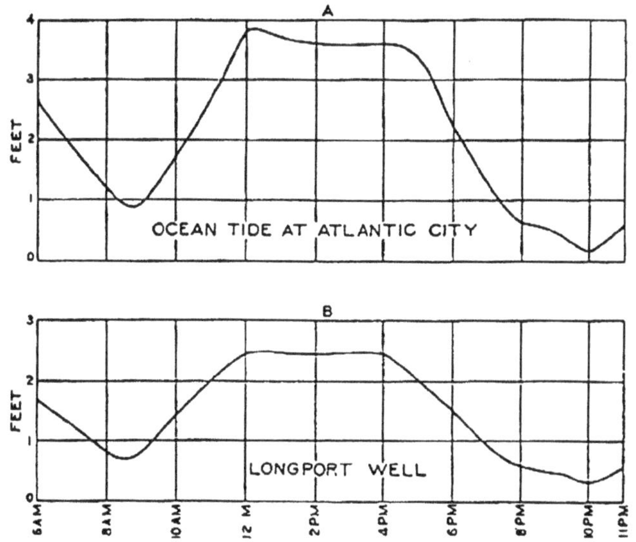

Figure 1.4: Ocean tides at Atlantic City. The in-phase response of well levels to ocean tides on January 22, 1926, in an 800-foot well near Longport, New Jersey, is evidence of solid-to-fluid coupling. A lag time would be evident if the well levels were responding to fluid flow (Meinzer, 1928, p. 274).

> gradient, thermal conductivity, and specific heat. I think a close approach to the solution of some of our problems is probably already worked out in the theory of heat conduction. Is this problem in radial flow worked out? Given a plate of given constant thickness and with constant thermal characteristics at a uniform initial temperature to compute the temperatures throughout the plate at any time after the introduction of a sink kept at 0 temperature? And a more valuable one from our standpoint: Given the same plate under the same conditions to compute the temperatures after the introduction of a sink into which heat flows at a uniform rate? I forgot to say that the plate may be considered to have infinite areal extent.

Lubin provided Theis the solution from the 1921 edition of Carslaw's treatise on heat conduction. He declined Theis's offer to be a coauthor with the words "from the standpoint of mathematics the work is not of fundamental importance" (Freeze, 1985, p. 442). In the resulting paper, Theis (1935) stated the analogy between groundwater storage and specific heat:

> In heat-conduction a specific amount of heat is lost concomitantly and instantaneously with fall in temperature. It appears probable,

> analogously, that in elastic artesian[4] aquifers a specific amount of water is discharged instantaneously from storage as the pressure falls.

Thus, Theis recognized that confined aquifers possess a property analogous to heat capacity, which he called the *coefficient of storage* in a subsequent paper (Theis, 1938). Jacob (1940) derived the transient flow equation for a horizontal aquifer "from scratch," as he put it, rather than appeal to the heat flow analogy. Jacob translated Theis's verbal definition of coefficient of storage $\mathcal{S}$ into a mathematical definition:

$$\mathcal{S} = \frac{1}{\rho_f A} \frac{\delta M_f}{\delta h} \tag{1.2}$$

where ρ_f is the density of water, A is the horizontal cross-sectional area of a vertical column of aquifer, δM_f is the change in the mass of water in a column of area A, and δh is the change in head. Eqn. 1.2 is a macroscopic definition, which gives no insight into the physical mechanisms behind storage capacity. Jacob, therefore, considered the micromechanics[5] of the release of water from aquifer compression and fluid expansion. With the assumption that the individual mineral grains are incompressible, Jacob derived the expression

$$\mathcal{S} = \rho_f g b \left(\frac{\phi}{K_f} + \frac{1}{K_v} \right) \tag{1.3}$$

where g is the acceleration of gravity, ϕ is the porosity, b is the aquifer thickness, K_f is the bulk modulus of water, and K_v is a modulus of compression of the aquifer. Jacob made no explicit specification of the particular modulus of compression represented by K_v; he did not restrict it to be a vertical modulus.

Jacob (1950) later arrived at the three-dimensional governing equation, which is the usual starting point for transient fluid-flow analyses for confined aquifers,

$$\frac{\partial h}{\partial t} = \frac{\rho_f g k}{\mu S_s} \nabla^2 h \tag{1.4}$$

where k is the permeability, μ is the fluid viscosity, and S_s is the standard hydrogeological specific storage (Hantush, 1960). The fluid-flow equation obtained from poroelastic theory contains an additional term associated with the time dependence of strain of the elementary volume under consideration.

[4] The word *confined* is today generally used in place of *artesian*.

[5] Micromechanics is the science of mechanics applied to granular materials at the grain and pore scale.

This additional term was a decade-long issue in hydrogeology in the 1960s until 1969, when Verruijt showed that the most general (linear) description of aquifer behavior is obtained from Biot's theory.

1.6 PETROLEUM ENGINEERING

Muskat wrote *Flow of Homogeneous Fluids through Porous Media* over a period of about 5 years in the 1930s with the encouragement of his employer, Gulf Oil in Pittsburgh. His book can be viewed as a treatise on applied potential theory. Muskat systematically developed solutions to Laplace's and Poisson's equations and the diffusion equation for boundary and initial conditions relevant to problems of groundwater flow in aquifers and fluid movement in hydrocarbon reservoirs. Although he was affiliated with an oil company, his book provides a balanced coverage of literature in both hydrogeology and petroleum engineering. He included many problems he had solved personally. Muskat considered storage effects to result primarily from fluid expansion, since rocks were thought to be incompressible at reservoir depths. Fluid expansion, however, was inadequate to account for the 500 million barrels of oil extracted with a pressure decline of 375 pounds per square inch from the East Texas Oil Field in the late 1920s and early 1930s. Muskat attributed the high production to be due to drive from small pockets of highly compressible gas or to "water drive" from the peripheries of the reservoir. Jacob (1940) suggested alternatively that the high production was due to the compressibility of the Woodbine sand and clays.

Poroelastic theory was used in petroleum engineering primarily to understand subsidence, estimate hydrocarbon volumes, and predict stresses around boreholes. The subsidence of the Goose Creek oil field described by Pratt and Johnson (1926) was the first conceptual realization of the coupling between large volumes of fluid extraction and large-scale mechanical deformation. Because the problem of large groundwater withdrawals from aquifers is identical, Geertsma, who was affiliated with Shell in the Netherlands, referred in his 1966 paper to subsidence in Mexico City and the Houston-Galveston region in Texas. The groundwater subsidence literature in the 1970s (e.g., Gambolati and Freeze, 1973; Gambolati et al., 1974) used numerical techniques more heavily than the analytical techniques introduced by Geertsma and used most recently by Segall (Segall, 1992; Segall et al., 1994). Segall's work connects hydrocarbon extraction with induced seismicity.

Poroelasticity research in the 1940s and 1950s was oriented primarily toward rock mechanics. Hughes and Cooke (1953) made laboratory measurements of pore compressibilities to correct for available pore space at reservoir depths. Laboratory measurements of poroelastic parameters continue (e.g., Laurent et al., 1993; Hart and Wang, 1995). Of particular importance in petroleum engineering and geophysics is the use of hydraulic fracturing as a technique for measuring earth stresses (Hubbert and Willis, 1957; Haimson and Fairhurst, 1969; Detournay and Cheng, 1993).

1.7 BIOT'S THEORY

Terzaghi's mathematical treatment was based on his *one-dimensional* laboratory experiments. The general *three-dimensional* theory of poroelasticity[6] was formulated by Biot in 1941 when he was a professor of mechanics at Columbia University. It appeared in the *Journal of Applied Physics* with the title "General theory of three-dimensional consolidation" (Biot, 1941a). It contained the constitutive equations that are the central subject of this book. Biot subsequently extended the theory to describe wave propagation in fluid-filled porous media (Biot, 1956a, 1956b).

In his 1941 paper Biot introduced a quantity he called the *variation in water content*, which he defined as "the increment of water volume per unit volume of soil" (Biot, 1941a). The increment of water content is the volume of the water *exchanged* by flow into or out of the control volume. In other words, Biot's increment of water content is the volume of water added to storage as used in the earlier hydrogeologic work of Theis or Jacob, although Biot's work appears to be independent of theirs. Biot's variable for a generic fluid is called the *increment of fluid content* ζ. No symbol has been used generally in the hydrogeologic literature for this quantity. Biot's (1941a) description of a hypothetical experiment in which a thin tube is used to extract water from a soil sample evokes the image of withdrawing water from a well:

> ... consider a sample of soil enclosed in a thin rubber bag so that the stress applied to the soil be zero. Let us drain the water from this soil through a thin tube passing through the walls of the bag. If a negative pressure $-p_o$ is applied to the tube a certain amount of water will be sucked out. This amount is given by $\zeta = -p_o/R$.

A negative value of ζ indicates the removal of water. The proportionality constant $1/R$ is called the *specific storage coefficient at constant stress*, because it is the ratio of the change in increment of fluid content ζ to the change in pore pressure p for a stress-free sample. The specific storage coefficient at constant stress is also called the *unconstrained specific storage coefficient*, or the *three-dimensional specific storage coefficient*.

Four poroelastic moduli, rather than the usual two moduli in standard elasticity, are necessary to relate the strains and increment of fluid content to

[6] Rendulic (1936) formulated a three-dimensional theory by substituting the Laplacian for the second spatial derivative in Eqn. 1.1. In three dimensions, this ad hoc generalization ignores an additional stress-coupling term.

Maurice Anthony Biot (1905–1985). Biot graduated with a bachelor's degree in philosophy (1927); degrees in mining engineering (1929) and electrical engineering (1930), and a D.Sc. (1931), all from Louvain University in Belgium, followed by a Ph.D. in aeronautical science (1932) from the California Institute of Technology under the supervision of Theodore von Kármán (Williams, 1983; Tolstoy, 1986; Cheng et al., 1998). Biot held academic positions at Harvard (1934–1935), Louvain (1935–1937), Columbia (1937–1946), and Brown (1946–1952), after which he became an independent researcher and consultant. Shell Development and Cornell Aeronautical Laboratory are two affiliations Biot used in his publications.

Biot worked on both sides of the poroelasticity-thermoelasticity coin, often citing the isomorphism between the two theories. Twenty-one of his papers were collected in a volume edited by Ivan Tolstoy and published by the Acoustical Society of America (Tolstoy, 1992). All but 2 of Biot's 175 papers have been collected into a CD-ROM by Thimus et al. (1998). Biot's work was in the nineteenth-century tradition of natural philosophy and mathematical physics. These roots are acknowledged in the preface of von Kármán and Biot's (1940) textbook *Mathematical Methods in Engineering* in their quote from Lord Kelvin and Peter Guthrie Tait's *Treatise on Natural Philosophy*, Part II: "Neither seeking nor avoiding mathematical exercitations we enter into problems solely with a view to possible usefulness for physical science." Equally apt historical quotations open each chapter of the text.

stresses and pore pressure. The constant $1/R$ is one of the two new constants. The other new constant $1/H$ is a *poroelastic expansion coefficient*, which relates the volumetric strain to pore pressure changes for conditions of constant stress. To the author's knowledge, the terminology *poroelastic expansion coefficient* appears to be new here, as this coefficient is often called *compressibility* (cf. Section 3.1).

Biot showed that Terzaghi's one-dimensional consolidation problem is a special case of his theory. McNamee and Gibson (1960a, 1960b) used Biot's theory to obtain analytical solutions for consolidation of a half space due to a strip or circular load. Other important applications of Biot's theory were to subsidence (Geertsma, 1966) and hydraulic fracturing (Haimson and Fairhurst, 1969). In 1969 in a paper entitled "Elastic storage of aquifers," Verruijt recognized the general applicability of Biot's theory to aquifer behavior. Verruijt opened his paper with the statement that parallel developments occurred in soil mechanics and groundwater hydrology.

Rice and Cleary's 1976 reformulation of Biot's linear poroelastic constitutive equations has been adopted widely for geophysical problems (Wang,

Specific Storage Coefficient of a Balloon. This simple experiment to measure the specific storage coefficient $1/R$ of a water-filled balloon (Fig. 1.7) provides observational meaning to the variables increment of fluid content and storage. A balloon is stretched over the end of a burette clamped to a meter stick and attached to a tall ring stand. The balloon and burette are filled with a volume V_f of water to an arbitrary height h_1 on the meter stick and volume V_1 on the burette. The volume of water in the balloon is $V_b = V_f - V_1$. Adding a measured volume of water ΔV_f from a beaker raises the water height to h_2 and increases the volume to V_2 in the burette. If the compressibility of water and the burette are assumed to be small, the volume of water ΔV_f is divided between the balloon and the burette: $\Delta V_f = \Delta V_b + (V_2 - V_1)$, where ΔV_b is the additional volume of water in the balloon. Because the increment of fluid content is the ratio $\Delta V_b / V_b$, $\zeta = [\Delta V_f - (V_2 - V_1)]/V_b$. The increase in pressure is $p = \rho_f g(h_2 - h_1)$. The specific storage coefficient $1/R$ is the ratio of ζ to p.

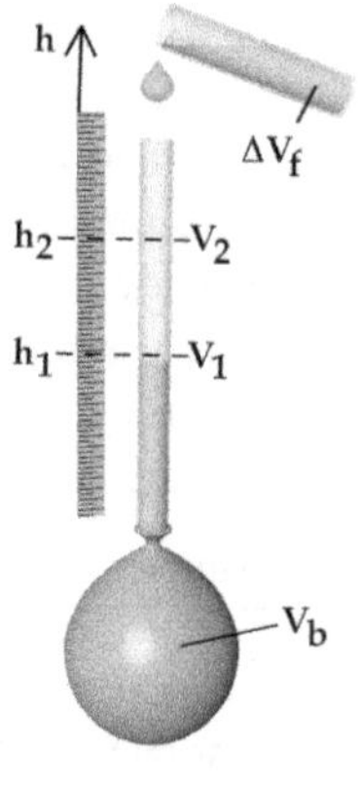

Figure 1.5: Apparatus to measure the specific storage coefficient of a water-filled balloon.

1993). Rice and Cleary chose constitutive parameters that emphasized the drained (constant pore pressure) and undrained (no flow) limits of long- and short-time behavior, respectively. Their perspective was from the field of applied mechanics, similar to that of Biot's, but they apparently were unaware of his much earlier contributions when they initially approached the problem. Rice and Cleary defined *fluid mass content* m_f to be the fluid mass per unit reference volume (see also Biot, 1973, p. 4930). The *change* in fluid mass content $\delta m_f = m_f - m_{f_o}$, where m_{f_o} is the fluid mass content in the reference state, is related to increment of fluid content ζ by

$$\zeta = \frac{\delta m_f}{\rho_{f_o}} \tag{1.5}$$

where ρ_{f_o} is the fluid density in the reference state. Fluid mass content is a state property, whereas the increment of fluid content must be viewed in the hydrogeologic sense of volume of fluid *transported* into or out of storage. Jacob (1940) defined storage in terms of fluid mass (Eqn. 1.2), but an advantage to using ζ as a primary variable is that it is dimensionless, like strain, and the constitutive equations do not have to include a density factor.

1.8 PREVIEW OF CONSTITUTIVE RELATIONS

The key concepts of Biot's 1941 poroelastic theory for an isotropic fluid-filled porous medium are contained in just two linear constitutive equations for the case of an isotropic applied stress field σ. In addition to σ, the other field quantities are the volumetric strain $\epsilon \equiv \delta V/V$, where V is the bulk volume, the increment of fluid content ζ, and the fluid pressure p. The volumetric strain $\delta V/V$ is taken to be positive in expansion and negative in contraction. Stress σ is positive if tensile and negative if compressive. Increment of fluid content ζ is positive for fluid added to the control volume and negative for fluid withdrawn from the control volume. Fluid pressure (pore pressure) p greater than atmospheric is positive. The constitutive equations express ϵ and ζ as a linear combination of σ and p:

$$\epsilon = a_{11}\sigma + a_{12}p \tag{1.6}$$

$$\zeta = a_{21}\sigma + a_{22}p \tag{1.7}$$

Generic coefficients a_{ij} are used in Eqns. 1.6 and 1.7 to emphasize the simple form of the constitutive equations. The first constitutive equation is a statement of the observation that changes in applied stress and pore pressure produce a fractional volume change. The second constitutive equation is a statement of the observation that changes in applied stress and pore pressure require fluid be added to or removed from storage.

Poroelastic constants are defined as ratios of field variables while maintaining various constraints on the elementary control volume. The physical meaning of each coefficient in Eqns. 1.6 and 1.7 is found by taking the ratio of the change in a dependent variable relative to the change in an independent variable, while holding the remaining independent variable constant:

$$a_{11} = \left.\frac{\delta\epsilon}{\delta\sigma}\right|_{p=0} \equiv \frac{1}{K}$$

$$a_{12} = \left.\frac{\delta\epsilon}{\delta p}\right|_{\sigma=0} \equiv \frac{1}{H}$$

$$a_{21} = \left.\frac{\delta\zeta}{\delta\sigma}\right|_{p=0} \equiv \frac{1}{H_1}$$

$$a_{22} = \left.\frac{\delta\zeta}{\delta p}\right|_{\sigma=0} \equiv \frac{1}{R} \tag{1.8}$$

The coefficient $1/K$ is obtained by measuring the volumetric strain due to changes in applied stress *while holding pore pressure constant.*[7] The state of constant pore pressure can be imagined to be enforced by inserting a tube into the rock and connecting it to a fluid reservoir at the same pressure. This state is called a *drained condition*, which is more general than the tube being vented to atmospheric pressure. Therefore, $1/K$ is the compressibility of the material measured under drained conditions, and K is the drained bulk modulus.

The coefficient $1/H$ is a property not encountered in ordinary elasticity. It describes how much the bulk volume changes due to a pore pressure change *while holding the applied stress constant.* By analogy with thermal expansion, it is called the *poroelastic expansion coefficient.* Energy considerations lead to the result that the coefficient $1/H$ is the same as $1/H_1$, that is, the linear transformation matrix is symmetric (see box).

The coefficient $1/R$ is a specific storage coefficient measured under conditions of constant applied stress; it is the ratio of the change in the volume of water added to storage per unit aquifer volume divided by the change in pore pressure. In this book the specific storage coefficient at constant stress is also called the *unconstrained* specific storage coefficient and is designated S_σ:

$$S_\sigma \equiv \frac{1}{R} \tag{1.9}$$

[7] Constitutive equations are generally written in terms of absolute quantities rather than their differentials. Each quantity is considered to be relative to a reference state. Also, the expressions $p = 0$ and $\delta p = 0$ are considered equivalent, so that the words *constant pore pressure* or *drained conditions* are associated with both equations.

Potential Energy and Reciprocity. Biot (1941a) assumed the existence of a potential energy density:

$$U = \frac{1}{2}(\sigma\epsilon + p\zeta) \tag{1.13}$$

The potential energy density is required to satisfy the conditions

$$\frac{\partial U}{\partial \epsilon} = \sigma \qquad \text{and} \qquad \frac{\partial U}{\partial \zeta} = p \tag{1.14}$$

The mixed partial derivatives $\partial^2 U/\partial\zeta\,\partial\epsilon$ and $\partial^2 U/\partial\epsilon\,\partial\zeta$ must be equal; hence,

$$\frac{\partial \sigma}{\partial \zeta} = \frac{\partial p}{\partial \epsilon} \tag{1.15}$$

Solving Eqns. 1.6 and 1.7 for σ and p gives

$$\sigma = \frac{a_{22}}{\Delta}\epsilon - \frac{a_{12}}{\Delta}\zeta \tag{1.16}$$

$$p = -\frac{a_{21}}{\Delta}\epsilon + \frac{a_{11}}{\Delta}\zeta \tag{1.17}$$

where $\Delta = a_{11}a_{22} - a_{12}a_{21}$. Substituting Eqns. 1.16 and 1.17 into Eqn. 1.15 leads to the result that $a_{12} = a_{21}$, that is

$$\frac{1}{H} = \frac{1}{H_1} \tag{1.18}$$

Eqn. 1.18 implies that the following reciprocity relation holds: the volume of fluid expelled at constant fluid pressure due to an increase in compressive stress is the same as the unconstrained volume expansion due to an increase in pore pressure.

Biot also introduced the coefficient $1/M$, which is the specific storage coefficient *at constant strain*. It is called the *constrained*[8] specific storage coefficient and designated S_ϵ. Micromechanical analysis (cf. Section 3.3) will show that the value of $1/R$ is determined by the compressibilities of the frame, the pores, the fluid, and the solid grains. Although $1/R$ has the same

[8] The term *constrained* is used in soil mechanics (e.g., Lambe and Whitman, 1979) to mean uniaxially constrained. In this book *constrained* means three-dimensionally constrained, unless a qualifying adjective is used.

units as compressibility and can be expressed in terms of different compressibilities, its physical meaning is that of a storage coefficient.[9]

The introduction of three coefficients—drained compressibility ($1/K$), poroelastic expansion coefficient ($1/H$), and unconstrained specific storage coefficient ($1/R$)—completely characterizes the poroelastic response for isotropic applied stress. These three coefficients are the three independent components of a symmetric 2×2 matrix:

$$\begin{pmatrix} \frac{1}{K} & \frac{1}{H} \\ \frac{1}{H} & \frac{1}{R} \end{pmatrix} \tag{1.10}$$

The drained compressibility and the unconstrained storage coefficient are the diagonal components. The poroelastic expansion coefficient is the off-diagonal component. The symmetry condition means that $1/H$ has the same value for the coupling between strain and fluid pressure at constant stress as it does for the coupling between increment of fluid content and stress at constant pore pressure. Using Eqn. 1.8 in Eqns. 1.6 and 1.7 yields

$$\epsilon \equiv \frac{\delta V}{V} = \frac{1}{K}\sigma + \frac{1}{H}p \tag{1.11}$$

$$\zeta = \frac{1}{H}\sigma + \frac{1}{R}p \tag{1.12}$$

Two additional coefficients—Skempton's coefficient B and constrained specific storage coefficient $S_\epsilon \equiv 1/M$—are now introduced and expressed in terms of the three already defined. These examples illustrate further that poroelastic constants are ratios of field variables with specified constraints on the elementary control volume.

Skempton's coefficient is defined to be the ratio of the induced pore pressure to the change in applied stress for *undrained* conditions—that is, no fluid is allowed to move into or out of the control volume:

$$B \equiv -\left.\frac{\delta p}{\delta \sigma}\right|_{\zeta=0} \tag{1.19}$$

[9] Bear (1972, p. 211) in his treatise on flow in porous media referred to Biot's theory as follows:

> ... assuming validity of Darcy's law and the generalized Hooke law [Eqns. 1.6 and 1.7], a theory was developed for flow in a consolidating medium, *without actually defining a storage coefficient* [emphasis added].

A negative sign is included in the definition because the sign convention for stress means that an increase in compressive stress, which induces a pore pressure increase, is a decrease in σ. The subscript equation $\zeta = 0$ is important because it expresses the undrained condition that no fluid is exchanged with the control volume. Substituting $\zeta = 0$ into Eqn. 1.12 shows that the induced pore pressure change is proportional to the applied stress: $\delta p = -(R/H)\delta\sigma$. Therefore,

$$B = \frac{R}{H} \tag{1.20}$$

If compressive stress is applied suddenly to a small volume of saturated porous material surrounded by an impermeable boundary, the induced pore pressure is B times the applied stress. Skempton's coefficient must lie between zero and one. Skempton's coefficient is a measure of how the applied stress is distributed between the skeletal framework and the fluid. It tends toward one for saturated soils, because the load is supported by the fluid. It tends toward zero for gas-filled pores because the load is supported by the framework.

The constrained specific storage coefficient, or specific storage coefficient at constant strain, is defined by

$$S_\epsilon \equiv \left.\frac{\delta\zeta}{\delta p}\right|_{\epsilon=0} \equiv \frac{1}{M} \tag{1.21}$$

where ϵ is the volumetric strain. The procedure for relating S_ϵ to the previously defined coefficients is to solve Eqn. 1.11 for σ and substitute it into Eqn. 1.12. The increment of fluid content is then expressed as a function of volumetric strain and pore pressure:

$$\zeta = \frac{K}{H}\epsilon + \left(\frac{1}{R} - \frac{K}{H^2}\right)p \tag{1.22}$$

Eqns. 1.21 and 1.22 show that

$$S_\epsilon = S_\sigma - \frac{K}{H^2} \tag{1.23}$$

Thus, the specific storage coefficient at constant strain is smaller than the specific storage coefficient at constant stress due to the constraint that the bulk volume remains constant. A schematic diagram comparing the constrained and unconstrained storage concepts is shown in Figure 1.6.

The ratio K/H is known as the *Biot-Willis coefficient* α:

$$\alpha \equiv \frac{K}{H} \tag{1.24}$$

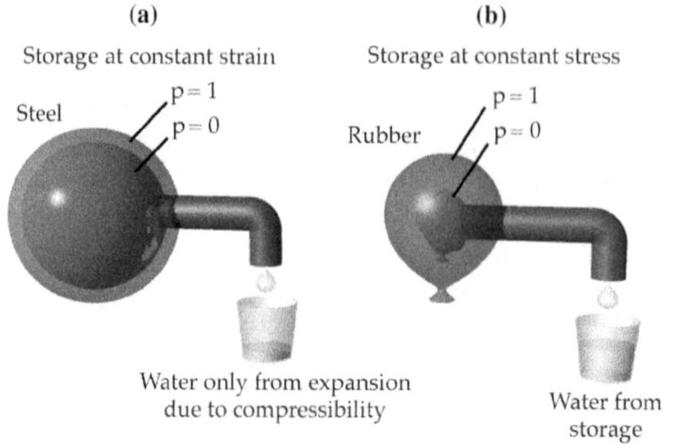

Figure 1.6: Constrained versus unconstrained storage. (a) Storage at constant strain is represented by the steel container. When the fluid pressure p decreases from one to zero, the only water released is from expansion of the water due to its compressibility, as the steel container does not change in volume. (b) Storage at constant stress is represented by the rubber balloon. When the fluid pressure p decreases from one to zero, the water released is due to both the expansion of the water because of its compressibility and the significant decrease in the volume of the balloon.

Eqn. 1.22 can then be rewritten as

$$\zeta = \alpha\epsilon + \frac{1}{M}p \tag{1.25}$$

Therefore, α is the ratio of volume of fluid that is added to storage divided by the change in bulk volume under the constraint that pore pressure remains constant.

In summary, three material constants—drained bulk modulus, poroelastic expansion coefficient, and unconstrained storage coefficient—characterize completely the linear, poroelastic response to volumetric deformation. Other constants, such as Skempton's coefficient and the constrained storage coefficient, can be derived from the three originally defined constants. A fourth independent constant, the shear modulus or the drained or undrained Poisson's ratio, is required for the complete poroelastic constitutive equations when shear stresses are present. Typical values of poroelastic constants for different rock types are given in Appendix C.

1.9 THERMOELASTIC ANALOGY

A complete mathematical analogy exists between poroelasticity and thermoelasticity (Biot, 1941a, 1956c; Rice and Cleary, 1976; Norris, 1992). Thermoelasticity combines the theory of heat conduction with constitutive equations that couple the stress and temperature fields. Temperature T corresponds to pore pressure, heat conduction corresponds to fluid flow, and entropy corresponds to fluid mass. In the thermoelastic equivalent of Eqn. 1.11, the volumetric strain is a function of mean stress and temperature:

$$\epsilon = \frac{\delta V}{V} = \frac{1}{K_t}\sigma + \alpha_t T \tag{1.26}$$

where $1/K_t$ is the isothermal compressibility and α_t is the volumetric thermal expansion coefficient [$^\circ K^{-1}$]. Identifying $1/K_t$ in Eqn. 1.26 as the isothermal ($T = 0$) compressibility is analogous to identifying $1/K$ in Eqn. 1.11 as the drained ($p = 0$) compressibility. Similarly, α_t is the stress-free thermal expansion coefficient, which corresponds to $1/H$, the stress-free poroelastic expansion coefficient.

The change in *specific* entropy δs (entropy per unit volume) for small changes from a reference equilibrium state at temperature T_o is given by the thermodynamic relationship

$$\delta s = \frac{h_Q}{T_o} \tag{1.27}$$

where h_Q is the quantity of heat added per unit volume. A change of entropy is proportional to heat transferred from storage. The analogue of δs in poroelasticity is the change of fluid mass content δm_f. The analogue of h_Q in poroelasticity is ζ, because ζ is a normalized volume of fluid added to or released from the porous material—just as h_Q is a normalized quantity of heat added to or released from the material. The variable m_f is a state variable of a poroelastic system, just as s is a state variable of a thermoelastic system. On the other hand, ζ and h_Q are not state variables because they are quantities that exist only when a change takes place within the system.

Biot (1956c) expressed the second thermoelastic constitutive equation as

$$\delta s = \alpha_t K_t \delta\epsilon + \frac{c_V}{T_o} T \tag{1.28}$$

where c_V is the specific heat at constant volume. Substituting Eqn. 1.27 into Eqn. 1.28 gives

$$h_Q = T_o \alpha_t K_t \epsilon + c_V T \tag{1.29}$$

Eqn. 1.29 can be compared with its poroelastic counterpart (Eqn. 1.25). The storage coefficient at constant strain, S_ϵ, is analogous to specific heat capacity at constant volume, c_V. Isothermal conditions ($T = 0$) in thermoelasticity are analogous to drained conditions ($p = 0$) in poroelasticity. Isentropic conditions ($\delta s = 0$) and adiabatic conditions ($h_Q = 0$) are equivalent by Eqn. 1.27, and hence both isentropic and adiabatic conditions are equivalent to undrained conditions ($\delta m_f = \zeta = 0$) in poroelasticity.

To summarize, a one-to-one analogy has been established for the linear constitutive equations of poroelasticity and thermoelasticity. The thermal analogue to stress-free poroelastic strain induced by pore pressure changes (cf. Eqn. 1.11) is stress-free thermal strain induced by temperature changes. Thus, the volumetric thermal expansion corresponds to $1/H$ in Eqn. 1.11. The other coupling coefficient is the specific heat, which is the amount of heat required to change the temperature of the body, defined either for constant volume or constant entropy (adiabatic) conditions. Constant temperature (isothermal) and insulated (adiabatic) boundary conditions correspond to constant pore pressure (drained) and no-flow (undrained) boundary conditions, respectively, in the poroelastic case.

The completion of the analogy between poroelasticity and thermoelasticity requires a consideration of time-dependent transport. The mathematical form of Fourier's law for heat transport in response to temperature gradients is identical to Darcy's law for groundwater flow in response to pressure gradients. Heat conduction and groundwater flow are both thermodynamically irreversible processes. However, they are assumed to occur slowly enough that the system passes through a continuous sequence of equilibrium states.

The thermoelasticity literature (e.g., Boley and Weiner, 1985) is a valuable resource for poroelasticity because it can reinforce concepts of analogous poroelastic behavior and also allow some solutions to be transferred directly. For example, the opening paragraph of Nowacki's (1986) treatise on thermoelasticity is a good statement of the basic coupling between deformation and temperature:

> A deformation of a body is inseparably connected with a change of its heat content and therefore with a change of the temperature distribution in the body. A deformation of a body which varies in time leads to temperature changes, and conversely. The internal energy of the body depends on both the temperature and the deformation. The science which deals with the investigation of the above coupled processes, is called thermoelasticity.

Biot's *increment of fluid content* is parallel to the term *change of heat content* used in thermoelasticity. In thermoelasticity, temperature changes produce thermal stresses, but stress changes do not significantly alter the temperature field in most materials. A similar uncoupling approximation exists in poroelasticity, in which the pore pressure field is calculated independently. Nowacki distinguishes between thermoelasticity and the theory of thermal stresses as being the difference between full coupling and the assumption that the influence of deformation on the temperature field can be neglected. In poroelasticity it is also useful to make a similar distinction between full coupling and the assumption that changes in fluid pressure affect strain, but not vice versa.

2

Linear Constitutive Equations

2.0 CHAPTER OVERVIEW

Biot's linear constitutive equations and their coefficients are described in this chapter. The four basic variables of poroelasticity can be grouped into two conjugate pairs: (1) stress and strain and (2) pore pressure and increment of fluid content (fluid volume added to storage per unit of control volume). Each of these variables is defined by its average value over a suitable representative elementary volume (REV), which consists of a sufficient volume of grains and void space such that a volume average of porosity approaches a stable limit. Stress and strain are symmetric second-rank tensors, and hence are represented by six scalar components. Pore pressure and increment of fluid content are scalars. A pair of dependent variables consists of one variable from each conjugate pair. The other two variables become the independent variables. Biot's constitutive relations consist of a set of seven linear equations. Each component of the two dependent variables is expressed as a linear combination of each component of the two independent variables. The coefficients of the linear transformation define different physical properties that characterize a particular fluid-filled porous medium. Four independent coefficients are required to characterize an isotropic medium. Only three independent coefficients are required if the stress field is isotropic; hence, one of the four poroelastic coefficients must be related to shear stress. The constitutive equations can be expressed in terms of an effective stress, which is a linear combination of normal stress and pore pressure.

2.1 KINEMATIC VARIABLES

Kinematics is the study of motion. Displacement is a basic kinematic variable because it is a vector whose tail is at the initial location of a point and whose head is at the final location of that point after motion has occurred. The two basic kinematic variables in poroelasticity are strain and increment of fluid content. Strain components are calculated from spatial derivatives of displacements. Increment of fluid content can also be regarded as a kinematic variable because it is a measure of the quantity of fluid moved into or out of a control volume.

2.1.1 Strain

The definition of the strain tensor for a fluid-filled porous material is the same as it is for an elastic solid (e.g., Means, 1976):

$$\epsilon_{xx} = \frac{\partial u}{\partial x} \tag{2.1}$$

$$\epsilon_{yy} = \frac{\partial v}{\partial y} \tag{2.2}$$

$$\epsilon_{zz} = \frac{\partial w}{\partial z} \tag{2.3}$$

$$\epsilon_{xy} = \frac{1}{2}\left(\frac{\partial u}{\partial y} + \frac{\partial v}{\partial x}\right) \tag{2.4}$$

$$\epsilon_{xz} = \frac{1}{2}\left(\frac{\partial u}{\partial z} + \frac{\partial w}{\partial x}\right) \tag{2.5}$$

$$\epsilon_{yz} = \frac{1}{2}\left(\frac{\partial v}{\partial z} + \frac{\partial w}{\partial y}\right) \tag{2.6}$$

where u, v, and w are the components of displacement in the x, y, and z directions, respectively. Eqns. 2.1–2.6 can be written in index notation:

$$\epsilon_{ij} = \frac{1}{2}\left(\frac{\partial u_i}{\partial x_j} + \frac{\partial u_j}{\partial x_i}\right) \tag{2.7}$$

where u_i is the displacement in the i-direction and each of the subscripts i and j takes on the values 1, 2, and 3 corresponding to the coordinate axes x, y, and z, respectively. The term *bulk strain* is sometimes used to emphasize that it is a measure of the deformation of the REV and not that of the solid or fluid phase within it. The longitudinal strains, ϵ_{11}, ϵ_{22}, and ϵ_{33}, are the relative-length changes parallel to the coordinate axes (Fig. 2.1a). Extensional strains are taken as positive, which is the usual convention in applied mechanics. The preponderance of the poroelastic literature also follows this convention, whereas the earth sciences take compression as positive. The volumetric strain (cf. Eqn. 1.11) is the sum of the three orthogonal longitudinal strains:

$$\epsilon = \frac{\delta V}{V} = \epsilon_{11} + \epsilon_{22} + \epsilon_{33} \tag{2.8}$$

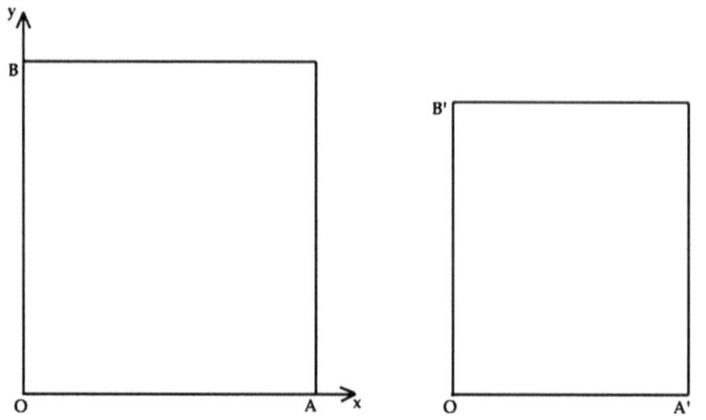

(a) Longitudinal Strain

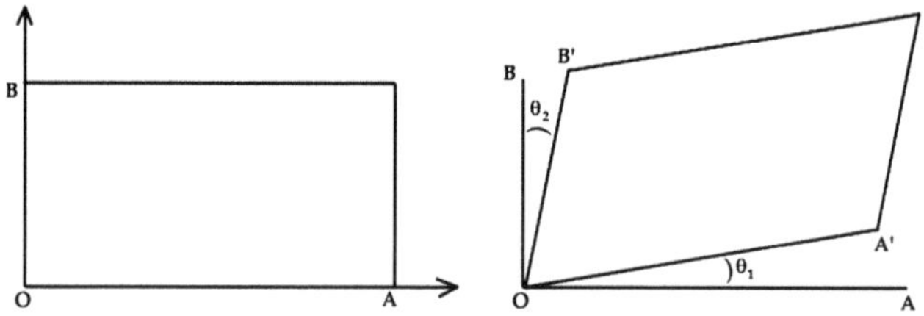

(b) Shear Strain

Figure 2.1: Strain in a representative elementary volume of porous material. The longitudinal strains are the lengths $(\overline{OA'} - \overline{OA})/\overline{OA}$ and $(\overline{OB'} - \overline{OB})/\overline{OB}$. For infinitesimal deformations the shear strain is $(1/2)(\theta_1 + \theta_2)$.

The shear strains ϵ_{ij} are the components in Eqn. 2.7 for which $i \neq j$. They measure the change from a right angle of lines parallel to the coordinate axes (Fig. 2.1b).

2.1.2 Increment of Fluid Content

The increment of fluid content ζ is the fluid volume imported into a control volume normalized by the control volume. Fluid exchange with a control volume can be due to deformation, change of fluid pressure, or the presence of a source or sink.

Biot and Willis (1957) expressed increment of fluid content as

$$\zeta = -\phi \vec{\nabla} \cdot (\vec{U}_f - \vec{U}_s) \tag{2.9}$$

where $\vec{U}_f$ and $\vec{U}_s$ are the average displacements of the fluid and solid, respectively. Eqn. 2.9 carries the assumption that the porosity is homogeneous (Biot, 1962). If ϕ varies spatially, it must be moved under the derivative sign. The physical interpretation of Eqn. 2.9 is shown in Figure 2.2. The porosity enters Eqn. 2.9 because it is the fraction of each face of the control volume

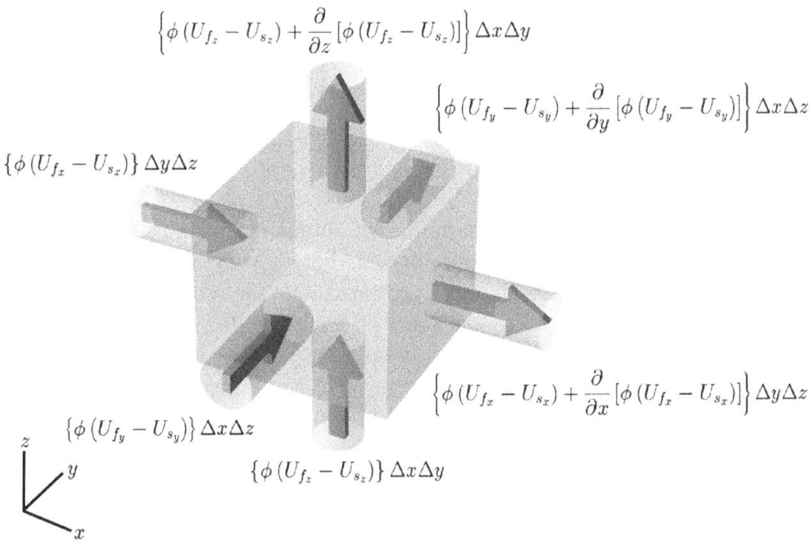

Figure 2.2: Schematic illustration of the increment of fluid content, ζ, as net fluid volume per unit bulk volume added to a control volume. Each arrow represents the average relative displacement vector $\vec{U}_f - \vec{U}_s$ at one of the six faces of the control volume. The cross-sectional area of each cylindrical tube is the porosity times the cross-sectional area of the face of the cube. Therefore, the volume of a cylindrical tube is the volume of water entering or exiting each face. The net volume of fluid exiting the control volume in the x-direction is $(\partial/\partial x)[\phi(U_{f_x} - U_{s_x})]\Delta y\Delta z$. Summing the results in the three coordinate directions leads to Eqn. 2.9.

occupied by fluid. For isotropy, the ratio of void area to the total area of a face is the same as the volumetric porosity ϕ. Because the divergence of fluid displacement relative to solid displacement represents the net loss of fluid volume from a unit control volume, the negative sign in Eqn. 2.9 expresses ζ as the net fluid volume imported into a unit control volume. Eqn. 2.9 represents continuity; the volume of fluid added to storage is equal to the net inflow of fluid.

An alternative micromechanical expression for the increment of fluid content is that it is the difference between the fractional change in pore volume and the fractional change in fluid volume (Berryman, 1992):

$$\zeta = \frac{\delta V_p - \delta V_f}{V} \tag{2.10}$$

where V_p is the pore volume, V_f is the fluid volume, and V is the reference volume. If the fluid were incompressible ($\delta V_f = 0$), the increment of fluid content ζ would be only the change in pore volume divided by the reference volume. If the pore volume were incompressible ($\delta V_p = 0$), the fluid volume added would be that due only to the compressibility of the fluid and $\zeta = (\phi\beta_f)p$, where β_f is the fluid compressibility.

2.2 DYNAMIC VARIABLES

Whereas kinematics is the description of motion, dynamics is the study of the causal forces that produce motion. Force is the most fundamental dynamic variable. Both stress and pore pressure—the dynamic variables in poroelasticity—are defined as the ratio of force divided by the area in the limit as the area goes to zero. An area element has magnitude and direction defined by the outward normal to the surface. Because both force and area are vector quantities, stress is represented by six components of a symmetric second-rank tensor. Symmetry is a consequence of rotational equilibrium of an REV. Fluid pressure is a scalar quantity because fluid pressure produces an equal force per unit area in all directions.

2.2.1 Stress

The tensor components σ_{ij} of the applied stress on an REV are shown in Figure 2.3. The stress components are *total stresses*—that is, σ_{ij} is the total force in the i-direction acting per unit area on the face whose normal is in the j-direction. Thus, an outwardly directed axial load of magnitude F on a sample face of area A produces an axial stress magnitude F/A. The area A is the area enclosed by the perimeter of the face and includes both solid grains and pores.

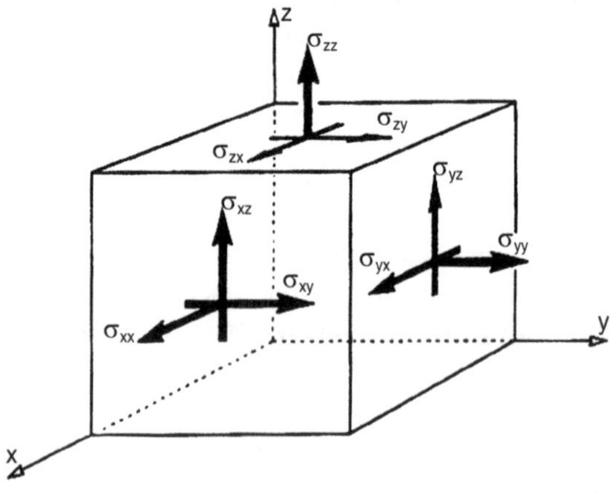

Figure 2.3: Stress tensor (after Verruijt, 1969, p. 338). The first subscript refers to the direction of the normal to the plane, and the second subscript refers to the direction of the force.

The *normal* stresses—σ_{xx}, σ_{yy}, and σ_{zz}—are those for which the force is parallel to the normal to the face. The sum of the three normal stresses, $\sigma_{xx} + \sigma_{yy} + \sigma_{zz}$, is called the *octahedral* stress; it is three times the mean normal stress $(\sigma_{xx} + \sigma_{yy} + \sigma_{zz})/3$. The sum of the three normal stresses is designated by σ_{kk} according to the convention that repeated subscripts signify summation. A particular normal stress may sometimes be designated by σ_{ii}, which is not intended to represent a summation. Using kk for the repeated subscripts or the context usually makes clear the intent of a repeated subscript.

The *shear* stresses—σ_{xy}, σ_{xz}, σ_{yz}—are those for which the force is perpendicular to the normal to the face. Rotational equilibrium (cf. Section 4.1) requires the stress tensor to be symmetric so that $\sigma_{ij} = \sigma_{ji}$.

2.2.2 Pore Pressure

Pore pressure p is the pressure of the fluid occupying the pore space. Its physical meaning is defined by the imaginary experiment of measuring the pressure in a fluid reservoir in static equilibrium with the pore. In practice, pore pressure is measured using a pressure transducer connected by a tube to a laboratory sample or located in a packed off section of a well.

2.3 CONSTITUTIVE RELATIONS

A poroelastic problem consists of four basic variables—stress (σ_{ij}), strain (ϵ_{ij}), pore pressure (p), and increment of fluid content (ζ). Stress and strain are symmetric tensor quantities with six components each. One of the mechanical quantities—stress or strain—is combined with one of the fluid quantities—pore pressure or increment of fluid content—to form a pair of independent variables. The remaining two variables then become the pair of dependent variables. Four permutations of independent and dependent variables are possible (Table 2.1). The set of constitutive equations in which stress is one dependent variable is called a *stiffness formulation* because the coefficients of the strain components are ratios of stress to strain. Similarly, the set of constitutive equations in which strain is one dependent variable is called a *compliance formulation*, because the coefficients of the stress components are ratios of strain to stress. There are two stiffness formulations and two compliance formulations depending on whether pore pressure or increment of fluid content is the other independent variable. The formulation is called a *pure stiffness formulation* when stress and pore pressure are paired as dependent variables, because they have the same physical units. The formulation is called a *mixed stiffness formulation* when stress is paired with increment of fluid content as dependent variables. Similarly, the formulation

TABLE 2.1.
Permutations of Independent and Dependent Poroelastic Variables

Case	Dependent-Variable Pair		Independent-Variable Pair	
Pure compliance	strain	increment of fluid content	stress	pore pressure
Mixed compliance	strain	pore pressure	stress	increment of fluid content
Pure stiffness	stress	pore pressure	strain	increment of fluid content
Mixed stiffness	stress	increment of fluid content	strain	pore pressure

is called a *pure compliance formulation* when strain and increment of fluid content are paired as dependent variables, because they have the same physical units. The formulation is called a *mixed compliance formulation* when strain is paired with pore pressure as dependent variables.

The complete set of poroelastic constitutive equations for a general anisotropic state of stress consists of seven linear equations—six for the tensor quantity (strain or stress) and one for the scalar quantity (increment of fluid content or pore pressure). The seven equations in a pure compliance formulation can be placed in three categories:

1. *Three equations in which the three normal strain components are related to three normal stress components and the pore pressure.* Three poroelastic moduli appear in each of these three equations. Two of the poroelastic moduli are equivalent to the two used in standard elasticity (e.g., a bulk modulus and a shear modulus). The third modulus describes the effect of pore pressure on strain.
2. *Three equations in which the three shear strain components are related to the corresponding shear stress components.* The absence of a dependence on pore pressure in these equations is due to the assumption that pore pressure changes do not induce shear strains. For an isotropic medium, the same shear modulus appears in all three equations. These shear strain equations are not needed when principal coordinates are selected because the shear strains and stresses vanish.
3. *One equation in which the increment of fluid content is related to the mean stress and pore pressure.* Two poroelastic moduli appear in this equation, including one that does not appear in any of the previous six equations.

The seven constitutive equations contain four independent poroelastic moduli. Different sets of four poroelastic moduli appear in the literature. Algebraic relationships exist among the different poroelastic moduli, just as they exist among Young's modulus, Poisson's ratio, bulk modulus, and shear modulus in standard linear elasticity. Many poroelastic moduli are introduced in the several sections that follow, because they arise naturally for different choices of independent and dependent variables in the constitutive equations. Chapter 3 considers additional interrelationships among the poroelastic moduli as well as their micromechanical relationships to fluid, pore, and solid-grain compressibilities.

2.4 PRINCIPAL COORDINATES

The stress and strain tensors each contain six components for an arbitrary orientation of the coordinate system. However, the shear stresses σ_{ij} and shear strains ϵ_{ij} $(i \neq j)$ are zero in *principal* coordinates. Therefore, the three constitutive equations relating the shear strain components to the three shear stress components are no longer necessary, and the seven constitutive equations reduce to four equations for four dependent variables each containing four independent variables:

$$\epsilon_1 = \frac{\sigma_1}{E} - \frac{\nu}{E}\sigma_2 - \frac{\nu}{E}\sigma_3 + \frac{p}{3H} \tag{2.11}$$

$$\epsilon_2 = -\frac{\nu}{E}\sigma_1 + \frac{\sigma_2}{E} - \frac{\nu}{E}\sigma_3 + \frac{p}{3H} \tag{2.12}$$

$$\epsilon_3 = -\frac{\nu}{E}\sigma_1 - \frac{\nu}{E}\sigma_2 + \frac{\sigma_3}{E} + \frac{p}{3H} \tag{2.13}$$

$$\zeta = \frac{1}{3H}(\sigma_1 + \sigma_2 + \sigma_3) + \frac{p}{R} \tag{2.14}$$

where ν is the drained Poisson's ratio, E is the drained Young's modulus, and $1/H$ and $1/R$ are Biot moduli introduced in Chapter 1. Single subscripts are used for ϵ and σ to represent the three normal strains and stresses, respectively, in principal coordinate directions. As noted in Chapter 1, the linear constitutive equations apply to small changes in each of the variables (strain, stress, pore pressure, and increment of fluid content) relative to a reference state. But the theory "may be applied to incremental variations in nonlinear systems" (Biot and Willis, 1957). For notational simplicity, the change in a quantity X is denoted by X itself rather than by δX. Thus, a drained condition is designated by $p = 0$.

The *poro* part of the poroelastic constitutive equations is found in the pore pressure term in Eqns. 2.11–2.13 and in Eqn. 2.14 in its entirety. Eqns. 2.11–2.13 are the standard constitutive equations of linear elasticity, but with the caveat that the elastic constants ν (Poisson's ratio) and E (Young's modulus) are measured under *drained* conditions ($p = 0$). Eqn. 1.11 is obtained after summing the normal strains and using the following standard linear elastic relationship:

$$K = \frac{E}{3(1 - 2\nu)} \tag{2.15}$$

Eqn. 2.15 holds for drained poroelastic constants; this same relationship holds for their undrained counterparts.

If all normal stresses are zero, each normal strain is directly proportional to the pore pressure, and $1/3H$ is the proportionality constant. The quantity $1/3H$ is mathematically analogous to the linear thermal expansion, because it is the ratio of linear strain to pore pressure change when the boundary stresses are held constant.

The physical meaning of $1/R$ in Eqn. 2.14 is found by taking the ratio of increment of fluid content to the pore pressure when the normal stresses are held constant. As discussed in Chapter 1, the quantity $1/R$ is the specific storage coefficient S_σ measured under constant stress conditions. The specific storage coefficient S_σ is also referred to as the three-dimensional specific storage. It is analogous to the specific heat at constant pressure in the theory of heat conduction and thermoelasticity.

The definition of Skempton's coefficient (cf. Eqn. 1.19) can be generalized to a state of nonisotropic stress by setting $\zeta = 0$ in Eqn. 2.14. The condition $\zeta = 0$ means that a control volume is undrained, because no fluid is transported into or out of the control volume. Eqn. 2.14 can then be solved to yield

$$B = \frac{R}{H} = -\left.\frac{p}{(\sigma_1 + \sigma_2 + \sigma_3)/3}\right|_{\zeta=0} \tag{2.16}$$

Eqn. 2.16 shows that the undrained pore pressure change is determined by the change in the mean normal stress $(\sigma_1 + \sigma_2 + \sigma_3)/3$.

2.5 ISOTROPIC STRESS AND STRAIN

The constitutive equations for isotropic stress and strain were previewed in Chapter 1. Only two constitutive equations containing three independent material properties are needed, because stress is represented by a scalar value σ and strain is represented by volumetric strain ϵ. A confining pressure P_c is represented by $\sigma = -P_c$. The negative sign is due to the sign convention that tensile (or less compressive) stress is positive. Physical interpretations are obtained for poroelastic coefficients by considering the ratio of a dependent to independent variable while holding the other independent variable constant (cf. Eqns. 2.11–2.14). Constant increment of fluid content means undrained conditions, constant pore pressure means drained conditions, constant stress means no external stress change, and constant strain means zero external displacement.

2.5.1 Pure Compliance Formulation

The sum of Eqns. 2.11–2.13 is equivalent to Eqn. 1.11, and Eqn. 2.14 is equivalent to Eqn. 1.12 by taking σ to be equal to the mean normal stress. Eqns. 1.11 and 1.12 can be written as follows:

$$\epsilon = \frac{1}{K}\sigma + \frac{\alpha}{K}p \tag{2.17}$$

$$\zeta = \frac{\alpha}{K}\sigma + \frac{\alpha}{KB}p \tag{2.18}$$

The constant $1/H$ in Eqn. 1.11 has been replaced by α/K (cf. Eqn. 1.24), and the storage coefficient $1/R$ in Eqn. 1.12 has been replaced by $\alpha/(KB)$ (cf. Eqn. 2.16). Because $1/R$ has been defined to be the unconstrained storage coefficient S_σ, the following equalities express the various relationships among these coefficients:

$$S_\sigma = \frac{1}{R} = \frac{\alpha}{KB} \tag{2.19}$$

Factoring out $1/K$ from Eqns. 2.17 and 2.18 leaves coefficients 1, α, and α/B.

2.5.2 Pure Stiffness Formulation

Inverting Eqns. 2.17 and 2.18 yields a stiffness formulation of the constitutive equations, because the coefficients are the ratio of a stress to a strain:

$$\sigma = K_u\epsilon - (K_uB)\zeta \tag{2.20}$$

$$p = -(K_uB)\epsilon + \frac{K_uB}{\alpha}\zeta \tag{2.21}$$

The physical meaning of the newly appearing constant K_u is that of undrained ($\zeta = 0$) bulk modulus. Its definition and relationship to coefficients in the compliance formulation are

$$K_u \equiv \left.\frac{\delta\sigma}{\delta\epsilon}\right|_{\zeta=0} = \frac{K}{1-\alpha B} \tag{2.22}$$

The constants K_u, B, and α form a set of three independent coefficients.

The idea that water wells are strain meters is obtained from Eqn. 2.21. If it is assumed that the pore pressure is an undrained response to volume strain

(Rojstaczer and Agnew, 1989, p. 12404; Roeloffs, 1988), the well sensitivity to strain is given by the ratio

$$\frac{w}{\epsilon} = \frac{1}{\rho_w g} \frac{p}{\epsilon}\bigg|_{\zeta=0} = -\frac{K_u B}{\rho_w g} \tag{2.23}$$

where w is the change in water level, ρ_w is the density of water, and g is the acceleration of gravity. Sensitive water wells have values of $-w/\epsilon$ between 1 and 4 millimeters per 10^{-9} of volume strain. The water well sensitivity can be expressed alternatively in terms of its stress response. As a stress meter, the ideal water well response depends on Skempton's coefficient divided by $\rho_w g$. Water well sensitivities between 0.025 and 0.1 mm per Pa correspond to values of Skempton's coefficient between 0.25 and 1. The validity of the undrained assumption depends on the rate of loading relative to the permeability of the formation. Also, well bore storage limits the sensitivity of a real well.

2.5.3 Mixed Stiffness Formulation

In Eqns. 2.17 and 2.18, stress and pore pressure are the independent variables, and strain and increment of fluid content are the dependent variables. Alternatively, strain and pore pressure can be the independent variables and stress and increment of fluid content the dependent variables. Solving for stress in Eqn. 2.17 yields

$$\sigma = K\epsilon - \alpha p \tag{2.24}$$

Substituting Eqn. 2.24 into Eqn. 2.18 yields

$$\zeta = \alpha\epsilon + \frac{\alpha}{K_u B} p \tag{2.25}$$

Eqn. 2.25 leads to a physical interpretation for the Biot-Willis coefficient α:

$$\alpha = \frac{\zeta}{\epsilon}\bigg|_{p=0} \tag{2.26}$$

In words, α is the ratio of increment of fluid content to volumetric strain while maintaining a drained (constant fluid pressure) condition. The constant fluid pressure condition means that the volume of fluid that goes into or out of storage is the change in pore volume only. The usual range for α is $\phi \leq \alpha \leq 1$ (Berryman, 1992).

Eqn. 2.25 also yields an expression for the *constrained specific storage coefficient*:

$$S_\epsilon \equiv \frac{1}{M} \equiv \left.\frac{\zeta}{p}\right|_{\epsilon=0} = \frac{\alpha}{K_u B} \tag{2.27}$$

Thus, dividing Eqn. 2.19 by Eqn. 2.27 shows that the ratio S_σ/S_ϵ of the unconstrained to constrained specific storage coefficients is the same as the ratio K_u/K of the undrained to drained bulk modulus. Because $K_u \geq K$, the unconstrained specific storage coefficient is larger than or equal to the constrained specific storage coefficient.

2.5.4 Mixed Compliance Formulation

The volumetric strain ϵ can be obtained as a linear function of σ and ζ by solving Eqn. 2.20. Alternatively, p can be eliminated in favor of ζ in Eqns. 2.17 and 2.18. By either route, the resulting equation is

$$\epsilon = \frac{\sigma}{K_u} + B\zeta \tag{2.28}$$

This equation formed the basis of Segall's (1989) analysis of subsidence, horizontal strain, and seismicity induced by fluid extraction from hydrocarbon reservoirs. In his words, "[Eqn. 2.28] states that the volume strain ϵ is composed of two parts: an elastic component under conditions of no pore-fluid flow σ/K_u, and a component resulting from changes in the amount of fluid in the pores of the rock $B\zeta$. Extraction of pore fluids results in localized contraction of the reservoir rocks, stressing and deforming the surrounding rock mass." Eqn. 2.28 also provides a second physical interpretation for Skempton's coefficient as the ratio of volumetric strain to increment of fluid content while maintaining constant mean stress:

$$B = \left.\frac{\epsilon}{\zeta}\right|_{\sigma=0} \tag{2.29}$$

2.6 NONPRINCIPAL COORDINATES

A full set of seven constitutive equations is necessary for nonprincipal coordinates. Three of the constitutive equations are the same as for principal coordinates, but normal strains and stresses in the coordinate directions are used in place of the principal strains and stresses. Three additional constitutive equations relate the shear strains to the shear stresses. The seventh equation for either pore pressure or increment of fluid content is also the same as for principal coordinates because shear stress does not affect either of these fluid variables, and because the sum of normal stresses is invariant to coordinate rotations. Permutations of dependent and independent variables bring out many different choices for the poroelastic coefficients.

2.6.1 Pure Compliance Formulation

The three principal stresses—σ_1, σ_2, and σ_3—in Eqns. 2.11–2.14 are replaced by the three normal stresses—σ_{xx}, σ_{yy}, and σ_{zz}—respectively. These equations are manipulated into a form that contains the sum of the normal stresses: $\sigma_{kk} = \sigma_{xx} + \sigma_{yy} + \sigma_{zz}$. Also, the shear modulus G is used in place of E based on the relationship $E = 2G(1+\nu)$ (Appendix B, Table B.1), and α/K is used in place of $1/H$. With these substitutions, Eqns. 2.11–2.13 become

$$\epsilon_{xx} = \frac{1}{2G}\left[\sigma_{xx} - \frac{\nu}{1+\nu}\sigma_{kk}\right] + \frac{\alpha}{3K}p \tag{2.30}$$

$$\epsilon_{yy} = \frac{1}{2G}\left[\sigma_{yy} - \frac{\nu}{1+\nu}\sigma_{kk}\right] + \frac{\alpha}{3K}p \tag{2.31}$$

$$\epsilon_{zz} = \frac{1}{2G}\left[\sigma_{zz} - \frac{\nu}{1+\nu}\sigma_{kk}\right] + \frac{\alpha}{3K}p \tag{2.32}$$

The three additional equations, which relate the shear strains to the shear stresses, are

$$\epsilon_{xy} = \frac{1}{2G}\sigma_{xy} \tag{2.33}$$

$$\epsilon_{yz} = \frac{1}{2G}\sigma_{yz} \tag{2.34}$$

$$\epsilon_{xz} = \frac{1}{2G}\sigma_{xz} \tag{2.35}$$

Eqns. 2.33–2.35 do not contain a pore pressure term because changes in pore pressure are assumed not to induce shear strain. Eqns. 2.30–2.35 can all be represented by a single equation in index notation:

$$\underbrace{\epsilon_{ij}}_{\textit{Total Strain}} = \underbrace{\frac{1}{2G}\left[\sigma_{ij} - \frac{\nu}{1+\nu}\sigma_{kk}\delta_{ij}\right]}_{\textit{Poroelastic Strain}} + \underbrace{\frac{\alpha}{3K}p\delta_{ij}}_{\textit{Free Strain}} \tag{2.36}$$

where the Kronecker delta in Eqn. 2.36 is defined by

$$\delta_{ij} = \begin{cases} 1 & \text{if } i = j \\ 0 & \text{if } i \neq j \end{cases} \tag{2.37}$$

The total strain in Eqn. 2.36 can be viewed as the sum of a poroelastic strain and a free strain. The total strain is the strain that would be measured by a strain gage at that point in the body. The free strain is the strain that would be measured if the REV about a point were unconstrained. The free normal strains in Eqn. 2.36 are equal in each coordinate direction, and no free shear strains are produced. Hence, an unconstrained body undergoes a uniform volumetric dilatation and appears as a scaled replica of itself when it is subjected to a uniform internal pore pressure change. However, a nonuniform pore pressure distribution or constraints imposed by boundary tractions induce poroelastic stresses σ_{ij} within the body owing to requirements of compatibility and force equilibrium (see Chapter 4). The poroelastic stress is also called the grain-to-grain or total stress in the poroelastic literature. Poroelastic stresses are analogous to thermal stresses induced by a nonuniform temperature distribution within a body. The poroelastic strain is the strain produced by the poroelastic stress. If applied boundary stresses are zero, nonzero poroelastic stresses and strains are due entirely to a nonuniform pore pressure field.[1] Poroelastic strains accommodate the differential free strains in adjacent REVs. Figure 2.4 schematically illustrates poroelastic shear stresses induced in a hydrocarbon reservoir as a consequence of a reduction in pore pressure due to fluid extraction.

Equation 2.36 can be written in different equivalent forms using the same relationships among the drained poroelastic moduli as that among the elastic moduli of a solid material (Appendix B, Table B.1). Eqn. 2.36 contains G, ν, and K, so one can be eliminated in favor of the other two. Eliminating Poisson's ratio ν yields

$$\epsilon_{ij} = \frac{1}{2G}\left[\sigma_{ij} - \left(\frac{3K-2G}{9K}\right)\sigma_{kk}\delta_{ij} + \frac{2G}{3K}\alpha p \delta_{ij}\right] \tag{2.38}$$

Eliminating bulk modulus K yields

$$\epsilon_{ij} = \frac{1}{2G}\left[\sigma_{ij} - \frac{\nu}{1+\nu}\sigma_{kk}\delta_{ij} + \frac{1-2\nu}{1+\nu}\alpha p \delta_{ij}\right] \tag{2.39}$$

The remaining constitutive equation, which relates increment of fluid content to mean stress and pore pressure, can be written in two equivalent forms based on Eqns. 2.14 and 2.18, respectively:

$$\zeta = \frac{1}{H}\frac{\sigma_{kk}}{3} + \frac{p}{R} \tag{2.40}$$

$$\zeta = \frac{\alpha}{K}\frac{\sigma_{kk}}{3} + \frac{\alpha}{KB}p \tag{2.41}$$

[1] An exception is that linear variations in pore pressure in rectangular coordinates do not induce poroelastic strains (Boley and Weiner, 1985, p. 244).

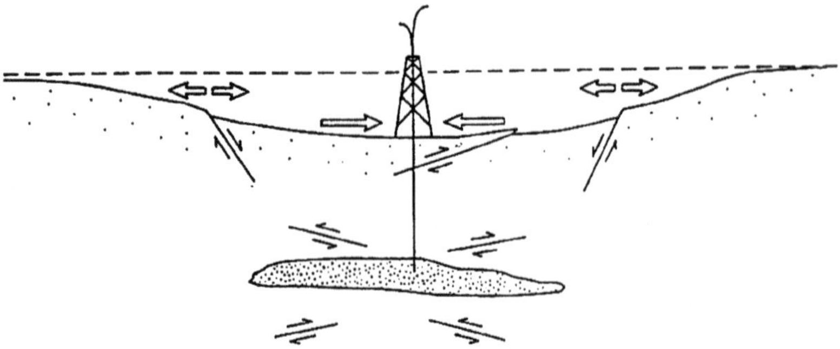

Figure 2.4: Poroelastic stresses arising from fluid extraction (Segall, 1989). As a result of compatibility and force equilibrium, the shrinkage of the reservoir due to pore-pressure reduction produces poroelastic stresses that pull the free surface into a smooth depression. Open arrows show horizontal strains at the surface. The poroelastic stress field includes shear stresses large enough to induce earthquakes in various parts of the reservoir.

Eqn. 2.40 follows from Eqn. 2.14 because the sum of the normal stresses is invariant with respect to coordinate rotations. Eqn. 2.41 generalizes Eqn. 2.18 by assuming that the same increment of fluid content results from a change in mean stress as from the same change in isotropic stress.

2.6.2 Mixed Stiffness Formulation

The constitutive equations can be expressed using stress components as the dependent variables. Solving Eqn 2.36 for σ_{ij} and choosing the coefficients in the stress-strain portion of the resulting equations to be shear modulus G and drained Lamé's constant λ yields

$$\sigma_{ij} = 2G\epsilon_{ij} + \lambda\epsilon_{kk}\delta_{ij} - \alpha p\delta_{ij} \tag{2.42}$$

where the standard relationships among drained moduli (Appendix B, Table B.1) have been applied. Alternatively, ν can be used in place of λ:

$$\sigma_{ij} = 2G\epsilon_{ij} + 2G\frac{\nu}{1-2\nu}\epsilon_{kk}\delta_{ij} - \alpha p\delta_{ij} \tag{2.43}$$

For future reference it is useful to write out Eqn. 2.43 explicitly for the six stress components:

$$\sigma_{xx} = 2G\epsilon_{xx} + 2G\frac{\nu}{1-2\nu}\epsilon_{kk} - \alpha p \tag{2.44}$$

$$\sigma_{yy} = 2G\epsilon_{yy} + 2G\frac{\nu}{1-2\nu}\epsilon_{kk} - \alpha p \tag{2.45}$$

$$\sigma_{zz} = 2G\epsilon_{zz} + 2G\frac{\nu}{1-2\nu}\epsilon_{kk} - \alpha p \tag{2.46}$$

$$\sigma_{xy} = 2G\epsilon_{xy} \tag{2.47}$$

$$\sigma_{xz} = 2G\epsilon_{xz} \tag{2.48}$$

$$\sigma_{yz} = 2G\epsilon_{yz} \tag{2.49}$$

The seventh constitutive equation is the same as Eqn. 2.25 where the volumetric strain ϵ is expressed as ϵ_{kk}:

$$\zeta = \alpha\epsilon_{kk} + \frac{\alpha}{K_u B} p \tag{2.50}$$

2.6.3 Mixed Compliance Formulation

The constitutive equations for strain components expressed in terms of stress components and increment of fluid content are obtained by solving Eqn. 2.41 for p and substituting the result into Eqn. 2.36:

$$\epsilon_{ij} = \frac{1}{2G}\left[\sigma_{ij} - \frac{\nu_u}{1+\nu_u}\sigma_{kk}\delta_{ij} + \frac{2GB}{3}\zeta\delta_{ij}\right] \tag{2.51}$$

The *undrained* Poisson's ratio ν_u is related to previously defined poroelastic constants by

$$\nu_u = \frac{3\nu + \alpha B(1-2\nu)}{3 - \alpha B(1-2\nu)} \tag{2.52}$$

Eqn. 2.52 is discussed further in Chapter 3. Segall (1985, Eqn. A3) used Eqn. 2.51 to model subsidence associated with large volumes of fluid extraction. Setting $\zeta = 0$ in Eqn. 2.51 shows that its strain-stress portion is the undrained equivalent of Eqn. 2.36. The shear modulus is assumed to have the same value for drained or undrained conditions.

2.6.4 Pure Stiffness Formulation

Solving Eqn. 2.51 for σ_{ij} yields

$$\sigma_{ij} = 2G\epsilon_{ij} + 2G\frac{\nu_u}{1-2\nu_u}\epsilon_{kk}\delta_{ij} - BK_u\zeta\delta_{ij} \tag{2.53}$$

where $K_u = 2G(1+\nu_u)/[3(1-2\nu_u)]$ is the undrained bulk modulus. The relationship between K_u, G, and ν_u is the same as the standard relationship among drained moduli (Appendix B, Table B.1). Eqn. 2.53 can also be expressed using the undrained Lamé constant $\lambda_u = 2G\nu_u/(1-2\nu_u)$:

$$\sigma_{ij} = 2G\epsilon_{ij} + \lambda_u \epsilon_{kk}\delta_{ij} - BK_u\zeta\delta_{ij} \tag{2.54}$$

A further variation of Eqn. 2.53 is obtained if Eqn. 2.27 is used to express BK_u as αM:

$$\sigma_{ij} = 2G\epsilon_{ij} + 2G\frac{\nu_u}{1-2\nu_u}\epsilon_{kk}\delta_{ij} - \alpha M\zeta\delta_{ij} \tag{2.55}$$

2.7 ANISOTROPY

The constitutive equations for an orthotropic material are presented for reference because many geologic materials are anisotropic (Biot, 1955; Carroll, 1979; Thompson and Willis, 1991; Cheng, 1997). Orthotropic symmetry reduces to transverse isotropy if properties are rotationally symmetric about one axis. The generalization of Eqns. 2.30–2.35 and Eqn. 2.40 can be expressed as

$$\epsilon_{xx} = \frac{\sigma_{xx}}{E_x} - \frac{\nu_{yx}}{E_y}\sigma_{yy} - \frac{\nu_{zx}}{E_z}\sigma_{zz} + \frac{\alpha_x}{3K}p \tag{2.56}$$

$$\epsilon_{yy} = -\frac{\nu_{yx}}{E_y}\sigma_{xx} + \frac{\sigma_{yy}}{E_y} - \frac{\nu_{zy}}{E_z}\sigma_{zz} + \frac{\alpha_y}{3K}p \tag{2.57}$$

$$\epsilon_{zz} = -\frac{\nu_{zx}}{E_z}\sigma_{xx} - \frac{\nu_{zy}}{E_z}\sigma_{yy} + \frac{\sigma_{zz}}{E_z} + \frac{\alpha_z}{3K}p \tag{2.58}$$

$$\epsilon_{xy} = \frac{1}{2G_{xy}}\sigma_{xy} \tag{2.59}$$

$$\epsilon_{yz} = \frac{1}{2G_{yz}}\sigma_{yz} \tag{2.60}$$

$$\epsilon_{xz} = \frac{1}{2G_{zx}}\sigma_{xz} \tag{2.61}$$

$$\zeta = \frac{1}{3K}(\alpha_x\sigma_{xx} + \alpha_y\sigma_{yy} + \alpha_z\sigma_{zz}) + \frac{p}{R} \tag{2.62}$$

where E_x, E_y, and E_z are the drained Young's moduli in the x, y, and z directions, respectively; G_{xy}, G_{yz}, and G_{zx} are shear moduli for coordinate planes xy, yz, and zx, respectively; ν_{yx}, ν_{zy}, and ν_{zx} are drained Poisson's ratios in which the second subscript is the direction of the compressive strain induced by a tensile stress in the direction of the first subscript; $\alpha_x/3K$, $\alpha_y/3K$, and $\alpha_z/3K$ are linear poroelastic expansion coefficients in the x, y, and z directions, respectively. Note that $\nu_{ij} \neq \nu_{ji}$ but that the compliance tensor is symmetric so that $\nu_{ij}/E_i = \nu_{ji}/E_j$.

For undrained conditions ($\zeta = 0$), Eqn. 2.62 can be solved for the induced pore pressure:

$$p^{(u)} = -\frac{R}{3K}(\alpha_x \sigma_{xx} + \alpha_y \sigma_{yy} + \alpha_z \sigma_{zz}) \tag{2.63}$$

Generalized Skempton's coefficients can be defined for orthotropic symmetry:

$$B_i \equiv -\left.\frac{p^{(u)}}{\sigma_{ii}}\right|_{\sigma_{jj}=0 \text{ for } j \neq i} = \frac{R}{K}\alpha_i \tag{2.64}$$

Eqn. 2.63 can then be expressed as

$$p^{(u)} = -\frac{1}{3}(B_x \sigma_{xx} + B_y \sigma_{yy} + B_z \sigma_{zz}) \tag{2.65}$$

The anisotropy of Skempton's coefficient means that deviatoric stresses can induce an undrained pore pressure change, an effect that does not occur under the assumption of isotropy.

Altogether, orthotropy requires 13 poroelastic constants—9 drained elastic constants, 3 poroelastic expansion coefficients, and 1 volumetric storage coefficient. The total number of poroelastic constants is reduced to 8 for transverse isotropy—5 drained elastic constants, 2 poroelastic expansion coefficients, and 1 volumetric storage coefficient.

2.8 EFFECTIVE STRESS

An increase in applied extensional stress expands a rock by about the same amount as an equal increase in pore pressure. Similarly, equal changes in applied compressive stress and pore pressure tend to offset each other. The idea that rock behavior is governed by the sum of stress and pore pressure is called the *law of effective stress*. It was first enunciated by Terzaghi, who studied both volumetric deformation and shear failure in soils. The effective principal stresses are approximated well by the differential principal stresses

σ_1+p, σ_2+p, and σ_3+p. These differential principal stresses are also called *Terzaghi effective stresses*. In Terzaghi's words (cf. Skempton, 1960):[2]

> The stresses in any point of a section through a mass of soil can be computed from the *total principal stresses* σ_1, σ_2, σ_3, which act in this point. If the voids of the soil are filled with water under a stress p, the total principal stresses consist of two parts. One part, p, acts in the water *and* in the solid in every direction with equal intensity. It is called the *neutral stress* (or the porewater pressure). The balance $\sigma_1' = \sigma_1 + p$, $\sigma_2' = \sigma_2 + p$ and $\sigma_3' = \sigma_3 + p$ represents an excess over the neutral stress p and it has its seat exclusively in the solid phase of the soil. This fraction of the total principal stresses will be called the *effective principal stresses*.

A precise definition for the effective mean stress for volumetric strain can be obtained from the Biot constitutive equations by rewriting Eqn. 2.17 as

$$\epsilon = \frac{1}{K}\left(\frac{\sigma_{kk}}{3} + \alpha p\right) \tag{2.66}$$

Eqn. 2.66 shows that volumetric strain is determined by the linear combination of mean stress and pore pressure

$$\frac{\sigma'_{kk}}{3} \equiv \frac{\sigma_{kk}}{3} + \alpha p \tag{2.67}$$

where $\sigma'_{kk}/3$ is defined to be the *effective mean stress*.

2.8.1 Experimental Example

Nur and Byerlee (1971) tested Eqn. 2.66 experimentally for Weber sandstone. The volumetric strain was measured on a copper-jacketed sample for different values of confining pressure P_c and a separately controlled pore pressure p. The volumetric strain was also measured as a function of confining pressure for a dry sample, which was considered equivalent to a completely drained ($p = 0$) sample. Therefore, the effective pressure for the dry sample was simply the confining pressure. The volumetric strain for a saturated sample under both a confining pressure and an elevated pore pressure is always less than that of the drained sample at the same confining pressure (Fig. 2.5*a*). Plotting the volumetric strain versus the Terzaghi effective pressure, $P_c - p$, overcompensated for the volume expansion effect of the pore pressure (Fig. 2.5*b*). However, plotting the volumetric strain versus the

[2] The notation and sign convention have been changed where necessary to be consistent with this book.

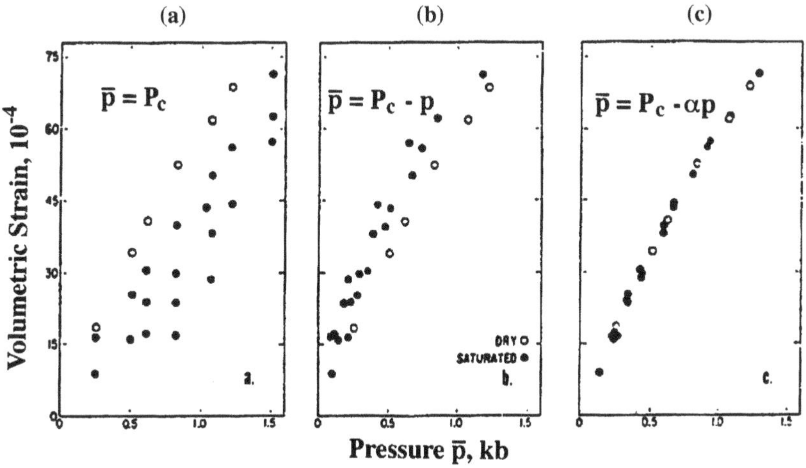

Figure 2.5: Volumetric strain in Weber sandstone (Nur and Byerlee, 1971). (a) Strain versus confining pressure. Solid circles represent data at elevated pore pressures. Open circles represent data for a dry sample in which $p = 0$. (b) Strain versus Terzaghi effective pressure $P_c - p$. Subtracting the pore pressure from the confining pressure "overcorrects" the volume strain data. (c) Strain versus effective pressure $P_c - \alpha p$. All the data overlap the dry sample data, because the effective pressure produces the same change in volumetric strain as does confining pressure with zero pore pressure.

effective pressure, $P_c - \alpha p$, brought the data at elevated pore pressures to the same value as a drained sample at $p = 0$ (Fig. 2.5*c*).

2.8.2 Constitutive Equations

The effective stress tensor components σ'_{ij} are defined by

$$\sigma'_{ij} = \sigma_{ij} + \alpha p \delta_{ij} \tag{2.68}$$

This definition reduces to Terzaghi's definition if $\alpha = 1$, which is equivalent to the unjacketed compressibility being negligible (cf. Section 3.1.2). Eqn. 2.36 can be expressed in terms of effective stresses σ'_{ij}:

$$2G\epsilon_{ij} = \sigma'_{ij} - \frac{\nu}{1+\nu}\sigma'_{kk}\delta_{ij} \tag{2.69}$$

In Eqn. 2.68 the pore pressure term is added only to normal stresses. Pore pressure is not added to shear stresses because shear stresses do not produce volumetric strains. Eqn. 2.69 has the same form as the constitutive equation used in standard, linear elasticity.

TABLE 2.2.
Effective Stress Coefficients for Different Poroelastic Quantities (Berryman, 1992)*

Physical Quantity	Effective Stress Coefficient	Single-Constituent Bounds	Two-Constituent Bounds
Total volume, V	$\alpha = 1 - \frac{K}{K_s'}$	$\phi \leq \alpha \leq 1$	$\alpha \leq 1$
Pore volume, V_p	$\beta = 1 - \frac{K_p}{K_\phi}$	$\alpha \leq \beta \leq 1$	$\phi \leq \beta$
Increment of fluid content, ζ	$\xi = \beta + \frac{K_p}{K_f} = \frac{1}{B}$	$\beta \leq \xi$	$\beta \leq \xi$
Porosity, ϕ	$\chi = \left(\frac{\beta - \phi}{\alpha - \phi}\right)\alpha$	$\chi = 1$	$\alpha \leq \chi$
Solid volume, $V_s = V - V_p$	$\sigma = \frac{\phi K_s'}{K_\phi} = \left(\frac{1-\beta}{1-\alpha}\right)\alpha$		

*The meaning of the variables β, ξ, χ, and σ apply only for this table and not elsewhere in the text.

2.8.3 Other Effective Stress Principles

The concept of an effective stress principle is a general one, but the particular linear combination of stress and pore pressure is different for bulk volume, pore volume, increment of fluid content, porosity, and solid volume. The coefficient of pore pressure in the effective stress for bulk volume is the Biot-Willis coefficient α. It is a measure of the efficiency with which pore pressure counteracts confining pressure to produce volumetric strain. Berryman's (1992, 1993) results for several other poroelastic quantities are given in Table 2.2 without derivation. In Table 2.2 the physical quantity in the first column remains unchanged if the ratio P_c/p is equal to the effective stress coefficient in the second column. The bulk moduli K_s', K_p, and K_ϕ, which appear in the expressions, are defined in Chapter 3. The third column gives bounds for the effective stress coefficient if the solid phase consists of a single constituent. The fourth column gives bounds for the effective stress coefficient if the solid phase consists of two constituents.

3

Poroelastic Constants

3.0 CHAPTER OVERVIEW

Poroelastic constants can be grouped into the following categories: (1) compressibility or its inverse, bulk modulus; (2) Poisson's ratio; (3) storage capacity; (4) poroelastic expansion coefficient; (5) undrained pore pressure buildup coefficient; and (6) shear modulus. The drained and undrained shear moduli are assumed to have a single value G. Several different constants within each group exist because a basic definition can include different constraints on the control volume. For example, bulk modulus, Poisson's ratio, and shear modulus can be drained or undrained; storage capacity can be constrained, unconstrained, or uniaxially constrained; undrained pore pressure buildup coefficients can be defined for isotropic loading (Skempton's coefficient) or for uniaxial loading; and compressibility can be at constant differential pressure or at constant pore pressure. In a complete set, at least one of the poroelastic constants must include a property related to a shear deformation. Biot and Willis (1957) suggested that the set $\{G, 1/K, 1/K_s', S_\gamma\}$, where K_s' is the unjacketed compressibility and S_γ is the unjacketed storage coefficient,[1] was the most convenient to measure. Detournay and Cheng (1993, p. 129) chose $\{G, \alpha, \nu, \nu_u\}$ as their basis set. Another complete set is $\{K, K_u, \nu, \nu_u\}$, consisting of the drained and undrained bulk moduli and the drained and undrained Poisson's ratios.

3.1 COMPRESSIBILITY

A compressibility, in general terms, is the ratio of a fractional volume change with respect to a pressure change. The volume can be bulk volume, pore volume, fluid volume, or solid grain volume; and the pressure can be confining pressure, pore pressure, or differential pressure (Zimmerman et al., 1986, Laurent et al., 1993). Two exceptions to this terminology are made in this

[1] Biot and Willis used the symbol γ for S_γ and the name *coefficient of fluid content*. Because γ is used in this book for the uniaxial loading efficiency, the S_γ notation chosen here emphasizes the nature of this coefficient as a measure of storage capacity.

book: (1) The ratio of increment of fluid content to change in pore pressure for different boundary conditions is designated a storage coefficient; and (2) the ratio of change in bulk volume to change in pore pressure for different boundary conditions is designated a poroelastic expansion coefficient.

Various compressibilities can be defined by introducing alternative constitutive equations. In particular, let V_p be the pore volume and define the differential pressure P_d to be

$$P_d \equiv P_c - p \tag{3.1}$$

where P_c is confining pressure. Then the following pair of constitutive equations introduces three compressibilities K'_s, K_p, and K_ϕ:

$$\frac{\delta V}{V} = -\frac{1}{K} P_d - \frac{1}{K'_s} p \tag{3.2}$$

$$\frac{\delta V_p}{V_p} = -\frac{1}{K_p} P_d - \frac{1}{K_\phi} p \tag{3.3}$$

K'_s, K_p, and K_ϕ are demonstrated to have physical meanings as the unjacketed bulk compressibility, drained pore compressibility, and unjacketed pore compressibility, respectively. These compressibilities are equally valid choices for poroelastic coefficients. Eqns. 3.2 and 3.3 can be rewritten using solid volume V_s and porosity ϕ as the dependent variables (Detournay and Cheng, 1993, Eqn. 26, p. 120):

$$\frac{\delta V_s}{V_s} = -\frac{1}{(1-\phi)K'_s} P_d - \frac{1}{1-\phi}\left(\frac{1}{K'_s} - \frac{\phi}{K_\phi}\right) p \tag{3.4}$$

$$\frac{\delta \phi}{1-\phi} = -\left(\frac{1}{K} - \frac{1}{1-\phi}\frac{1}{K'_s}\right) P_d + \frac{\phi}{1-\phi}\left(\frac{1}{K'_s} - \frac{1}{K_\phi}\right) p \tag{3.5}$$

3.1.1 Drained Bulk Compressibility

The drained compressibility, $1/K$, was already defined to be one of the basic poroelastic constants in Eqn. 2.17.

3.1.2 Unjacketed Bulk Compressibility

In an unjacketed experiment, a laboratory sample is immersed in confining fluid (Fig. 3.1). The confining fluid permeates the pore spaces throughout the interior of the rock sample. Therefore, a change in confining pressure produces an equal change in pore pressure (i.e., $P_c = p$, and hence

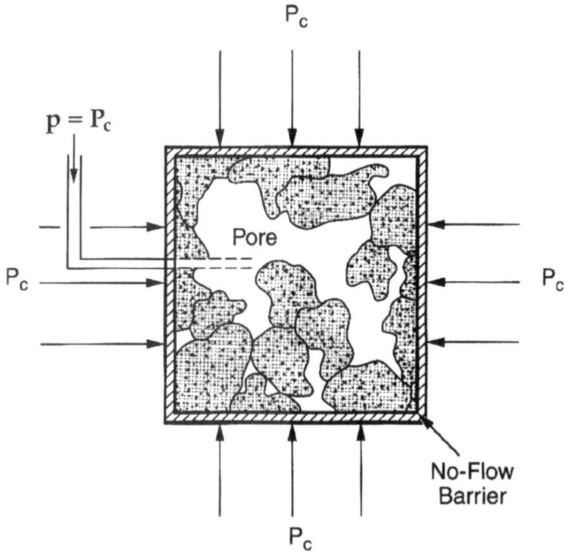

Figure 3.1: Unjacketed compressibility. Change in confining pressure and change in pore pressure are equal ($p = P_c$), which is the case for an unjacketed sample.

$P_d = P_c - p = 0$). Thus, relative to a reference state, an unjacketed condition is one in which the differential pressure is held constant. Eqn. 2.17 for volumetric strain can be written using P_d and p as the independent variables, because P_d is a linear combination of P_c and p:

$$\begin{aligned}\frac{\delta V}{V} &= -\frac{P_d}{K} + \left(\frac{1}{H} - \frac{1}{K}\right)p \\ &= -\frac{P_d}{K} + \left(\frac{\alpha}{K} - \frac{1}{K}\right)p \qquad (3.6)\end{aligned}$$

Eqn. 3.6 can be viewed as Eqn. 2.17 with a change of basis vectors from the pair of independent variables (P_c, p) to (P_d, p). The coefficient of p in Eqn. 3.6 is the unjacketed compressibility, $1/K_s'$:

$$\frac{1}{K_s'} \equiv -\frac{1}{V}\frac{\delta V}{\delta p}\bigg|_{P_d=0} \qquad (3.7)$$

Comparing Eqn. 3.6 with Eqn. 3.2 shows that

$$\frac{1}{K_s'} = \frac{1}{K} - \frac{1}{H} = \frac{1}{K} - \frac{\alpha}{K} \qquad (3.8)$$

Changing pore pressure at constant differential pressure is generally pictured to change bulk volume primarily through its effect on the solid grains, because an increase in pore pressure counteracts an increase in applied stress, leaving the grain-to-grain stresses unchanged. Eqn. 3.4 shows that K_s' can be interpreted as the bulk modulus of the solid phase. However, the unjacketed modulus K_s' is exactly the solid-grain modulus that would be measured on an individual mineral grain only if all the solid grains are the same mineral. For example, a clayey sandstone does not satisfy this condition (Brown and Korringa, 1975; Berryman and Milton, 1992). The prime symbol is therefore used in Eqn. 3.7 to distinguish it from the solid-grain modulus that would be measured on individual grains. Although a distinction exists, the unjacketed bulk modulus K_s' is often called the solid-grain modulus, and K_s might be substituted for K_s'.

3.1.3 Drained Pore Compressibility

The drained pore compressibility is defined as the ratio of fractional change in pore volume to change in confining pressure while maintaining constant pore pressure:

$$\frac{1}{K_p} \equiv -\frac{1}{V_p}\frac{\delta V_p}{\delta P_c}\bigg|_{p=0} \tag{3.9}$$

The drained pore compressibility can be expressed in terms of the drained bulk compressibility and unjacketed compressibility. For a fully saturated porous medium, the porosity is related to the pore volume and fluid volume in a representative elementary volume (REV) by $\phi = V_p/V = V_f/V$. These relationships can be inserted into the definition for the increment of fluid content (Eqn. 2.10) to obtain

$$\zeta = \phi\frac{\delta V_p}{V_p} - \phi\frac{\delta V_f}{V_f} \tag{3.10}$$

The fractional change in fluid volume is related to the change in fluid pressure through the fluid compressibility:

$$\frac{1}{K_f} \equiv -\frac{1}{V_f}\frac{\delta V_f}{\delta p} \tag{3.11}$$

Substituting Eqn. 3.10 and $\sigma = -P_c$ into Eqn. 2.18, using Eqn. 3.11, and solving for $\delta V_p/V_p$ give

$$\frac{\delta V_p}{V_p} = -\frac{\alpha}{\phi K}P_c + \frac{1}{\phi}\left(\frac{\alpha}{KB} - \frac{\phi}{K_f}\right)p \tag{3.12}$$

Applying the definition in Eqn. 3.9 to Eqn. 3.12 gives

$$\frac{1}{K_p} = \frac{\alpha}{\phi K} \tag{3.13}$$

3.1.4 Unjacketed Pore Compressibility

The unjacketed pore compressibility is the pore volume counterpart to the unjacketed bulk compressibility. The unjacketed pore compressibility is a measure of the change in pore volume with change in pore pressure when the confining pressure is required to track with the pore pressure. The unjacketed pore compressibility is defined by[2]

$$\frac{1}{K_\phi} \equiv -\frac{1}{V_p}\frac{\delta V_p}{\delta p}\bigg|_{P_d=0} \tag{3.14}$$

The negative sign in Eqn. 3.14 is consistent with other definitions of compressibility. Berge and Berryman (1995) showed that negative values of K_ϕ are physically realizable. Substituting $P_c = P_d + p$ into Eqn. 3.12 yields

$$\frac{\delta V_p}{V_p} = -\frac{\alpha}{\phi K}P_d + \frac{1}{\phi}\left(\frac{\alpha}{KB} - \frac{\phi}{K_f} - \frac{\alpha}{K}\right)p \tag{3.15}$$

Applying the definition in Eqn. 3.14 to Eqn 3.15 gives

$$\frac{1}{K_\phi} = -\frac{1}{\phi}\left[\frac{\alpha}{KB} - \frac{\phi}{K_f} - \frac{\alpha}{K}\right] \tag{3.16}$$

$$= -\frac{1}{\phi}\left[\left(\frac{1}{K} - \frac{1}{K_s'}\right)\left(\frac{1}{B} - 1\right) - \frac{\phi}{K_f}\right] \tag{3.17}$$

If the porosity remains constant for equal changes in pore pressure and confining pressure ($P_d = 0$), Eqn. 3.5 shows that the unjacketed pore compressibility, $1/K_\phi$, must equal the unjacketed bulk compressibility, $1/K_s'$. Porosity remains constant for a solid phase composed of a single constituent, because the material is a scaled replica of itself when the differential pressure remains constant (Carroll, 1980; Berryman, 1992). But in general, $1/K_\phi$ and

[2] The symbol K_ϕ is used to represent the unjacketed pore compressibility (Brown and Korringa, 1975; Berge and Berryman, 1995). Previously, Rice and Cleary (1976) used the symbol K_s'' for the unjacketed pore compressibility. Their expression for K_s'' was corrected later by Green and Wang (1986). The symbol K_{V_p} is the most consistent notationally because the subscript ϕ suggests a compressibility associated with porosity rather than with pore volume. Detournay and Cheng (1993, Eqn. 26b, p. 120) used the symbol K_ϕ for the coefficient of P_d in Eqn. 3.5.

$1/K_s'$ are different (Brown and Korringa, 1975; Rice and Cleary, 1976). Indirect experimental evidence (Green and Wang, 1986; Hart and Wang, 1995; Berge, 1998) suggests that $1/K_\phi$ is closer to the fluid compressibility at low effective stress for clayey sandstones. For $B = 1$, the result $K_\phi = K_f$ can be obtained from Eqn. 3.17.

The constitutive equation for V_p (Eqn. 3.3) can be written in terms of an effective stress $P_c - \beta p$ as

$$\frac{\delta V_p}{V_p} = -\frac{1}{K_p}(P_c - \beta p) \tag{3.18}$$

where

$$\beta \equiv 1 - \frac{K_p}{K_\phi} \tag{3.19}$$

Eqn. 3.19 was previously listed in Table 2.2 as the effective stress coefficient for pore volume. The coefficient β is analogous to the Biot-Willis coefficient α, which is the effective stress coefficient for bulk volume.

3.1.5 Additional Compressibilities

Zimmerman et al. (1986) defined four compressibilities as coefficients in the following constitutive equations for bulk strain and pore strain (Fig. 3.2):

$$\frac{\delta V}{V} = -C_{bc}P_c + C_{bp}p \tag{3.20}$$

$$\frac{\delta V_p}{V_p} = -C_{pc}P_c + C_{pp}p \tag{3.21}$$

The first subscript refers to the type of volume strain (b is bulk volumetric strain and p is pore strain), and the second subscript refers to the type

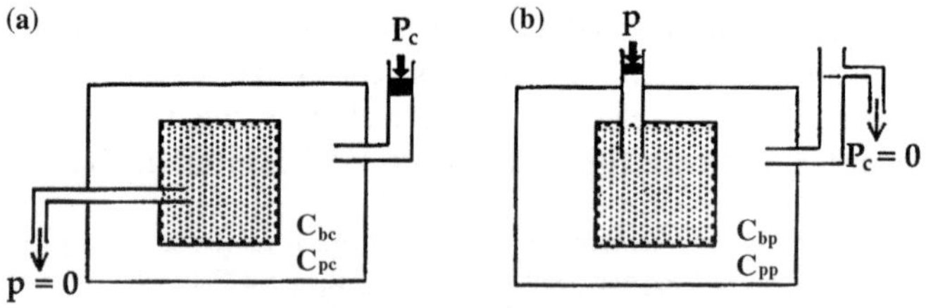

Figure 3.2: Confining and pore-pressure changes associated with various compressibilities (after Kümpel, 1991). (a) C_{bc} and C_{pc} are associated with $P_c \neq 0$ and $p = 0$. (b) C_{bp} and C_{pp} are associated with $p \neq 0$ and $P_c = 0$.

of pressure change (c is confining pressure and p is pore pressure). The coefficients C_{bp} and C_{pp} are the bulk volume expansion coefficient and pore volume expansion coefficient, respectively, in the terminology of this book. Zimmerman et al.'s (1986) compressibilities are related to previously defined compressibilities:

$$C_{bc} = \frac{1}{K} \tag{3.22}$$

$$C_{bp} = \frac{\alpha}{K} \tag{3.23}$$

$$C_{pc} = \frac{1}{K_p} \tag{3.24}$$

$$C_{pp} = \frac{\beta}{K_p} \tag{3.25}$$

The following relationships are also easily derived (Kümpel, 1991, Eqns. 24–26, p. 789):

$$\frac{1}{K_s'} = C_{bc} - C_{bp} \tag{3.26}$$

$$\frac{1}{K_\phi} = C_{pc} - C_{pp} \tag{3.27}$$

$$C_{pc} = \frac{1}{\phi}\left(\frac{1}{K} - \frac{1}{K_s'}\right) \tag{3.28}$$

$$C_{pp} = \frac{1}{\phi}\left(\frac{1}{K} - \frac{1}{K_s'}\right) - \frac{1}{K_\phi} \tag{3.29}$$

3.2 DRAINED VERSUS UNDRAINED MODULI

The importance of poroelastic coupling in a physical situation depends on the rate of pore fluid movement relative to the rate of change of stress conditions. The drained and undrained cases are the limiting cases of slow and fast loading, respectively. Relatively slow loading leaves the fluid pressure unchanged in an REV because fluid flow has adequate time to equilibrate with an external boundary. On the other hand, little fluid flows into or out of an REV if the loading is rapid. Different poroelastic moduli characterize the drained and undrained responses, and the definition of an elastic modulus, such as bulk modulus, Poisson's ratio, or Lamé's constant, must include a drained or undrained constraint in addition to the constraints for stress and strain on the boundaries of the REV.

3.2.1 Bulk Modulus

The relationship between the drained and undrained bulk moduli was given in Eqn. 2.22. Because α and B are less than one, K_u is greater than K. The bulk porous material is stiffer for undrained conditions because the fluid resists compression as well as the skeletal frame. The induced pore pressure opposes the applied stress that produced it.

3.2.2 Poisson's Ratio

Poisson's ratio in solid mechanics is defined as the negative of the ratio of lateral to longitudinal strain under axial loading. In poroelasticity, as was the case for bulk modulus, a drained or undrained condition must be stated explicitly in the definition. The drained Poisson's ratio for an axial stress σ_{ii} is defined by

$$\nu = -\frac{\epsilon_{jj}}{\epsilon_{ii}}\bigg|_{p=0;\,\sigma_{jj}=0} \qquad (i \neq j) \tag{3.30}$$

The undrained Poisson's ratio is defined by

$$\nu_u = -\frac{\epsilon_{jj}}{\epsilon_{ii}}\bigg|_{\zeta=0;\,\sigma_{jj}=0} \qquad (i \neq j) \tag{3.31}$$

The undrained Poisson's ratio can be expressed in terms of the drained Poisson's ratio by substituting $p = -B(\sigma_{kk}/3)$ into Eqn. 2.36 and solving for the strain ratio in Eqn. 3.31 for an axial stress in one direction (e.g., $\sigma_{11} \neq 0$ and $\sigma_{22} = \sigma_{33} = 0$):

$$\nu_u = \frac{3\nu + \alpha B(1 - 2\nu)}{3 - \alpha B(1 - 2\nu)} \tag{3.32}$$

Solving Eqn. 3.32 for $\alpha B(1 - 2\nu)$ gives a relation useful in the derivation of the plane strain equations (cf. Section 7.1):

$$\frac{\alpha B(1 - 2\nu)}{3} = \frac{\nu_u - \nu}{1 + \nu_u} \tag{3.33}$$

The undrained Poisson's ratio is larger than the drained value, because an increase in fluid pressure (1) decreases the vertical strain and (2) increases the unconstrained lateral strain. Inverting Eqn. 3.32 yields

$$\nu = \frac{3\nu_u - \alpha B(1 + \nu_u)}{3 - 2\alpha B(1 + \nu_u)} \tag{3.34}$$

The undrained Poisson's ratio (Eqn. 3.32) and undrained bulk modulus (or undrained compressibility) (Eqn. 2.22) are a sufficient set for any problem characterized by undrained conditions, just as the drained Poisson's ratio and drained bulk modulus are a sufficient set for drained conditions. The drained Poisson's ratio, undrained Poisson's ratio, drained compressibility, and undrained compressibility form a complete set of poroelastic constants. Alternatively, the Biot-Willis parameter α and Skempton's coefficient B can be chosen as the two poroelastic constants in addition to the drained constants K and ν, because they are the parameters used in Eqns. 2.22 and 3.32 that relate the drained and undrained bulk moduli and Poisson's ratios.

3.3 STORAGE CAPACITY

Storage capacity is a measure of the amount of fluid that must be added to or removed from an REV to effect a given fluid pressure change. Storage capacity is sensitive to mechanical constraints on the REV itself, which leads to the definition of several different storage coefficients. In addition, a minor confusion can arise because the hydrogeologic definition of specific storage is in terms of head rather than pressure, which requires the hydrogeologic storage coefficient to be divided by the factor $\rho_f g$ to yield its poroelastic equivalent. The symbols that follow are the ones chosen for the poroelastic case. The three commonly used measures of storage capacity are the uniaxial or one-dimensional specific storage coefficient, S; the unconstrained or three-dimensional specific storage coefficient, S_σ; and the constrained specific storage coefficient, S_ϵ. Each storage coefficient, like any poroelastic constant, is defined in terms of a representative elementary volume. Their values can, of course, vary spatially.

3.3.1 Unconstrained Specific Storage

The unconstrained specific storage has already been discussed as one of Biot's primary poroelastic coefficients (cf. Eqns. 1.9 and 2.19). An important micromechanical relationship can be found by substituting $S_\sigma = \alpha/KB$ (Eqn. 2.19) into Eqn. 3.16 and solving for S_σ:

$$S_\sigma = \left(\frac{1}{K} - \frac{1}{K_s'}\right) + \phi\left(\frac{1}{K_f} - \frac{1}{K_\phi}\right) \tag{3.35}$$

For incompressible solid grains and pores, Eqn. 3.35 reduces to $S_\sigma = 1/K + \phi/K_f$, which resembles the Jacob equation for one-dimensional storage but with bulk compressibility in place of vertical compressibility.

The hydrogeologic three-dimensional specific storage S' is defined to be the volume of water released from storage per unit control volume per unit decline in head *while holding the mean applied stress constant* (Van der Kamp and Gale, 1983; Green and Wang, 1990; Kümpel, 1991; Wang, 1993). Hence, S' is simply proportional to the unconstrained specific storage S_σ:

$$S' = (\rho_f g) S_\sigma \tag{3.36}$$

3.3.2 Constrained Specific Storage

A second specific storage coefficient, S_ϵ, is the reciprocal of Biot's modulus, M (cf. Eqn.1.21). It is the volume of fluid released from storage per unit control volume per unit pressure decline *while holding the control volume constant*:

$$S_\epsilon = \frac{1}{M} \equiv \left.\frac{\delta\zeta}{\delta p}\right|_{\epsilon=0} \tag{3.37}$$

Constraining the REV to be at constant volume means that S_ϵ is smaller than S_σ, because the volume of fluid released from storage is due mainly to the fluid compressibility. Changes in pore and solid-grain volume are generally secondary in magnitude. Several different expressions for S_ϵ are presented:

1. In Chapter 1, S_ϵ was shown to be equal to $S_\sigma - K/H^2$ (Eqn. 1.23). Because $\alpha = K/H$ (Eqn. 1.24),

$$S_\epsilon = S_\sigma - \frac{\alpha^2}{K} \tag{3.38}$$

2. Another expression given previously (Eqn. 2.27) is that $S_\epsilon = \alpha/K_u B$. Because $S_\sigma = \alpha/KB$ (Eqn. 2.19) (see also Palciauskas and Domenico, 1989, p. 210),

$$\frac{S_\epsilon}{S_\sigma} = \frac{K}{K_u} \tag{3.39}$$

3. Because $K_u = K/(1 - \alpha B)$ (Eqn. 2.22),

$$S_\epsilon = S_\sigma(1 - \alpha B) \tag{3.40}$$

   Also, from Eqns. 3.38 and 3.39,

$$S_\epsilon = \frac{\alpha^2}{K_u - K} \tag{3.41}$$

4. The constrained storage coefficient S_ϵ can be expressed in terms of compressibilities using Eqns. 3.8, 3.35, and 3.38:

$$S_\epsilon = \frac{1}{K_s'}\left(1 - \frac{K}{K_s'}\right) + \phi\left(\frac{1}{K_f} - \frac{1}{K_\phi}\right) \tag{3.42}$$

   Assuming that $K_s' = K_\phi = \infty$ reduces Eqn. 3.42 to $S_\epsilon = \phi/K_f$. As $1/K$ approaches $1/K_s'$, which would be expected at greater depths in the earth, S_σ also approaches ϕ/K_f, an approximation for storage coefficient that is used often in petroleum engineering. Thus, in this limit the constrained and unconstrained storage coefficients approach each other.

5. Finally, M, the reciprocal of S_ϵ, can be expressed (Detournay and Cheng, 1993, Eqn. 52b, p. 129) as

$$M = \frac{2G(\nu_u - \nu)}{\alpha^2(1 - 2\nu_u)(1 - 2\nu)} \tag{3.43}$$

3.3.3 Uniaxial Specific Storage

The hydrogeologic definition of specific storage is the volume of water released per unit decline of head per unit bulk volume while maintaining the REV in a state of zero lateral strain and constant vertical stress (Fig. 3.3):

$$S_s \equiv (\rho_f g)\frac{\delta\zeta}{\delta p}\bigg|_{\epsilon_{xx}=\epsilon_{yy}=0;\ \sigma_{zz}=0} \tag{3.44}$$

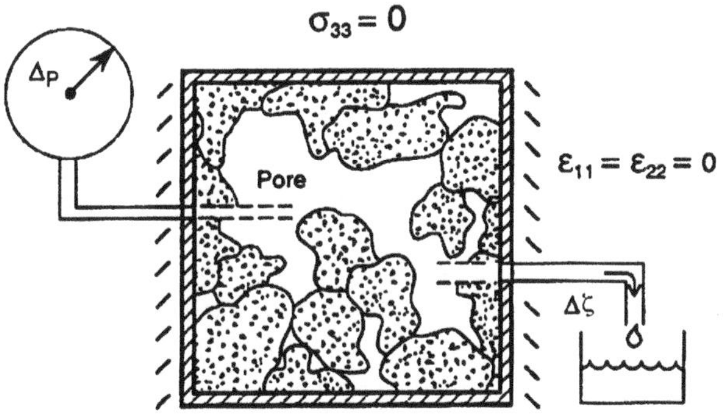

Figure 3.3: Schematic illustration of the definition of uniaxial specific storage.

The equivalent storage parameter in terms of pressure is

$$S \equiv \frac{S_s}{\rho_f g} \tag{3.45}$$

The two strain and one stress conditions in Eqn. 3.44 are three constraints among the four constitutive equations that determine the ratio $\delta\zeta/\delta p$. First, the constraints $\epsilon_{xx} = \epsilon_{yy} = 0$ and $\sigma_{zz} = 0$ are inserted in Eqns. 2.30 and 2.31 to obtain expressions for σ_{xx} and σ_{yy}. Then the normal stresses are summed to yield

$$\sigma_{kk} = -4\eta p \tag{3.46}$$

where

$$\eta = \frac{1-2\nu}{2(1-\nu)}\alpha \tag{3.47}$$

The dimensionless variable η has been called the *poroelastic stress coefficient* by Detournay and Cheng (1993, p. 129). Eqn. 3.46 shows that the change in mean stress is a scalar multiple of the change in pore pressure for the special case of zero horizontal strain and constant vertical stress (Van der Kamp and Gale, 1983). This result means that the stress field *uncouples* from the pore pressure field (cf. Chapter 6).

An expression is obtained for uniaxial specific storage by substituting Eqn. 3.46 into Eqn. 2.41, using Eqn. 2.19, and taking the ratio of ζ to p:

$$S = S_\sigma\left(1 - \frac{4\eta B}{3}\right) \tag{3.48}$$

Thus, for conditions of uniaxial strain and constant vertical stress,

$$\zeta = Sp \tag{3.49}$$

Eqn. 3.48 shows that the one-dimensional or uniaxial specific storage is smaller than the three-dimensional or unconstrained specific storage. This result is expected on physical grounds, because the lateral constraint on the REV restricts the amount of pore volume decrease when pressure is reduced, which in turn reduces the amount of fluid released from storage.

The relationship $S_\epsilon = S_\sigma(1 - \alpha B)$ (Eqn. 3.40) shows that $S_\epsilon < S_\sigma$. Eqn. 3.48 shows that $S < S_\sigma$. It can be demonstrated that $\alpha B > 4\eta B/3$ if $\nu > -1$; this constraint for ν is well known in the theory of elasticity (e.g., Jaeger and Cook, 1976, p. 112). Therefore, the magnitudes of the three different storage coefficients are in the following order:

$$S_\sigma \geq S \geq S_\epsilon \tag{3.50}$$

The constant-stress storage coefficient is greatest because the least restraint is placed on the framework as fluid pressure is decreased. The constant-strain storage coefficient is smallest because the volume is held constant, which means that the volume of fluid released is due primarily to fluid compressibility. The uniaxial specific storage, S, lies in between because only lateral strain is held constant. These three different storage coefficients illustrate not only the general nature of storage—release of fluid due to fluid pressure decline—but also the importance of the defining stress or strain boundary conditions on the control volume. The distinction between uniaxial and unconstrained specific storage is analogous to the distinction between vertical compressibility and bulk compressibility.

The generalization of Jacob's expression (Eqn. 1.3) for the one-dimensional specific storage in terms of the vertical compressibility and fluid compressibility is obtained by inserting $B = \alpha/KS_\sigma$ (cf. Eqn. 2.19) and Eqns. 3.8, 3.47, and 3.35 into Eqn. 3.48 and simplifying:

$$S = \frac{\alpha^2}{K_v} + \phi\left(\frac{1}{K_f} - \frac{1}{K_\phi}\right) \tag{3.51}$$

The poroelastic form of Jacob's equation is recovered from Eqn. 3.51 under the assumptions that the solid grains and pores are incompressible ($1/K_s' = 1/K_\phi = 0$):

$$S = \frac{1}{K_v} + \phi\frac{1}{K_f} \tag{3.52}$$

The ratio of K/K_s' is between 0.2 and 0.3 for Berea sandstone and Indiana limestone at an effective stress of 10 MPa (Appendix C, Table C.1) in which case the incompressible grain assumption leads to an error of approximately 35 to 50% in the calculated value of S.

The uniaxial specific storage can be expressed in terms of S_ϵ times a factor containing ν_u and ν (Detournay and Cheng, 1993, Eqn. 53, p. 129):

$$S = S_\epsilon \frac{1-\nu_u}{1-\nu}\frac{1-2\nu}{1-2\nu_u} \tag{3.53}$$

$$= S_\epsilon \frac{K_v^{(u)}}{K_v} \tag{3.54}$$

where

$$\frac{1}{K_v} \equiv \frac{1}{3K}\frac{1+\nu}{1-\nu} = \left(K + \frac{4G}{3}\right)^{-1} \tag{3.55}$$

and

$$\frac{1}{K_v^{(u)}} \equiv \frac{1}{3K_u}\frac{1+\nu_u}{1-\nu_u} \tag{3.56}$$

K_v and $K_v^{(u)}$ are often called the drained and undrained vertical incompressibility, respectively (cf. Appendix B, Table B.1). More generally, however, $1/K_v$ and $1/K_v^{(u)}$ are one-dimensional or uniaxial compressibilities that are not restricted to apply only for the vertical direction. They are the ratio of longitudinal strain to longitudinal stress for any direction under the assumption that the orthogonal strains are zero; hence, the longitudinal strain and volumetric strain are equal. Undrained conditions are thought to apply during the propagation of a compressional wave whose wave velocity is then given by $[K_v^{(u)}/\rho]^{1/2}$, where ρ is the bulk density of the material (Gassmann, 1951).

An equation similar to Eqn. 3.54 is

$$S = S_\sigma \frac{1-\nu_u}{1+\nu_u}\frac{1+\nu}{1-\nu} \tag{3.57}$$

$$= S_\sigma \frac{K}{K_u}\frac{K_v^{(u)}}{K_v} \tag{3.58}$$

In the limit that $\nu_u = \nu$, the three specific storage coefficients, S, S_σ, and S_ϵ are equal.

In Section 3.6.1 (cf. Eqn. 3.84), the uniaxial specific storage is expressed in a form similar to $S_\sigma = \alpha/KB$ (Eqn. 2.19) and $S_\epsilon = \alpha/K_u B$ (Eqn. 2.27) as

$$S = \frac{\alpha}{K_v \gamma} \tag{3.59}$$

where γ, the loading efficiency, can be thought of as a uniaxial Skempton's coefficient. Eqns. 2.19, 2.27, and 3.59 express the similar relationships among various storage coefficients, compressibilities, and undrained pore pressure buildup coefficients.

3.3.4 Unjacketed Specific Storage

Biot and Willis (1957) introduced a coefficient for an unjacketed test, which they called the *coefficient of fluid content*. It is the amount of fluid added per unit bulk volume per unit change in fluid pressure under conditions of constant differential pressure:

$$S_\gamma \equiv \left.\frac{\delta\zeta}{\delta p}\right|_{P_d=0} \tag{3.60}$$

Like the other storage coefficients in this section, the definition of S_γ is the ratio of the increment of fluid content to the change in pore pressure. Because the boundary condition is that the REV be maintained at constant differential pressure, it is called the *unjacketed storage coefficient* in this book. Substituting Eqns. 3.3 and 3.11 into Eqn. 3.10 yields

$$S_\gamma = \frac{\phi}{K_f} - \frac{\phi}{K_\phi} \tag{3.61}$$

If $1/K_\phi = 0$, then $S_\gamma = \phi/K_f$, because no change in pore volume means that the only fluid added to or withdrawn from the REV is due to self-compression of the fluid. Another expression for S_γ is (Detournay and Cheng, 1993, Eqn. 50, p. 127):

$$S_\gamma = S_\epsilon - \frac{\alpha(1-\alpha)}{K} \tag{3.62}$$

3.4 HYDRAULIC DIFFUSIVITY

The consolidation coefficient c in Eqn. 1.1 is shown in Chapter 4 to be also the diffusivity governing the diffusion of increment of fluid content (or, equivalently, fluid mass per unit volume of porous material). The derivation shows c to be directly proportional to the permeability and inversely proportional to the uniaxial specific storage. Higher permeabilities allow faster movement of fluid, whereas higher storage capacity requires more fluid to be moved. Three hydraulic diffusivities corresponding to S_σ, S_ϵ, and S can be defined:

$$c_\sigma \equiv \frac{k}{\mu S_\sigma} \tag{3.63}$$

$$c_\epsilon \equiv \frac{k}{\mu S_\epsilon} \tag{3.64}$$

$$c \equiv \frac{k}{\mu S} \tag{3.65}$$

The value of c is intermediate between c_σ and c_ϵ. The relative magnitude of the hydraulic diffusivities is $c_\sigma \leq c \leq c_\epsilon$ because $S_\sigma \geq S \geq S_\epsilon$. Hydraulic diffusivity values calculated from typical corresponding values of permeability and uniaxial specific storage are shown in Table 3.1.

TABLE 3.1.
Typical Values for Permeability, Specific Storage, and Hydraulic Diffusivity for Several Rock Types*

Permeability, k[m^2]	Specific Storage, S_s[m^{-1}]	Specific Storage, $S = S_s/(\rho_f g)$ [Pa^{-1}]	Hydraulic Diffusivity, c [m^2/s]	Rock Type
10^{-12}	10^{-5}	10^{-9}	1	Sand or sandstone
10^{-15}	10^{-6}	10^{-10}	10^{-2}	Sandstone or limestone
10^{-18}	10^{-7}	10^{-11}	10^{-4}	Granite

*Specific storage is given both for changes of pressure and for changes of head. Conversions shown assume that the density of water is 1000 kg/m^3, viscosity of water is 10^{-3} Pa·s, and acceleration of gravity is 10 m/s^2.

3.5 POROELASTIC EXPANSION COEFFICIENTS

Pore pressure in poroelasticity plays the same role as temperature in thermoelasticity. A volume increase in an unconstrained body accompanies an increase in pore pressure just as it would for an increase in temperature. Therefore, in keeping with this analogy, the physical property governing the volume increase in poroelasticity is designated the *volumetric poroelastic expansion coefficient*. An additional poroelastic expansion coefficient, Geertsma's parameter, can be defined for the volume increase due to a pore pressure increase in a uniaxially constrained body. This coefficient appears in the governing equation for volumetric strain (Eqn. 4.20).

3.5.1 Biot's Volumetric Expansion Coefficient

Biot's constant $1/H$ in Eqn. 1.11 is the poroelastic expansion constant governing the volumetric strain due to a change in pore pressure taking place at constant mean stress. As a result of symmetry, $1/H$ also appears in Eqn. 1.12 as the ratio of ζ to change in mean stress at constant pore pressure. Therefore, it also can be thought of as a type of storage coefficient, but one associated with fluid added to storage due to a change in mean stress rather than fluid pressure.

Rearranging Eqn. 3.7 shows that the poroelastic expansion coefficient is the difference between the bulk compressibility and the unjacketed compressibility:

$$\frac{1}{H} = \frac{\alpha}{K} = \frac{1}{K} - \frac{1}{K_s'} \tag{3.66}$$

For incompressible grains, $1/H = 1/K$. Finally, Eqn. 3.66 gives an expression for α:

$$\alpha = 1 - \frac{K}{K_s'}. \tag{3.67}$$

The Biot-Willis parameter α is the ratio of increment of fluid content to change in bulk volume when the pore fluid remains at constant pressure (cf. Eqn. 2.26). It would be exactly one if all of the bulk strain were due to pore volume change (i.e., the solid phase is incompressible). It is less than one for a compressible solid phase because the change in bulk volume is greater than the change in pore volume by the amount of the change in the solid volume.

Another expression for the poroelastic expansion coefficient is given by combining Eqn. 3.66 with a simple rearrangement of Eqn. 3.13:

$$\frac{1}{H} = \frac{\phi}{K_p} \tag{3.68}$$

3.5.2 Geertsma's Uniaxial Expansion Coefficient

A poroelastic expansion coefficient analogous to $1/H$ can be defined for a uniaxially constrained REV. For $\epsilon_{xx} = \epsilon_{yy} = 0$, Eqn. 2.46 becomes

$$\sigma_{zz}\Big|_{\epsilon_{xx}=\epsilon_{yy}=0} = \frac{2G(1-\nu)}{1-\nu}\epsilon_{zz} - \alpha p \tag{3.69}$$

Solving for ϵ_{zz} gives

$$\epsilon_{zz} = \frac{1}{K_v}\sigma_{zz}\Big|_{\epsilon_{xx}=\epsilon_{yy}=0} + \frac{\alpha}{K_v}p \tag{3.70}$$

Therefore, for zero vertical stress, the vertical and hence volumetric strain is proportional to the pore pressure change p. The proportionality constant is α/K_v, which was introduced by Geertsma (1966) as the coefficient c_m:

$$c_m \equiv \frac{\alpha}{K_v} \tag{3.71}$$

The coefficient c_m is designated here as *Geertsma's uniaxial poroelastic expansion coefficient,*[3] because it represents the ratio of axial strain to pore

[3] Geertsma's uniaxial poroelastic coefficient is sometimes called a compressibility because it has the units of compressibility. However, the term *expansion coefficient* is chosen on the basis of the thermoelastic analogy to poroelasticity.

pressure change while maintaining a condition of uniaxial strain. Geertsma's coefficient can also be expressed in terms of the poroelastic stress coefficient η (cf. Eqn. 3.47):

$$c_m = \frac{\eta}{G} \tag{3.72}$$

3.6 COEFFICIENTS OF UNDRAINED PORE-PRESSURE BUILDUP

The sudden application of a compressive stress induces a pore pressure increase. Two widely used pore pressure buildup coefficients are Skempton's coefficient and loading efficiency.

3.6.1 Skempton's Coefficient

In the undrained limit, the induced pore pressure due to a sudden change in the mean stress is given by Skempton's coefficient (cf. Eqns. 1.19 and 2.16). A laboratory determination of Skempton's coefficient requires measuring the change in fluid pressure after changing the applied stress (confining pressure) while sealing the fluid completely within the sample (Fig. 3.4). A pore

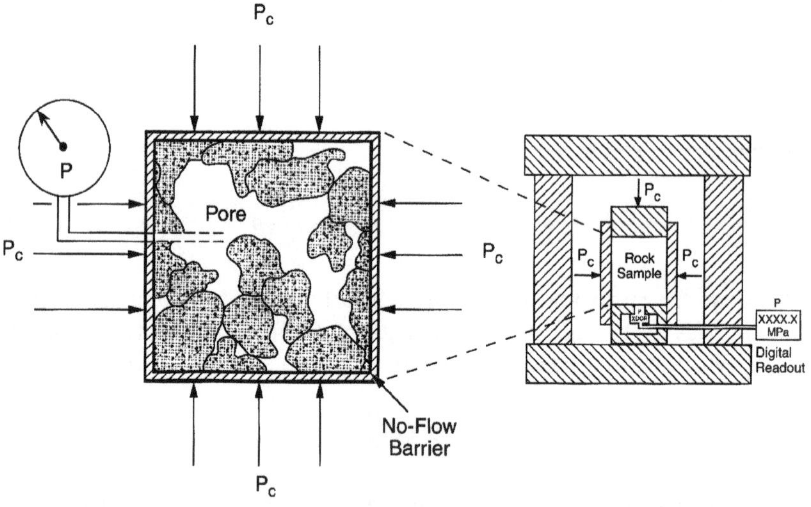

Figure 3.4: Skempton's coefficient measured in a pressure vessel (after Green and Wang, 1986). The ratio of pore pressure to confining pressure is obtained under undrained conditions by completely surrounding the sample with an impermeable jacket.

pressure transducer is placed in direct contact with the sample to eliminate corrections for fluid compression in tubing (Green and Wang, 1986; Berge et al., 1993).

An expression for Skempton's coefficient in terms of the different compressibilities can be obtained by solving Eqn. 3.17 for B:

$$B = \frac{1/K - 1/K_s'}{1/K - 1/K_s' + \phi(1/K_f - 1/K_\phi)} \tag{3.73}$$

Skempton's coefficient and the various storage coefficients depend on the porosity and fluid compressibility. If air fills the pores ($1/K_f \rightarrow \infty$), then Skempton's coefficient is nearly zero, whereas if water fills the pores, Skempton's coefficient is typically between 0.5 and 1.0 for fluid-saturated rock and close to 1.0 for fluid-saturated soil.

Another useful expression for Skempton's coefficient is obtained by solving Eqn. 2.22 for B and substituting Eqn. 3.67 for α:

$$B = \frac{1 - K/K_u}{1 - K/K_s'} \tag{3.74}$$

Also, Eqn. 3.33 yields

$$B = \frac{3(\nu_u - \nu)}{\alpha(1 + \nu_u)(1 - 2\nu)} \tag{3.75}$$

3.6.2 Loading Efficiency

A second pore pressure buildup coefficient is the loading efficiency, γ, which is an aquifer parameter that can be inferred from well response to barometric or ocean tide loading (Domenico and Schwartz, 1998). Loading efficiency is an undrained poroelastic coefficient that gives the pore pressure increase due to a compressive axial stress increase while maintaining the REV in a state of uniaxial strain:

$$\gamma \equiv -\frac{\delta p}{\delta \sigma_{zz}}\bigg|_{\epsilon_{xx}=\epsilon_{yy}=\zeta=0}. \tag{3.76}$$

The definition includes a negative sign because a positive (tensile) increase in axial stress produces a decrease in pore pressure. For undrained, uniaxial strain conditions, setting $\epsilon_{xx} = \epsilon_{yy} = 0$ in Eqn. 2.53 yields the horizontal stresses

$$\sigma_{xx}\big|_{\epsilon_{xx}=\epsilon_{yy}=\zeta=0} = \sigma_{yy}\big|_{\epsilon_{xx}=\epsilon_{yy}=\zeta=0} = \frac{\nu_u}{1-\nu_u}\sigma_{zz} \tag{3.77}$$

Eqn. 3.77 shows that the ratio of lateral stress to axial stress is $\nu_u/(1-\nu_u)$ for undrained loading. Eqn. 3.77 is the same as the expression for drained loading except that undrained Poisson's ratio replaces drained Poisson's ratio. The mean stress is obtained by summing the axial and lateral stresses:

$$\left.\frac{\sigma_{kk}}{3}\right|_{\epsilon_{xx}=\epsilon_{yy}=\zeta=0} = \frac{1}{3}\frac{1+\nu_u}{1-\nu_u}\sigma_{zz}$$

The loading efficiency can, therefore, be expressed in terms of Skempton's coefficient and undrained Poisson's ratio as

$$\gamma = \frac{B(1+\nu_u)}{3(1-\nu_u)} \tag{3.78}$$

Thus, the loading efficiency is determined by Skempton's coefficient times a factor representing the uniaxial strain constraint. This factor also can be expressed as a ratio of undrained bulk modulus to undrained vertical incompressibility $K_v^{(u)}$ (cf. Appendix B, Table B.1) so that Eqn. 3.78 becomes

$$\gamma = B\frac{K_u}{K_v^{(u)}} \tag{3.79}$$

where

$$K_v^{(u)} = G\frac{2-2\nu_u}{1-2\nu_u} \tag{3.80}$$

The difference between drained and undrained vertical compressibility is shown in Chapter 6 (cf. Eqn. 6.45) to be

$$\frac{1}{K_v} - \frac{1}{K_v^{(u)}} = \frac{\alpha}{K_v}\gamma \tag{3.81}$$

An equation similar to Eqn. 3.74 can be obtained from Eqn. 3.81:

$$\gamma = \frac{1-K_v/K_v^{(u)}}{1-K/K_s'} \tag{3.82}$$

Another expression for γ, which can be obtained after some algebra, is

$$\gamma = \frac{\eta}{GS} \tag{3.83}$$

Using Eqn. 3.72 in Eqn. 3.83 gives

$$\gamma = \frac{c_m}{S} = \frac{\alpha}{K_v S} \tag{3.84}$$

Finally, applying Eqn. 3.54 for S in Eqn. 3.84 gives

$$\gamma = \frac{\alpha}{K_v^{(u)} S_\epsilon} \tag{3.85}$$

Example: *Tidal Efficiency*

The uniaxial strain condition is often assumed to apply to the effect of ocean tidal loading on an aquifer close to the shoreline (cf. Fig. 1.4). The loading efficiency is then equal to the tidal efficiency, T.E., defined by Jacob (1940) to be the ratio of ocean water level changes divided by water level changes in a well:

$$\text{T.E.} = \gamma \tag{3.86}$$

A micromechanical expression for γ in the limit of incompressible grains and pores can be obtained from Eqns. 3.52 and 3.84 (Domenico and Schwartz, 1998, pp. 160–162; Domenico, 1983, pp. 283–287):

$$\text{T.E.} = \frac{1/K_v}{1/K_v + \phi/K_f} = \frac{K_f}{K_f + \phi K_v} \tag{3.87}$$

Therefore, the specific storage can be estimated from a measurement of T.E. if ϕ and K_f are known, because Eqn. 3.87 can be solved for K_v and substituted into Eqn. 3.52.

Example: *Barometric Efficiency*

Barometric efficiency, B.E., is defined to be the negative of $\rho_f g$ times the ratio of the change in water level δh in a well to the change in atmospheric pressure δP_{atm}:

$$\text{B.E.} = -(\rho_f g)\frac{\delta h}{\delta P_{\text{atm}}} \tag{3.88}$$

The atmospheric load not only exerts an external load on the surface but also directly pushes down the water level in an open well by an amount $\delta P_{\text{atm}}/(\rho_f g)$. The net change in water level is the decrease due to the depression of the water in the well plus the increase due to the compression of the aquifer from the load. Therefore, the barometric efficiency for an open well is

$$\text{B.E.} = 1 - \gamma \tag{3.89}$$

An example time series of the barometric loading effect is shown in Figure 3.5. In the limit of incompressible grains and pores, Eqns. 3.87

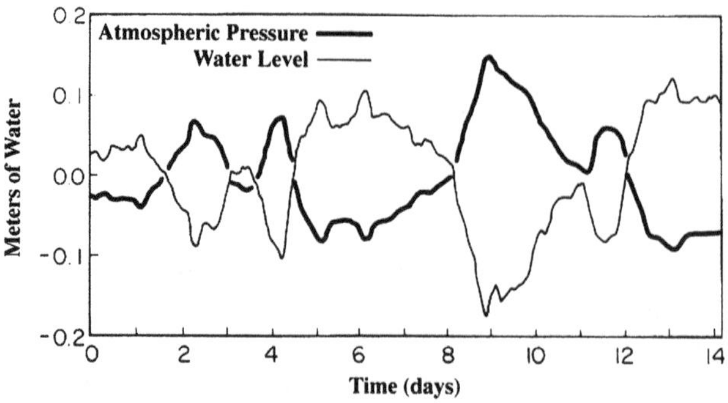

Figure 3.5: Atmospheric pressure versus water level in a well relative to a datum (after Furbish, 1991). The well is located at Maxey Flats near Morehead, Kentucky, in a thin, fractured sandstone. The static-confined barometric efficiency of this well is about 0.73.

and 3.89 give

$$\text{B.E.} = \frac{\phi/K_f}{1/K_v + \phi/K_f} = \frac{\phi K_v}{K_f + \phi K_v} \tag{3.90}$$

Finally, assuming incompressible grains only ($K_s' = \infty$ in Eqn. 3.82) gives

$$\text{B.E.} = \frac{K_v}{K_v^{(u)}} \tag{3.91}$$

As was the case for tidal efficiency, the specific storage can be estimated from a measurement of B.E. if ϕ and K_f are known. Independent measurements of both T.E. and B.E. in a well, in theory, yield S_s and ϕ.

3.7 INCOMPRESSIBLE OR HIGHLY COMPRESSIBLE CONSTITUENTS

In this chapter REV-scale micromechanical relations have been derived that express macroscopic poroelastic constants in Biot's constitutive equations in terms of compressibilities of the solid, pore, and fluid constituents of the porous material. Simplified expressions are tabulated in Table 3.2 for special cases in which the constituent compressibilities are extremely low or high (e.g., a highly compressible gas as the pore fluid). These expressions can be used to estimate the macroscopic poroelastic moduli when the assumptions are appropriate. In Table 3.2 a poroelastic material with an incompressible solid constituent is defined by both $K/K_s' \ll 1$ and $K/K_\phi \ll 1$, and one

TABLE 3.2.
Poroelastic Constants for Highly Compressible or Incompressible Constituents (Detournay and Cheng, 1993, p. 122)

Case	α	K_u	B	M
Incompressible solid ($K/K_s' \ll 1$; $K/K_\phi \ll 1$)	1	$K\left(1+\frac{K_f}{\phi K}\right)$	$\frac{K_f}{\phi K + K_f}$	$\frac{K_f}{\phi}$
Incompressible solid and infinitely incompressible fluid (above and $K_f \to \infty$)	1	∞	1	∞
Highly compressible fluid ($K_f/K \ll 1$)	$1-\frac{K}{K_s'}$	$K\left(1+\frac{\alpha^2 K_f}{\phi K}\right)$	$\frac{\alpha K_f}{\phi K}$	$\frac{K_f}{\phi}$
Infinitely compressible fluid ($K_f = 0$)	$1-\frac{K}{K_s'}$	K	0	0

containing an incompressible fluid is defined by $K/K_f \ll 1$. An infinitely incompressible fluid is defined by $K_f \to \infty$, and an infinitely compressible fluid is defined by $K_f = 0$.

For the case of an incompressible solid and infinitely incompressible fluid, the poroelastic moduli take on their upper-bound values: $\alpha = B = 1$, $K_u = M = \infty$, and $\nu_u = 0.5$. Porous materials with these constituent values display the strongest possible degree of poroelastic coupling.

For an infinitely compressible fluid, Skempton's coefficient is zero, and the fluid pressure of the porous material is uncoupled from changes in stress. All storage coefficients are infinite because no finite amount of fluid added or removed will change the pore pressure. The fact that the material has a nonzero poroelastic expansion coefficient is moot because the pore pressure cannot be changed. Overall, the porous material behaves like an elastic material without fluid—that is, $K_u \to K$ and $\nu_u \to \nu$.

3.8 LABORATORY DATA

A complete set of poroelastic constants for an isotropic porous material requires the measurement of four independent coefficients. Few direct determinations have been made of all the poroelastic moduli on the same rock under the same pore pressure and confining pressure conditions. Estimates, however, can be made from two drained moduli (e.g., E and ν), K_s, K_f and ϕ (Appendix C, Table C.1). These parameters are sufficient to compute all the remaining coefficients with the assumption that $K_s = K_s' = K_\phi$; the

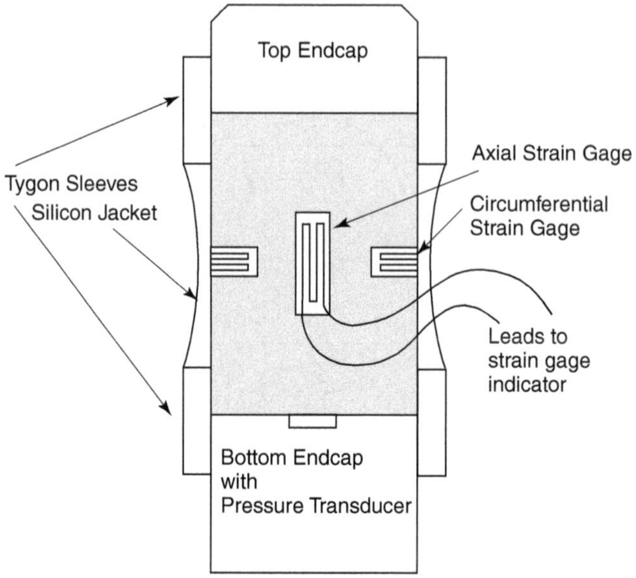

Figure 3.6: Sample assembly for the measurement of undrained poroelastic constants K_u, E_u, ν_u, and B (Hart and Wang, 1995). Strain gages are bonded directly on the sample before flexible jacketing material encases the assembly to exclude confining fluid from the interior of the sample.

solid-grain modulus is taken from handbooks of mineral properties. A particular fluid compressibility is always assumed in reporting undrained moduli. The Berea sandstone and Indiana limestone values by Hart and Wang (1995) in Table C.1 are based on measurements of eight different poroelastic moduli measured at an effective stress of 10 MPa. The eight measured moduli can be grouped into three sets: three drained measurements (K, E and ν), four undrained measurements (K_u, E_u, ν_u and B), and one unjacketed measurement (K_s'). The sample configuration for the undrained measurements is shown in Figure 3.6. A best-fit set was obtained by least-squares minimization. The four independent moduli were chosen to be K, E, K_u, and K_s'. The first two moduli characterize the drained state. The second two moduli allow computation of α and B in conjunction with K.

4

Governing Equations

4.0 CHAPTER OVERVIEW

The governing equations for fluid flow in a fully saturated, poroelastic medium are formulated in this chapter. The basic variables of poroelasticity are six stress components, three displacement components, pore pressure, and increment of fluid content. The six strain components are defined in terms of the three displacement components (cf. Eqn. 2.7). The eleven unknowns are determined from (1) seven constitutive equations—six for strain (or stress) components, and one for pore pressure (or increment of fluid content); (2) three force equilibrium equations—one for each coordinate direction; and (3) an inhomogeneous pressure diffusion equation, which contains the first time derivative of stress or strain, obtained by combining Darcy's law with the requirement of fluid continuity (conservation of fluid mass).

The mechanical problem is *elastostatic*, which means that static equilibrium obtains for each instant of time—that is, mechanical equilibrium is assumed to be attained instantaneously. In reality, a finite amount of time is required for a dynamic wave to transmit stress changes across the problem domain, but the wave propagation term is ignored in the quasistatic approximation. If a stress or fluid pressure change is suddenly applied to a poroelastic body, displacements and pore pressure within each representative elementary volume (REV) adjust instantaneously to accommodate this change to maintain a state of internal force equilibrium. Subsequent time-dependent fluid diffusion occurs as a result of the delaying effects of finite permeability and storage.

Coupling occurs among the equations because pore pressure (or increment of fluid content) appears in the force equilibrium equations, and because mean stress (or volumetric strain) appears in the fluid-flow equation. These coupled equations form the core of the theory of poroelasticity. The richness and variety of poroelastic phenomena arise from interactions between the mechanical requirement of force equilibrium and the fluid-flow requirement of continuity. The linear poroelastic problem requires that the mechanical equilibrium equations and the fluid continuity equation be satisfied simultaneously, and that initial and boundary conditions are specified.

The governing equations are, in general, difficult to solve. Methods developed in the standard theory of elasticity are complicated by the addition of the coupled pore pressure field. Some general methods of solution are described in Chapter 5 for unbounded domains. Simplifications occur for restricted geometries. Problems containing conditions of uniaxial strain, plane strain, radial symmetry, and axisymmetry are discussed in Chapters 6, 7, 8, and 9, respectively. Finite element and boundary element methods are outlined in Chapter 10.

4.1 FORCE EQUILIBRIUM EQUATIONS

The elastostatic governing equations are obtained from the requirement that an REV is in a state of force equilibrium at every instant in time. Specifically, an REV subjected to stresses on its faces must be in static translational and rotational equilibrium. Rotational equilibrium implies that the stress tensor is symmetric. For example, the clockwise and counterclockwise moments about the z-direction are equal to the shear forces times the lever arms (cf. Fig. 2.3):

$$\sigma_{yx}(\Delta x \Delta z)\Delta y = \sigma_{xy}(\Delta y \Delta z)\Delta x \tag{4.1}$$

Performing the same analysis for each coordinate direction yields the three symmetry equations:

$$\sigma_{xy} = \sigma_{yx} \tag{4.2}$$

$$\sigma_{yz} = \sigma_{zy} \tag{4.3}$$

$$\sigma_{zx} = \sigma_{xz} \tag{4.4}$$

The translational force balance equations, including a body force per unit bulk volume, $\vec{\mathbf{F}}$, can also be obtained with the help of Figure 2.3. Usually, the body force is the gravitational force per unit bulk volume, $\vec{\mathbf{F}} = (0, 0, -\rho g)$, where the z-axis is the vertical direction and is positive upward. The net force in the $+x$-direction must sum to zero:

$$\begin{aligned}&[\sigma_{xx}(x+\Delta x, y, z) - \sigma_{xx}(x, y, z)]\Delta y \Delta z + [\sigma_{yx}(x, y+\Delta y, z)\\&\quad - \sigma_{yx}(x, y, z)]\Delta x \Delta z + [\sigma_{zx}(x, y, z+\Delta z) - \sigma_{zx}(x, y, z)]\Delta x \Delta y\\&\quad + F_x(x, y, z)\Delta x \Delta y \Delta z = 0\end{aligned} \tag{4.5}$$

Each term in brackets can be approximated linearly by the derivative of the stress component times the distance across the REV:

$$\sigma_{xx}(x+\Delta x, y, z) - \sigma_{xx}(x, y, z) = \frac{\partial \sigma_{xx}}{\partial x}\Delta x \tag{4.6}$$

Performing this analysis for each coordinate direction yields the following three force equilibrium equations:

$$\frac{\partial \sigma_{xx}}{\partial x} + \frac{\partial \sigma_{yx}}{\partial y} + \frac{\partial \sigma_{zx}}{\partial z} + F_x = 0 \tag{4.7}$$

$$\frac{\partial \sigma_{xy}}{\partial x} + \frac{\partial \sigma_{yy}}{\partial y} + \frac{\partial \sigma_{zy}}{\partial z} + F_y = 0 \tag{4.8}$$

$$\frac{\partial \sigma_{xz}}{\partial x} + \frac{\partial \sigma_{yz}}{\partial y} + \frac{\partial \sigma_{zz}}{\partial z} + F_z = 0 \tag{4.9}$$

The three translational equilibrium equations (Eqns. 4.7–4.9) are written in compact notation as

$$\frac{\partial \sigma_{ji}}{\partial x_j} = -F_i \tag{4.10}$$

where the convention of summation of repeated indices is used. The symmetry of the stress tensor (Eqns. 4.2–4.4) is expressed by

$$\sigma_{ij} = \sigma_{ji} \tag{4.11}$$

4.2 DISPLACEMENT FORMULATION

Partial differential equations representing mechanical equilibrium are obtained by substituting the constitutive relations into the force equilibrium equations. Governing equations using displacement as the primary mechanical variable are derived in this section. Strains are recovered readily as various derivatives of displacement. A stress formulation is derived in Section 4.3. The governing equations must also contain either fluid pressure or increment of fluid content, depending on which variable is chosen as the coupling variable in the strain-stress constitutive equation (Eqn. 2.42 or 2.53, respectively). Equations for both displacement pressure and displacement increment of fluid content will be presented.

Three coupled partial differential equations containing the displacement components are obtained from the force balance equation in each of the three coordinate directions. The steps are first to substitute the constitutive equations for the stress components (Eqns. 2.44–2.49) into Eqns. 4.7–4.9, and then to apply the definitions of the strain components in terms of derivatives of displacements (Eqns. 2.1–2.6). Eqns. 4.7–4.9 become

$$G\nabla^2 u + \frac{G}{1-2\nu}\left[\frac{\partial^2 u}{\partial x^2} + \frac{\partial^2 v}{\partial x \partial y} + \frac{\partial^2 w}{\partial x \partial z}\right] = \alpha \frac{\partial p}{\partial x} - F_x \tag{4.12}$$

$$G\nabla^2 v + \frac{G}{1-2\nu}\left[\frac{\partial^2 u}{\partial y \partial x} + \frac{\partial^2 v}{\partial y^2} + \frac{\partial^2 w}{\partial y \partial z}\right] = \alpha \frac{\partial p}{\partial y} - F_y \tag{4.13}$$

$$G\nabla^2 w + \frac{G}{1-2\nu}\left[\frac{\partial^2 u}{\partial z \partial x} + \frac{\partial^2 v}{\partial z \partial y} + \frac{\partial^2 w}{\partial z^2}\right] = \alpha \frac{\partial p}{\partial z} - F_z \tag{4.14}$$

All three equations can be written compactly using index notation:

$$G\nabla^2 u_i + \frac{G}{1-2\nu}\frac{\partial^2 u_k}{\partial x_i \partial x_k} = \alpha \frac{\partial p}{\partial x_i} - F_i \tag{4.15}$$

The notation is that displacements in the x, y, and z directions are designated by u, v, and w, respectively—that is, $u_1 = u$, $u_2 = v$, and $u_3 = w$. Eqn. 4.15 would be identical to its counterpart in standard elasticity except for the additional pressure gradient term on the right-hand side. The fluid pressure gradient term is equivalent mathematically to a body force.

A concise governing equation for volumetric strain can be obtained from Eqn. 4.15. First, note that the volumetric strain can be expressed as the divergence of displacement:

$$\epsilon = \epsilon_{xx} + \epsilon_{yy} + \epsilon_{zz} = \frac{\partial u}{\partial x} + \frac{\partial v}{\partial y} + \frac{\partial w}{\partial z} \tag{4.16}$$

Eqn. 4.16 is written in index notation as

$$\epsilon = \epsilon_{kk} = \frac{\partial u_k}{\partial x_k} \tag{4.17}$$

Eqn. 4.15 can then be expressed as

$$G\nabla^2 u_i + \frac{G}{1-2\nu}\frac{\partial \epsilon}{\partial x_i} = \alpha \frac{\partial p}{\partial x_i} - F_i \tag{4.18}$$

Taking the derivative of Eqn. 4.18 with respect to x_i gives

$$G\nabla^2 \epsilon_{ii} + \frac{G}{1-2\nu}\frac{\partial^2 \epsilon}{\partial {x_i}^2} = \alpha \frac{\partial^2 p}{\partial {x_i}^2} - \frac{\partial F_i}{\partial x_i} \tag{4.19}$$

where $\epsilon_{ii} = \partial u_i / \partial x_i$ is the longitudinal strain parallel to the i-th coordinate direction (no implied summation). Summing the three equations represented by Eqn. 4.19 yields

$$\nabla^2(\epsilon - c_m p) = -\frac{1}{K_v}\vec{\nabla} \cdot \vec{F} \tag{4.20}$$

where Geertsma's parameter $c_m \equiv \alpha/K_v$ was introduced previously in Eqn. 3.71. In the absence of body-force gradients, the Laplacian of the volumetric strain is proportional to the Laplacian of the pore pressure. The Laplacian is a measure of the difference between the value of a function at a point and its local average in an infinitesimal neighborhood about the point (Kraut, 1967, p. 317). Therefore, in the absence of body-force gradients, deviations of mean volumetric strain from its local average are proportional to deviations of pore pressure from its local average.

The governing equations for displacement can be derived in the same manner as just described, but using increment of fluid content as the fluid variable in place of pore pressure. One form is

$$G\nabla^2 u_i + \frac{G}{1-2\nu_u}\frac{\partial^2 u_k}{\partial x_i \partial x_k} = BK_u \frac{\partial \zeta}{\partial x_i} - F_i \tag{4.21}$$

Relations among the poroelastic moduli can be used to express Eqn. 4.21 in other forms:

$$G\nabla^2 u_i + \frac{G}{1-2\nu_u}\frac{\partial^2 u_k}{\partial x_i \partial x_k} = \alpha M \frac{\partial \zeta}{\partial x_i} - F_i \tag{4.22}$$

$$G\nabla^2 u_i + (\lambda_u + G)\frac{\partial^2 u_k}{\partial x_i \partial x_k} = \frac{\lambda_u - \lambda}{\alpha}\frac{\partial \zeta}{\partial x_i} - F_i \tag{4.23}$$

where λ_u is the undrained Lamé's constant. Eqn. 4.23 is obtained from Eqn. 4.21 by using the relationship

$$BK_u = \frac{\lambda_u - \lambda}{\alpha} \tag{4.24}$$

Eqn. 4.24 is obtained by noting that $\lambda_u - \lambda = K_u - K = K_u - K_u(1 - \alpha B) = K_u \alpha B$, where the first equality is derived from the fact that both $K - \lambda$ and $K_u - \lambda_u$ are equal to $2G/3$ (Appendix B, Table B.1). The term containing the gradient of the increment of fluid content is equivalent mathematically to a body force (cf. Eqn. 4.15). This body-force analogy is discussed further in Chapter 5.

Drained moduli appear in Eqn. 4.15 because pressure is the coupling term, whereas undrained moduli appear in Eqns. 4.21–4.23 because increment of fluid content is the coupling term. A similar observation was noted previously when the corresponding constitutive equations (Eqns. 2.42 and 2.53) were presented. Starting with Eqn. 4.21 yields the governing equation equivalent to Eqn. 4.20 but with increment of fluid content and undrained constants,

$$\nabla^2(\epsilon - \gamma\zeta) = -\frac{1}{K_v^{(u)}}\vec{\nabla} \cdot \vec{F} \tag{4.25}$$

where γ is the loading efficiency (cf. Eqn. 3.78). If the body-force gradient is zero, the loading efficiency is the constant of proportionality between the Laplacian of volumetric strain and the Laplacian of increment of fluid content. Loading efficiency and Geertsma's parameter c_m play similar roles in Eqns. 4.20 and 4.25, the former when increment of fluid content is the fluid-coupling parameter and the latter when pore pressure is the fluid-coupling variable. Both loading efficiency and Geertsma's parameter are poroelastic parameters defined for uniaxial strain constraints on the elementary volume.

4.3 BELTRAMI-MICHELL EQUATIONS

The stress formulation consists of governing equations, known as the Beltrami-Michell equations, for the six stress components. They are derived by taking different spatial derivatives of Eqns. 4.12– 4.14. It is first useful to obtain the stress equivalent of Eqn. 4.20 by substituting into it the constitutive relationship Eqn. 2.17,

$$\nabla^2[\sigma_{kk} + 4\eta p] = -\frac{1+\nu}{1-\nu}\vec{\nabla} \cdot \vec{F} \tag{4.26}$$

where η has been defined previously (cf. Eqn. 3.47). In the absence of body-force gradients, the Laplacian of the mean stress is proportional to the Laplacian of the pore pressure—that is, deviations of mean stress from its local average are proportional to deviations of pore pressure from its local average. Eqns. 4.20, 4.25, and 4.26 are equivalent equations. The first form is in terms of the variable pair ϵ_{kk} and p, the second is in terms of the variable pair ϵ_{kk} and ζ, and the third is in terms of the variable pair σ_{kk} and p.

Three coupled governing equations for the normal stress components are obtained by substituting Eqn. 2.36 into Eqn. 4.19 and eliminating $\nabla^2\sigma_{kk}$ by solving Eqn. 4.26. Three more equations for the shear stresses are obtained by taking mixed derivatives of the displacement equations. For example, one equation is obtained by taking the partial derivative with respect to y of Eqn. 4.12 and adding the partial derivative with respect to x of Eqn. 4.13. The six Beltrami-Michell stress equations can be summarized in index notation:

$$\nabla^2\sigma_{ij} + \frac{1}{1+\nu}\frac{\partial^2\sigma_{kk}}{\partial x_i \partial x_j} + 2\eta\left[\frac{1-\nu}{1+\nu}\frac{\partial^2 p}{\partial x_i \partial x_j} + \delta_{ij}\nabla^2 p\right]$$
$$= -\frac{\nu}{1-\nu}\delta_{ij}\vec{\nabla}\cdot\vec{F} - \frac{\partial F_i}{\partial x_j} - \frac{\partial F_j}{\partial x_i} \quad (4.27)$$

The Beltrami-Michell equations can also be expressed in terms of effective stresses:

$$\nabla^2\sigma'_{ij} + \frac{1}{1+\nu}\frac{\partial^2\sigma'_{kk}}{\partial x_i \partial x_j} - \alpha\left[2\frac{\partial^2 p}{\partial x_i \partial x_j} + \frac{\nu}{1-\nu}\delta_{ij}\nabla^2 p\right]$$
$$= -\frac{\nu}{1-\nu}\delta_{ij}\vec{\nabla}\cdot\vec{F} - \frac{\partial F_i}{\partial x_j} - \frac{\partial F_j}{\partial x_i} \quad (4.28)$$

4.4 STRAIN COMPATIBILITY EQUATIONS

An alternative derivation for the Beltrami-Michell equations is to substitute constitutive equations and force equilibrium into six *strain compatibility equations*. Consequently, the Beltrami-Michell equations are sometimes referred to as compatibility equations expressed in stress components. The strain compatibility equations are derived by taking second spatial derivatives of strain components (Eqns. 2.1–2.6). For example, one compatibility equation is obtained by first taking additional derivatives of the equations for ϵ_{xx}, ϵ_{yy}, and ϵ_{xy} to obtain the following three equations:

$$\frac{\partial^2\epsilon_{xx}}{\partial y^2} = \frac{\partial^3 u}{\partial y^2 \partial x} \quad (4.29)$$

$$\frac{\partial^2\epsilon_{yy}}{\partial x^2} = \frac{\partial^3 v}{\partial x^2 \partial y} \quad (4.30)$$

$$2\frac{\partial^2\epsilon_{xy}}{\partial x \partial y} = \frac{\partial^3 u}{\partial y^2 \partial x} + \frac{\partial^3 v}{\partial x^2 \partial y} \quad (4.31)$$

Substituting the right-hand sides of Eqns. 4.29 and 4.30 into Eqn. 4.31 gives

$$2\frac{\partial^2\epsilon_{xy}}{\partial x \partial y} = \frac{\partial^2\epsilon_{xx}}{\partial y^2} + \frac{\partial^2\epsilon_{yy}}{\partial x^2} \quad (4.32)$$

Two similar equations are formed by permuting the indices cyclically, that is, $x \to y$, $y \to z$, and $z \to x$. These equations are

$$2\frac{\partial^2 \epsilon_{yz}}{\partial y \partial z} = \frac{\partial^2 \epsilon_{yy}}{\partial z^2} + \frac{\partial^2 \epsilon_{zz}}{\partial y^2} \tag{4.33}$$

$$2\frac{\partial^2 \epsilon_{zx}}{\partial z \partial x} = \frac{\partial^2 \epsilon_{zz}}{\partial x^2} + \frac{\partial^2 \epsilon_{xx}}{\partial z^2} \tag{4.34}$$

Three other strain compatibility equations are obtained via a different set of differentiations. For example, one equation is obtained by first noting the following four identities:

$$\frac{\partial^2 \epsilon_{xx}}{\partial y \partial z} = \frac{\partial^3 u}{\partial x \partial y \partial z} \tag{4.35}$$

$$2\frac{\partial}{\partial x}\frac{\partial \epsilon_{xy}}{\partial z} = \frac{\partial^3 u}{\partial x \partial y \partial z} + \frac{\partial^3 v}{\partial x^2 \partial z} \tag{4.36}$$

$$2\frac{\partial}{\partial x}\frac{\partial \epsilon_{yz}}{\partial x} = \frac{\partial^3 v}{\partial x^2 \partial z} + \frac{\partial^3 w}{\partial x^2 \partial y} \tag{4.37}$$

$$2\frac{\partial}{\partial x}\frac{\partial \epsilon_{xz}}{\partial y} = \frac{\partial^3 u}{\partial x \partial y \partial z} + \frac{\partial^3 w}{\partial x^2 \partial y} \tag{4.38}$$

Multiplying Eqn. 4.35 by two, subtracting Eqn. 4.36, adding Eqn. 4.37, and subtracting Eqn. 4.38 yield

$$\frac{\partial^2 \epsilon_{xx}}{\partial y \partial z} = \frac{\partial}{\partial x}\left(-\frac{\partial \epsilon_{yz}}{\partial x} + \frac{\partial \epsilon_{xz}}{\partial y} + \frac{\partial \epsilon_{xy}}{\partial z}\right) \tag{4.39}$$

Cyclic permutation gives the remaining two strain compatibility equations:

$$\frac{\partial^2 \epsilon_{yy}}{\partial z \partial x} = \frac{\partial}{\partial y}\left(\frac{\partial \epsilon_{yz}}{\partial x} - \frac{\partial \epsilon_{zx}}{\partial y} + \frac{\partial \epsilon_{yx}}{\partial z}\right) \tag{4.40}$$

$$\frac{\partial^2 \epsilon_{zz}}{\partial x \partial y} = \frac{\partial}{\partial z}\left(\frac{\partial \epsilon_{zy}}{\partial x} + \frac{\partial \epsilon_{zx}}{\partial y} - \frac{\partial \epsilon_{xy}}{\partial z}\right) \tag{4.41}$$

The Beltrami-Michell equations are recovered after substituting the constitutive relations and the force equilibrium equations into the strain compatibility equations.

4.5 MECHANICAL BOUNDARY AND INITIAL CONDITIONS

The complete specification of a poroelastic problem requires sufficient boundary conditions to constrain the solution of the governing equations within the problem domain. Boundary conditions are often difficult to specify for geologic problems. The three main types of mechanical boundary conditions are specified displacement, specified stress, and specified functional relationship between stress and displacement.

Typical stress boundary conditions are specified loads, specified regional stresses at infinity, or a stress-free boundary at the earth's surface. The most common displacement boundaries are zero displacement boundaries, both to represent a rigid restraint and to represent symmetry planes. Infinite lateral dimensions are usually represented by zero displacement boundaries at large distances. A more complicated boundary condition is a surface in contact with an elastic body. This condition is represented by an elastic constraint equation in which the traction is proportional to displacement (Boley and Weiner, 1985, p. 65).

The mechanical initial conditions—stresses or displacements—must also be specified for the entire problem domain. The initial conditions must satisfy the elastic equilibrium equations.

4.6 DARCY'S LAW

In 1856, Darcy conducted a series of experiments in which he varied the amount of water flowing downward through a vertical sand column (Fig. 4.1). Darcy deduced an empirical relationship for one-dimensional flow,

$$q_z = -K\frac{dh}{dz} \tag{4.42}$$

where q_z, the specific discharge, is the volume of fluid crossing a unit area per unit time; h, the head, is the height above a reference datum attained by a column of water in equilibrium with a point in the column; and K, the hydraulic conductivity, is the constant of proportionality. The context should keep clear when K is being used for hydraulic conductivity versus the framework bulk modulus. In three dimensions, Darcy's law is expressed in component form as

$$q_i = -K\frac{\partial h}{\partial x_i} \tag{4.43}$$

where q_i is the component of flux in the direction of the coordinate x_i.

Hubbert (1940, 1956) showed that Darcy's head is the potential energy of the fluid per unit weight. The force $\vec{\mathbf{E}}$ acting on a unit mass of fluid is

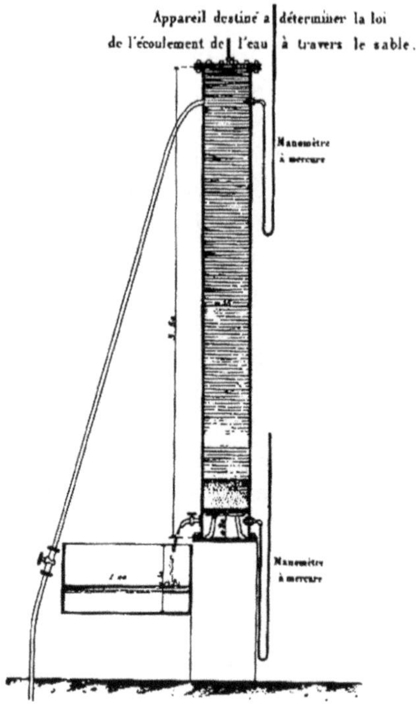

Figure 4.1: Darcy's (1856) sand column experiment. A steady-state flow was set up in the column, and heights of mercury in the manometers were recorded as equivalent columns of water relative to a datum.

the resultant of the gravitational force and the force exerted by the pressure gradient,

$$\vec{\mathbf{E}} = \vec{g} - \frac{1}{\rho_f}\vec{\nabla}p \tag{4.44}$$

where $\vec{g}$ is the acceleration of gravity. If ρ_f is either constant or a function only of p, Hubbert showed that the force field $\vec{\mathbf{E}}$ can be obtained as the negative of the gradient of a potential Φ,

$$\vec{\mathbf{E}} = -\vec{\nabla}\Phi \tag{4.45}$$

where

$$\Phi = gz + \int_0^p \frac{dp}{\rho_f} \tag{4.46}$$

The reference potential in Eqn. 4.46 is taken to be zero at $z = 0$, and pressure is taken to be gauge pressure (absolute pressure minus atmospheric pressure). For an incompressible and homogeneous fluid, Eqn. 4.46 reduces to

$$\Phi = gz + \frac{p}{\rho_f} \tag{4.47}$$

In terms of Darcy's experiment, the pressure p is that exerted by a column of water of height $h - z$ (z-axis directed upward), that is, $p = \rho_f g(h - z)$. Therefore, substituting this expression for p into Eqn. 4.47 shows that

$$\Phi = gh \tag{4.48}$$

and thus Darcy's head can be expressed as the sum of an elevation head z and a pressure head $p/\rho_f g$:

$$h = z + \frac{p}{\rho_f g} \tag{4.49}$$

The physical significance of head is that it is the potential energy per unit weight of fluid, whereas Φ is the potential energy per unit mass of fluid. Similarly, the gradient of head is a force per unit weight of fluid, whereas the gradient of Φ is a force per unit mass of fluid.

The use of hydraulic conductivity in Darcy's law (Eqn. 4.42) assumes that the pore fluid is water. The value of K would be different if oil were used in place of water, because density and viscosity are different for oil and water. The value of K is directly proportional to fluid density and inversely proportional to fluid viscosity μ. Accounting for these fluid properties leads to the three-dimensional form of Darcy's law,

$$\vec{\mathbf{q}} = -\frac{k}{\mu}\vec{\nabla}(p + \rho_f g z) \tag{4.50}$$

$$= -\frac{k}{\mu}\vec{\nabla}p - \frac{k\rho_f g}{\mu}\hat{\mathbf{k}} \tag{4.51}$$

where

$$K = \frac{\rho_f g k}{\mu} \tag{4.52}$$

is the relationship between the hydraulic conductivity and intrinsic permeability k, and the unit vector in the z-direction is represented by $\hat{\mathbf{k}}$. Eqn. 4.50 contains two important physical properties—permeability and viscosity—that are intrinsic to the rock and fluid, respectively. Permeability is a measure of

TABLE 4.1.
Permeability vs. Hydraulic Conductivity and Rock Types That Might Take on Each Value*

Permeability [m^2]	Permeability [Darcy]	Hydraulic Conductivity [m/s]	Representative Rock Type
10^{-12}	1	10^{-5}	Sand or sandstone
10^{-15}	10^{-3}	10^{-8}	Sandstone or limestone
10^{-18}	10^{-6}	10^{-11}	Granite or shale

*Conversions shown assume that the viscosity of water is 10^{-3} Pa·s, density of water is 1000 kg/m^3, and acceleration of gravity is 10 m/s^2.

the geometry of the pore structure of the porous medium and has units of area (e.g., m^2). Viscosity is a measure of the fluid resistance to shear and has units of stress multiplied by time (e.g., Pa·s). The ratio k/μ is sometimes called *mobility* in the petroleum engineering literature. The approximate conversion between hydraulic conductivity and permeability to water is given in Table 4.1.

4.6.1 Excess Pressure

A typical poroelastic problem in geomechanics is to find the fluid pressure or displacement response due to an applied load or to fluid extraction. If the fluid is static in the initial state, it is useful to decompose the total pressure into the sum of hydrostatic pressure and an excess pressure. Hydrostatic pressure is the fluid pressure due to the weight of the overlying water column. Excess pore pressure is the total fluid pressure minus hydrostatic pressure (Fig. 4.2). Thus, the total pressure can be expressed as

$$\begin{aligned} p &= p^{\text{hydro}} + p^{\text{ex}} \\ &= \rho_f g(z_o - z) + p^{\text{ex}} \end{aligned} \tag{4.53}$$

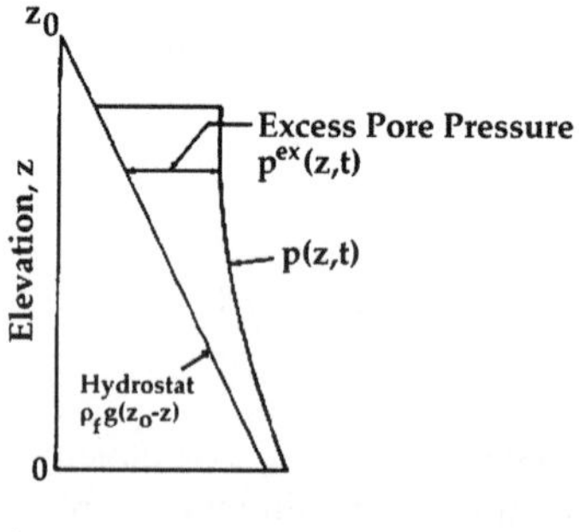

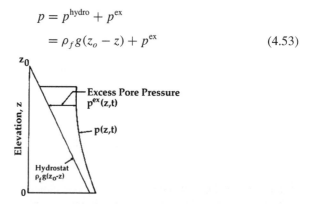

Figure 4.2: Definition of excess pore pressure as the difference between total fluid pressure and hydrostatic pressure (after Lambe and Whitman, 1979).

where z_o is the height of the static water column relative to datum, and $z_o - z$ is the height of the hydrostatic column above the measurement point z. Dividing Eqn. 4.53 by $\rho_f g$ gives the pressure head. Adding the elevation head z gives the total head:

$$h = z_o + \frac{p^{\mathrm{ex}}}{\rho_f g} \tag{4.54}$$

Thus, the excess pressure is the same as Darcy's head within an additive constant as long as the height of the static water column is independent of time. Because the datum for head is arbitrary, the excess pressure is proportional to head, and Darcy's law can be written as

$$\vec{\mathbf{q}} = -\frac{k}{\mu}\vec{\nabla} p \tag{4.55}$$

where the superscript *ex* has been dropped because the variable p in the constitutive equations is the excess pressure relative to a hydrostatic reference state.

4.6.2 Seepage Velocity

Darcy's law is an energy transport equation. Groundwater flow is a quasistatic process, meaning that the system passes continuously through a sequence of equilibrium states. It is irreversible because the potential energy of the fluid is dissipated as frictional heat throughout the porous medium. Inertial terms are ignored in the equation of motion of the fluid. The frictional force between the fluid and pore walls balances the driving force due to the potential gradient. The frictional dissipation of energy occurs as a result of relative motion between fluid and solid grains. Therefore, the seepage or average linear velocity, $\vec{v} = \vec{q}/\phi$, must be interpreted as the *relative* velocity between the fluid and solid:

$$\vec{v} = \frac{1}{\phi}\vec{q} = \dot{\vec{U}}_f - \dot{\vec{U}}_s \tag{4.56}$$

4.7 FLUID CONTINUITY

The continuity constraint can be expressed in terms of the specific discharge by inserting Eqn. 4.56 into the time derivative of Eqn. 2.9:

$$\frac{\partial \zeta}{\partial t} = -\vec{\nabla} \cdot \vec{q} \tag{4.57}$$

Eqn. 4.57 can be generalized to include a quantity of fluid added directly into the REV from an external source,

$$\frac{\partial \zeta}{\partial t} + \vec{\nabla} \cdot \vec{q} = Q \tag{4.58}$$

where $Q(x, y, z)$, volume of fluid per unit bulk volume per unit time, is a distribution function of fluid source (positive when fluid is added) within the REV. Eqn. 4.58 can be expressed using vector notation, index notation, and explicit notation:

$$\frac{\partial \zeta}{\partial t} + \mathrm{div}\vec{q} = Q \tag{4.59}$$

$$\frac{\partial \zeta}{\partial t} + \frac{\partial q_k}{\partial x_k} = Q \tag{4.60}$$

$$\frac{\partial \zeta}{\partial t} + \frac{\partial q_x}{\partial x} + \frac{\partial q_y}{\partial y} + \frac{\partial q_z}{\partial z} = Q \tag{4.61}$$

4.7.1 Pore Pressure—Mean Stress Equation

The partial differential equation governing fluid flow is obtained by substituting Darcy's law (Eqn. 4.55) into the continuity equation (Eqn. 4.61) with the assumption that the physical properties k and μ do not vary spatially:

$$\frac{\partial \zeta}{\partial t} - \frac{k}{\mu}\nabla^2 p = Q \tag{4.62}$$

Otherwise an additional term containing the spatial derivatives of these quantities must be included. Substituting for ζ from the constitutive equation (Eqn. 2.18) gives

$$\frac{\alpha}{KB}\left[\frac{B}{3}\frac{\partial \sigma_{kk}}{\partial t} + \frac{\partial p}{\partial t}\right] - \frac{k}{\mu}\nabla^2 p = Q \tag{4.63}$$

Eqn. 4.63 is an *inhomogeneous* diffusion equation for pore pressure even when no explicit fluid sources Q are present, because it contains a term proportional to changes of mean stress with time. The time-dependent mean stress term is equivalent mathematically to a fluid source and analogous to the fluid pressure gradient term in the mechanical equilibrium equation (Eqn. 4.15) being equivalent to a body force.

The combination of poroelastic parameters, α/KB, in Eqn. 4.63 is the specific storage at constant stress, S_σ (cf. Eqn. 2.19). Therefore,

$$S_\sigma\left[\frac{B}{3}\frac{\partial \sigma_{kk}}{\partial t}+\frac{\partial p}{\partial t}\right]=\frac{k}{\mu}\nabla^2 p+Q \tag{4.64}$$

4.7.2 Pore Pressure—Volumetric Strain Equation

As was the case for the constitutive equations, either pore pressure or increment of fluid content can be chosen to be the fluid variable, and stress or strain can be chosen to be the mechanical variable (see Table 2.1) in the fluid diffusion equation. Eqn. 4.64 used pore pressure and mean stress as the fluid and mechanical variables, respectively. The permutation considered in this section is pore pressure and volumetric strain.

Eqn. 2.24 is substituted into Eqn. 4.64 to yield

$$\alpha\frac{\partial \epsilon_{kk}}{\partial t}+S_\epsilon\frac{\partial p}{\partial t}=\frac{k}{\mu}\nabla^2 p+Q \tag{4.65}$$

where S_ϵ is the specific storage at constant strain (cf. Eqn. 3.40). Eqn. 4.65 is, again, an inhomogeneous diffusion equation for pore pressure. The pore pressure field is coupled with the time rate of change of the volumetric strain. This term behaves mathematically like a fluid source.

4.7.3 Hydrogeologic Transient Flow Equation

The assumptions of uniaxial strain and constant vertical stress led to the simplified constitutive relation $\zeta = Sp$ (Eqn. 3.49), which when substituted into Eqn. 4.62 yields the standard transient flow equation used in hydrogeology:

$$S\frac{\partial p}{\partial t}=\frac{k}{\mu}\nabla^2 p+Q \tag{4.66}$$

Eqns. 4.64 and 4.65 are similar to Eqn. 4.66, but with two important differences. First, the unconstrained or constrained specific storage appears in place of the one-dimensional specific storage. Second, the equations contain an additional term, $S_\sigma B\dot{\sigma}_{kk}/3$ or $\alpha\dot{\epsilon}$, where the dot notation signifies the time derivative. This term couples the time dependence of the stress or strain field into the governing equation for fluid flow.

The assumptions of uniaxial strain and constant vertical stress are not satisfied rigorously in every volume element of an aquifer when two- or three-dimensional flow is occurring, because the flow itself distorts the strain field. The complete poroelastic behavior associated with flow to a well is described in more detail in Chapter 9.

4.7.4 Equation for Increment of Fluid Content

Pore pressure can be eliminated from the governing equation for fluid flow (Eqn. 4.64) in favor of increment of fluid content. The constitutive relation for ζ (Eqn. 2.41) is solved for mean stress, $\sigma_{kk}/3$, and substituted into Eqns. 4.26 and 4.63. Adding the two equations produces a diffusion equation for ζ,

$$\frac{\partial \zeta}{\partial t} = \frac{k}{\mu S}\nabla^2 \zeta + Q + \frac{k}{\mu}\frac{c_m}{S} F_{k,k} \tag{4.67}$$

where $F_{k,k} \equiv \partial F_k/\partial x_k$. An alternative expression for the the coefficient of $F_{k,k}$ has been presented by Rudnicki (1986a, Eqn. 11, p. 385):

$$\frac{\partial \zeta}{\partial t} = \frac{k}{\mu S}\nabla^2 \zeta + Q + \frac{k}{\mu}\frac{\lambda_u - \lambda}{\alpha(\lambda_u + 2G)} F_{k,k} \tag{4.68}$$

Two features are noteworthy about Eqn. 4.67. First, the diffusion equation for ζ is uncoupled from the displacement, strain, and stress fields. If no fluid sources Q or body forces $\vec{F}$ are present, the increment of fluid content satisfies a homogeneous diffusion equation. Second, the diffusivity for ζ is $c = k/(\mu S)$ (Eqn. 3.65). The diffusivity for ζ is identical to the hydraulic diffusivity used in the diffusion equation in hydrogeology (Eqn. 4.66) and to the consolidation coefficient for uniaxial strain conditions in soil mechanics.

If the boundary conditions for fluid flow can be posed in terms of ζ, the problem can be solved independently and the result used in Eqn. 4.21 to solve for the displacements. Because ζ is a measure of fluid volume flowing into or out of an REV, it is the natural variable for no-flow boundaries.

The increment of fluid content, ζ, is proportional to $(\sigma_{kk}/3) + (1/B)p$ (Eqn. 2.41), which is the effective stress for ζ (cf. Table 2.2). Thus, a second interpretation of Eqn. 4.67 is that the effective stress for ζ satisfies the diffusion equation. Because the increment of fluid content is proportional to the fluid mass content (fluid mass per unit volume), a third interpretation of Eqn. 4.67 is that the fluid mass content satisfies the diffusion equation.

4.8 COMPARISON OF DIFFUSION EQUATIONS

Eqns. 4.64 and 4.65 are equivalent governing equations for fluid flow in a poroelastic medium. The term containing the time rate of change of mean stress or mean strain is paired with the specific storage at constant stress or constant strain, respectively, in the term containing the time rate of change of pore pressure. If $\dot{\sigma}_{kk} = 0$, then the diffusive pressure front travels at a speed governed by $c_\sigma = k/(\mu S_\sigma)$. If, on the other hand, $\dot{\epsilon}_{kk} = 0$, then the diffusive pressure front travels at a speed governed by $c_\epsilon = k/(\mu S_\epsilon)$, which is greater than c_σ because less water must be moved from or to storage to change the pressure relative to a state of constant stress. Thus, the significance of S_σ and S_ϵ as the coefficients of $\partial p/\partial t$ in Eqns. 4.64 and 4.65 is that the rate of pore-pressure diffusion is different for conditions of constant stress versus constant strain.

A uniform stress or strain condition can exist only approximately throughout a problem domain when transient flow occurs, because the changing flow field induces temporal and spatial deviations. Therefore, any simplification of Eqns. 4.64 or 4.65 to a homogeneous diffusion equation must be justified on the grounds that a condition of constant stress or constant strain is maintained approximately throughout the problem domain. A similar conclusion applies for the assumed conditions of uniaxial strain and constant vertical stress that lead to the homogeneous pore-pressure diffusion equation that is commonly applied in hydrogeology. In theory, these conditions are satisfied exactly for a one-dimensional laboratory consolidation experiment in which the walls of the apparatus are stiff compared with the soil. In a pumping test, however, in which a well is pumped at a uniform rate in a horizontal aquifer, volume elements near the well undergo lateral as well as vertical deformation, and hence do not satisfy the uniaxial strain assumption. As with constant stress or constant strain conditions, the uniaxial strain approximation is justified only insofar as transient flow does not induce significant deviations.

4.9 FLUID BOUNDARY AND INITIAL CONDITIONS

For the poroelastic formulation developed in this chapter, fluid boundary and initial conditions must be posed in terms of excess, rather than total, pore pressure. In a consolidation experiment, for example, the initial excess pore pressure is the undrained increase in pore pressure induced by a suddenly applied vertical load. A common pore-pressure boundary condition is specified pressure. Drained boundaries ($p = 0$) are a special case. A fluid load, such as a reservoir lake or a pressurized borehole, leads to a specified stress boundary condition for the mechanical problem and a specified pore-pressure condition for the fluid-flow problem. A common specified flow condition is a no-flow boundary ($\zeta = 0$) to represent an impermeable boundary or a symmetry plane. Infinite lateral dimensions are usually represented by no-flow boundaries at large distances.

The pore pressure initial condition for poroelastic problems is often based on a suddenly applied boundary load, which induces the starting distribution of pore pressure. Because mechanical stresses are transmitted throughout the body instantaneously (at the speed of sound), which is much more rapid than fluid movement, the fluid can be assumed to be undrained in every REV.

The mechanical equilibrium equations can then be solved for the induced pore-pressure distribution, using the sudden load as part of the mechanical boundary conditions and taking $\zeta = 0$ everywhere in the interior. For example, the initial pore pressure field is $-B\sigma_{kk}/3$ for a rectangular prism subjected suddenly to normal stresses, and it is $-\gamma\sigma_{zz}$ for a sudden vertical load and fixed lateral boundaries.

4.10 UNCOUPLING OF STRESS OR STRAIN FROM PORE PRESSURE

In thermoelasticity, the phrase *uncoupled quasistatic theory* means that the mechanical coupling term in the temperature diffusion equation is omitted (Boley and Weiner, 1985, p. 41). Similarly, the poroelastic meaning is that the mechanical coupling term in the fluid pressure diffusion equation is omitted. The uncoupled assumption, in fact, includes "one-way" poroelastic coupling; namely, changes in the fluid pressure field do produce stresses and strains, but changes in the stress or strain fields are assumed not to affect fluid pressures. The simplification that results for an uncoupled problem is that the transient fluid-flow equation in pressure can be solved independently of the time-dependent strain or stress field.[1] The resulting pore-pressure field can then be inserted into the mechanical equilibrium equations as a parametric function of time. The mechanical problem is then solved separately as a standard elastostatic problem.

The inhomogeneous pore-pressure diffusion equations (4.64 and 4.65) are mathematically uncoupled from the mechanical equilibrium equations in four special circumstances: (1) steady state, (2) a state of uniaxial strain and constant vertical stress, (3) a highly compressible fluid, and (4) an irrotational displacement field in an infinite or semi-infinite domain without body forces.

4.10.1 Steady State

The mechanical coupling term disappears from Eqn. 4.64 for steady state conditions. The resulting Poisson's equation is solved first for the pressure field, which is then inserted into the mechanical equilibrium equations. Both the pressure and displacement fields are independent of time.

4.10.2 Uniaxial Strain and Constant Vertical Stress

No stress or strain term appears in the pressure diffusion equation representing fluid continuity when an REV is in a state of uniaxial strain and constant

[1] Note that the equation for fluid flow using increment of fluid content for the fluid variable is always uncoupled from the force equilibrium equations.

vertical stress (Eqn. 4.66). This simplification occurs because the mean stress in this case is proportional to the pore pressure (Eqn. 3.46). The hydraulic diffusivity is given by $c = k/(\mu S)$. Thus, one-dimensional poroelastic (consolidation) problems for which these boundary conditions apply are mathematically uncoupled, because the external boundary conditions of uniaxial strain and constant vertical stress apply everywhere in the problem domain.

4.10.3 Highly Compressible Fluid

The problems of fluid flow and mechanics are uncoupled when a highly compressible fluid (e.g., air) fills the pore space. Building up pressure in a highly compressible gas requires that the space it occupies undergo a large volume decrease. Therefore, a change in the mean applied stress on an elementary volume of porous material will change the gas pressure only slightly if the fluid compressibility is much greater than the frame compressibility. This reasoning is quantified by using the result that Skempton's coefficient $B \simeq \alpha K_f/(\phi K)$ when $K_f/K \ll 1$ (cf. Table 3.2). Therefore, the coupling term $B\dot{\sigma}_{kk}/3$ in Eqn. 4.64 approaches zero as K_f/K approaches zero.

Because $S_\sigma \simeq \phi/K_f$ for a highly compressible fluid (Table 3.2), Eqn. 4.64 simplifies to the mathematically uncoupled pressure diffusion equation:

$$\frac{\phi}{K_f}\frac{\partial p}{\partial t} = \frac{k}{\mu}\nabla^2 p + Q \qquad \text{for } K_f/K \ll 1 \tag{4.69}$$

For a highly compressible fluid, $S_\epsilon = 1/M \simeq \phi/K_f$ (see Table 3.2). The storage coefficients S_ϵ and S_σ are the same because almost all the fluid volume change from storage is due to the compressibility of the gas, with only a negligible fraction due to pore volume change. Therefore, the time-dependent strain term in Eqn. 4.65 is negligible, and Eqn. 4.65 also reduces to Eqn. 4.69.

4.10.4 Irrotational Displacement Field in an Unbounded Domain

Finally, the mechanical problem uncouples from the pore-pressure equation for an irrotational displacement field $\vec{u}$ in an infinite or semi-infinite problem domain in the absence of a body force. The vector field $\vec{u}$ is said to be irrotational if $\vec{\nabla} \times \vec{u} = 0$, where

$$\vec{\nabla} \times \vec{u} = \begin{vmatrix} \hat{i} & \hat{j} & \hat{k} \\ \dfrac{\partial}{\partial x_1} & \dfrac{\partial}{\partial x_2} & \dfrac{\partial}{\partial x_3} \\ u_1 & u_2 & u_3 \end{vmatrix} \tag{4.70}$$

Thus, the component in the $\hat{k}$-direction is $\partial u_2/\partial x_1 - \partial u_1/\partial x_2$. Physically, $\vec{\nabla} \times \vec{u}$ is related to the degree of circulation of the displacement field. The velocity field of stirred coffee (Fig 4.3a), for example, displays circulation, whereas the irrotational velocity field of radial flow away from an injection well (Fig. 4.3b) does not. Another example of an irrotational vector field is the one-dimensional vector field $\vec{u} = (0, 0, u_3(z))$, because each partial derivative $\partial u_i / \partial x_j$ $(i \neq j)$ is zero.

Because $\partial u_i/\partial x_j = \partial u_j/\partial x_i$ for an irrotational displacement field, the following equation holds:

$$\begin{aligned} \nabla^2 u_i &= \frac{\partial^2 u_i}{\partial x_1^2} + \frac{\partial^2 u_i}{\partial x_2^2} + \frac{\partial^2 u_i}{\partial x_3^2} \\ &= \frac{\partial}{\partial x_i}\left(\frac{\partial u_1}{\partial x_1} + \frac{\partial u_2}{\partial x_2} + \frac{\partial u_3}{\partial x_3}\right) \\ &= \frac{\partial \epsilon}{\partial x_i} \end{aligned} \tag{4.71}$$

In the absence of a body force, Eqn. 4.18 becomes

$$\frac{\partial \epsilon}{\partial x_i} = c_m \frac{\partial p}{\partial x_i} \tag{4.72}$$

Integrating Eqn. 4.72 gives

$$\epsilon = c_m p + g(t) \tag{4.73}$$

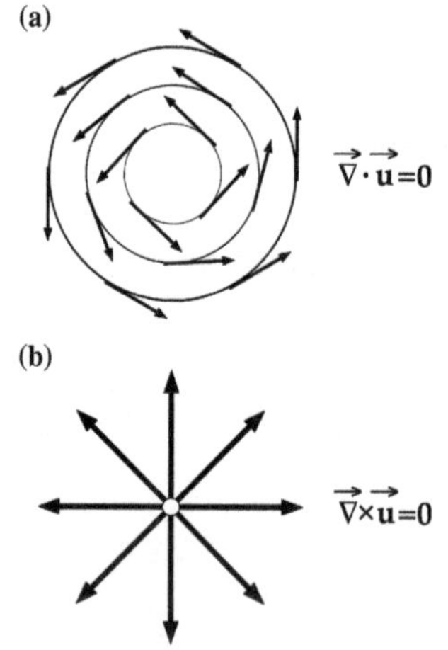

Figure 4.3: Comparison of (a) velocity field with no divergence and (b) an irrotational velocity field (after Kraut, 1967, p. 91). The curl of the velocity field in a stirred cup of coffee is nonzero, whereas it is zero in the radial velocity field. Particle tracks in the irrotational velocity field contain no curvature.

where $g(t)$ is an integration constant that can, in general, be a function of time. Substituting the constitutive equation for ϵ (Eqn. 2.17) into Eqn. 4.73 yields

$$\sigma_{kk} + 4\eta p = \frac{2G(1+\nu)}{1-2\nu} g(t) \tag{4.74}$$

$$= 3Kg(t) \tag{4.75}$$

Next, solving Eqn. 4.75 for σ_{kk}, substituting into Eqn. 4.64, and using the relationship between S and S_σ (cf. Eqn. 3.48) yields

$$S\frac{\partial p}{\partial t} = \frac{k}{\mu}\nabla^2 p - \alpha\frac{dg}{dt} + Q \tag{4.76}$$

or, equivalently,

$$\frac{\partial p}{\partial t} - c\nabla^2 p = -\frac{\alpha}{S}\frac{dg}{dt} + \frac{Q}{S} \tag{4.77}$$

where $c = k/(\mu S)$ is the consolidation coefficient or uniaxial hydraulic diffusivity.

For an infinite or semi-infinite problem domain, a physically realistic solution must satisfy the boundary conditions that ϵ, σ_{kk}, and p vanish at infinity, which requires $g(t)$ and dg/dt to be zero. Therefore, for an irrotational displacement field in an unbounded problem domain, $\epsilon - c_m p$ equals 0, and $\sigma_{kk} + 4\eta p$ equals 0. Thus, the mean stress or volumetric strain is proportional to the pore pressure, and the stress or strain uncouples from the pore pressure. A further consequence of this condition is that the increment of fluid content ζ equals Sp. These same relationships were obtained for conditions of uniaxial strain and constant vertical stress (cf. Section 3.3.3).

4.11 FORCE EQUILIBRIUM FOR THE SOLID MATRIX

Eqns. 4.7–4.9 were obtained by analyzing the total stresses (applied force per unit area of solid grains and pore fluid) exerted on a unit bulk volume of porous medium. In this section, equivalent results are derived by applying the force balance requirement just to the solid skeleton of the porous medium. The grain-to-grain stress that acts only in the solid matrix (Biot and Willis, 1957, p. 598) was called the Terzaghi effective stress T_{ij} by Detournay and Cheng (1993, p. 119):

$$T_{ij} \equiv \sigma_{ij} + p\delta_{ij} \tag{4.78}$$

The Terzaghi effective stress is the force carried by the skeletal frame divided by the total area of the surface (Lambe and Whitman, 1979, p. 241). The action of an increase in pore pressure is to decrease compressive grain-to-grain normal stresses due to the applied stresses by exactly the magnitude of the pore pressure. The pore pressure is assumed not to affect the grain-to-grain shear stresses. The Terzaghi effective stress differs from the previously defined effective stress (Eqn. 2.68) that removed pressure as an explicit variable from the constitutive equations. Taking $\partial T_{ij}/\partial x_j$ and substituting Eqns. 4.10 for $\partial \sigma_{ij}/\partial x_j$ gives

$$\frac{\partial T_{ij}}{\partial x_j} = \frac{\partial p}{\partial x_i} - F_i \tag{4.79}$$

where the body force $\vec{F} = -\rho g \vec{\nabla} z$ is the total weight per unit volume of solids and fluid in an REV.

The right-hand side of Eqn. 4.79 is now shown to equal the negative of the resultant of forces per unit volume acting on the solid fraction of the porous material. These forces are gravity, fluid pressure gradients, and seepage force (Bear, 1972, pp. 184–185):

1. *Gravity*. The weight of the solid per unit volume is

$$\vec{F}^{\text{gravity}} \equiv -\rho_s g(1-\phi)\vec{\nabla} z \tag{4.80}$$

$$= -\rho g \vec{\nabla} z + \phi \rho_f g \vec{\nabla} z \tag{4.81}$$

   The negative sign is due to the convention that the positive z-axis is upward. The second equality is obtained from $\rho = (1-\phi)\rho_s + \phi\rho_f$, where ρ is the bulk density of the porous material.
2. *Fluid Pressure Gradient*. The net force in the $+x$-direction exerted by a fluid pressure gradient on the solid particles is $-(1-\phi)\times(\partial p/\partial x)\Delta x \Delta y \Delta z$. A similar consideration in the other coordinate directions gives the resultant force per unit volume due to a fluid pressure gradient acting on the solid framework:

$$\vec{F}^{\text{pressure}} \equiv -(1-\phi)\vec{\nabla} p \tag{4.82}$$

3. *Seepage Force*. Seepage force is the drag force exerted by the relative motion of pore fluid on solid grains as it flows through the porous medium. In an isotropic medium, seepage force acts in the direction of fluid flow. Examples of seepage force action are sand produced from oil or water wells in unconsolidated formations, piping in earth dams (Lambe and Whitman, 1979, p. 478), the loss of strength of soil when a quick condition exists, and slope failure due to groundwater flow

(Iverson and Major, 1986; Orange and Breen, 1992). The force acting on a unit mass of fluid is $-\vec{\nabla}\Phi$ (Eqn. 4.45). The moving fluid exerts an equal drag force on the solid grains. Multiplying by $\phi\rho_f$ converts the force per unit fluid mass to force per unit volume of porous medium:

$$\vec{F}^{\text{seepage}} \equiv -\phi\rho_f\vec{\nabla}\Phi \tag{4.83}$$

$$= -\phi\rho_f g\vec{\nabla}z - \phi\vec{\nabla}p \tag{4.84}$$

where the second equality is obtained by substituting Eqn. 4.48 for Φ.

The resultant force $\vec{F}^{\text{solids}}$ acting on the solid grains per unit volume of porous material is

$$\vec{F}^{\text{solids}} = \vec{F}^{\text{gravity}} + \vec{F}^{\text{pressure}} + \vec{F}^{\text{seepage}} \tag{4.85}$$

$$= -\rho g\vec{\nabla}z - \vec{\nabla}p \tag{4.86}$$

$$= -(1-\phi)(\rho_s - \rho_f)g\vec{\nabla}z - \rho_f\vec{\nabla}\Phi \tag{4.87}$$

Eqn. 4.86 can be interpreted as the sum of the total weight of the porous material (solids + fluid) and the net applied forces on the faces of the unit porous volume due to the pressure gradient. Substituting Eqn. 4.86 in Eqn. 4.79 shows that

$$\frac{\partial T_{ij}}{\partial x_j} = -F_i^{\text{solids}} \tag{4.88}$$

Eqn. 4.88 expresses force balance for the solid framework.

Alternatively, the first term in Eqn. 4.87 can be interpreted as the submerged weight of the solid grains in a unit volume of porous material—that is, the weight of the solid grains $(1-\phi)\rho_s g$ minus the upward buoyant force equal to the weight of the displaced water $(1-\phi)\rho_f g$:

$$\vec{F}^{\text{sub}} = -(1-\phi)(\rho_s - \rho_f)g\vec{\nabla}z \tag{4.89}$$

In other words, the total force on the solid framework per unit volume of porous material is the submerged weight of the solids plus the seepage force per unit fluid volume:

$$\vec{F}^{\text{solids}} = \vec{F}^{\text{sub}} - \rho_f\vec{\nabla}\Phi \tag{4.90}$$

Substituting Eqn. 4.90 into Eqn. 4.88 yields

$$\frac{\partial T_{ij}}{\partial x_j} = -F_i^{\text{sub}} + \rho_f\frac{\partial\Phi}{\partial x_i} \tag{4.91}$$

Eqns. 4.10 and 4.91 show that force equilibrium can be expressed as (1) total stresses + total gravitational body forces = 0 or (2) solid grain-to-grain stresses + submerged weights + seepage forces = 0, respectively.

Example: *Quick Condition (Lambe and Whitman, 1979, p. 263)*

A vertical column of sand flows upward when the upward seepage force exceeds the submerged weight of the sand. Therefore, the critical seepage force for liquefaction satisfies (cf. Eqn. 4.90):

$$\rho_f \frac{d\Phi^{\text{liq}}}{dz} = -(1-\phi)(\rho_s - \rho_f)g \tag{4.92}$$

Eqn. 4.92 corresponds to a critical hydraulic gradient of $-(1-\phi)(\rho_s - \rho_f)/\rho_f$ (i.e., the ratio of the submerged density of the solid to the fluid density). The magnitude of the critical hydraulic gradient for saturated soils is approximately one.

Example: *Seepage Force and Slope Stability (Iverson and Major, 1986)*

Coulomb failure of a slope due to seepage force can be analyzed using a coordinate system oriented parallel and perpendicular to the hillside (Fig. 4.4). If the hillside is assumed to extend infinitely in the x and z directions, and if the flow field is uniform, the only spatial dependence of stress is with y. Force equilibrium on the solid frame (Eqn. 4.91) consists of one equation repre-

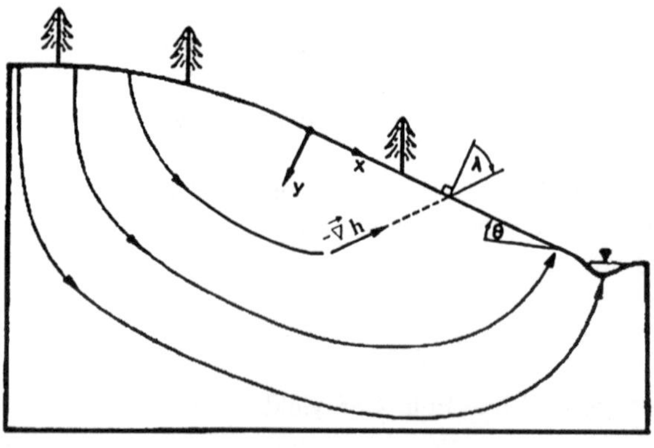

Figure 4.4: Influence of groundwater flow on slope stability (Iverson and Major, 1986). The seepage force and gravity can produce Coulomb failure.

senting the x-direction force balance and one representing the y-direction balance,

$$\frac{dT_{xy}}{dy} = -(\rho - \rho_f)g\sin\theta + \rho_f g\frac{\partial h}{\partial x} \tag{4.93}$$

$$\frac{dT_{yy}}{dy} = -(\rho - \rho_f)g\cos\theta + \rho_f g\frac{\partial h}{\partial y} \tag{4.94}$$

where T_{xy} is the shear stress on a plane normal to the y-direction due to a force in the x direction, T_{yy} is the Terzaghi normal stress in the y-direction, and θ is the slope angle. The Coulomb failure criterion is

$$T_{xy} = -T_{yy}\tan\phi_C \tag{4.95}$$

where $\tan\phi_C$ is the coefficient of friction, and the negative sign accounts for the convention that stresses are positive in extension. The components of the hydraulic gradient are

$$\frac{\partial h}{\partial x} = -|\vec{\nabla} h|\sin\lambda \tag{4.96}$$

$$\frac{\partial h}{\partial y} = |\vec{\nabla} h|\cos\lambda \tag{4.97}$$

where λ is the angle between the normal to the slope and the seepage force. Substituting Eqns. 4.93 and 4.94 into the derivative of Eqn. 4.95 with respect to y and using Eqns. 4.96 and 4.97 lead to the result

$$|\vec{\nabla} h|^{\text{crit}} = -\frac{\rho - \rho_f}{\rho_f}\frac{\sin(\theta - \phi_C)}{\sin(\lambda + \phi_C)} \tag{4.98}$$

where $|\vec{\nabla} h|^{\text{crit}}$ is the critical hydraulic gradient. For $\theta = 25°$, $\lambda = 5°$, $\phi_C = 30°$, and $\rho/\rho_f = 2$, the critical hydraulic gradient is 0.15. Liquefaction, on the other hand, requires that the upward component of the seepage force exceed the submerged weight:

$$|\vec{\nabla} h|^{\text{liq}} = \frac{\rho - \rho_f}{\rho_f}\frac{1}{\cos(\lambda + \theta)} \tag{4.99}$$

The liquefaction gradient is 1.15.

5

Unbounded Problem Domains

5.0 CHAPTER OVERVIEW

Elastostatic Green's functions for an infinite problem domain are obtained using classical techniques from potential theory. The displacement field due to a point fluid source is shown to be mathematically equivalent to a center of dilatation (three orthogonal double-force pairs). The equivalence is used to obtain the solution for a point source in a half space from the solution for a center of dilatation in a half space with a traction-free surface. Poroelastic displacements at the free surface for pressure distributions or increment-of-fluid distributions shaped as a cylinder, disk, or slab can be obtained from the mathematically analogous gravitational problem.

Time-dependent Green's functions for an instantaneous point source of fluid are readily obtained, because the increment of fluid content satisfies the heat-diffusion equation. The time-dependent solution for the sudden injection of a fluid dipole is then obtained by taking the derivative of the solution for the sudden injection of a point fluid source in the direction of the dipole moment. In the fluid diffusion equation, a dipole fluid source distribution is shown to be mathematically equivalent to a time-dependent body-force distribution whose direction is constant. The equivalence is used to obtain the solution for a sudden point force as the superposition of the solution for the equivalent dipole plus the undrained point-force solution.

5.1 POROELASTIC DISPLACEMENT POTENTIAL IN INFINITE DOMAIN

Eqns. 4.15 and 4.21 are nonhomogeneous partial differential equations. Therefore, the displacement can be obtained as the sum of a particular solution of the nonhomogeneous equation in infinite space and a complementary solution of the homogeneous equation satisfying modified boundary conditions. It is readily verified that a particular solution can be expressed as

$$u_i = \frac{\partial \Phi}{\partial x_i} \tag{5.1}$$

if Φ[1] satisfies

$$\nabla^2\Phi = c_m p \tag{5.2}$$

in the case of Eqn. 4.15, and satisfies

$$\nabla^2\Phi = \gamma\zeta \tag{5.3}$$

in the case of Eqn. 4.21. Eqn. 5.1 expresses the displacement as the gradient of Φ, and hence Φ is called a displacement potential. Note that the Laplacian of Φ is the volumetric strain because

$$\nabla^2\Phi = \frac{\partial u_k}{\partial x_k} = \epsilon_{kk} = \epsilon \tag{5.4}$$

Because the poroelastic displacement potential Φ satisfies Poisson's equation, the pressure p on the right-hand side of Eqn. 5.2, or the increment of fluid content ζ on the right-hand side of Eqn. 5.3, is analogous to a charge density or mass anomaly in electrical and gravitational potential theory, respectively. Therefore, a solution (e.g., Jackson, 1999, p. 30) of Eqn. 5.3 is

$$\Phi(x_1, x_2, x_3) = -\frac{\gamma}{4\pi}\int\int\int_D \frac{\zeta(\xi_1, \xi_2, \xi_3)}{|\vec{x} - \vec{\xi}|} d\xi_1 d\xi_2 d\xi_3 \tag{5.5}$$

where $|\vec{x} - \vec{\xi}| = \sqrt{(x_1 - \xi_1)^2 + (x_2 - \xi_2)^2 + (x_3 - \xi_3)^2}$ is the distance from the measurement point (x_1, x_2, x_3) to the integration point (ξ_1, ξ_2, ξ_3) in the problem domain D. Similarly, the displacement field for a general distribution of pressure $p(x, y, z)$ is

$$\Phi(x_1, x_2, x_3) = -\frac{c_m}{4\pi}\int\int\int_D \frac{p(\xi_1, \xi_2, \xi_3)}{|\vec{x} - \vec{\xi}|} d\xi_1 d\xi_2 d\xi_3 \tag{5.6}$$

The displacement components u_i are then found as the gradient of the potential. The displacements go to zero as $|\vec{x} - \vec{\xi}|$ goes to infinity. The stress components follow directly from the definition of strain components in terms of derivatives of displacement (Eqn. 2.7) and the constitutive relations (Boley and Weiner, 1985, p. 80):

$$\sigma_{ij} = 2G\left(\frac{\partial^2\Phi}{\partial x_i \partial x_j} - \delta_{ij}\nabla^2\Phi\right) \tag{5.7}$$

[1] The symbol Φ in this chapter signifies the poroelastic displacement potential and not Hubbert's force potential.

5.2 UNIFORM FLUID INJECTION IN A SPHERE

The potential due to a spherically shaped distribution of either increment of fluid content or pressure can be solved by superposition of the potential due to a uniform distribution in a spherical shell. The potential, at a point $\vec{x}$ at a distance R relative to the center of a spherical shell of radius a and thickness Δa injected uniformly with fluid volume per unit volume ζ_o, is obtained by carrying out the integration in Eqn. 5.5 over circular rings of latitude. The distance ξ_R from all points on a circular ring to $\vec{x}$ is constant (Fig. 5.1) and hence

$$\Phi^{\text{shell}}(\vec{x}) = -\frac{\gamma \zeta_o}{2} a^2 \Delta a \int_0^\pi \frac{\sin\phi}{\xi_R} d\phi \tag{5.8}$$

where

$$\xi_R = \sqrt{a^2 - 2aR\cos\phi + R^2} \tag{5.9}$$

(cf. MacMillan, 1930, p. 37). In Fig. 5.1, a point on the spherical shell is represented in spherical coordinates by (a, θ, ϕ), where a is the radius of the shell, θ is the azimuth, and ϕ is the colatitude with respect to a "north pole" in the direction of $\vec{x}$. Differentiating Eqn. 5.9 with respect to ϕ gives

$$\frac{\sin\phi}{\xi_R} d\phi = \frac{d\xi_R}{aR} \tag{5.10}$$

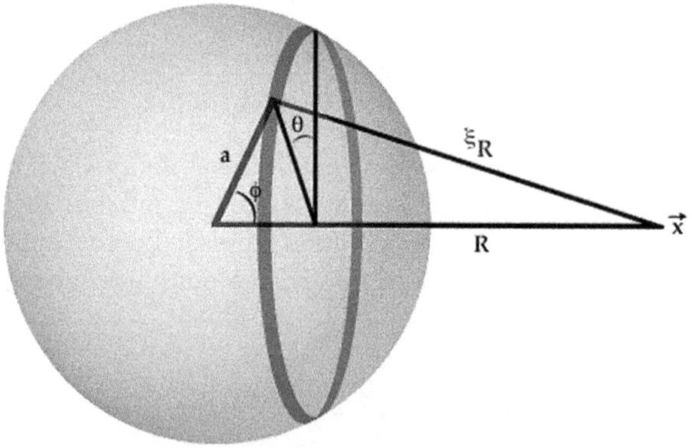

Figure 5.1: Geometry for obtaining displacement potential due to uniform fluid injection into a spherical shell (after MacMillan, 1930, p. 36). The "north pole" is in the $\vec{x}$ direction.

Hence,

$$\Phi^{\text{shell}} = -\frac{\gamma\zeta_o}{2}\frac{a\Delta a}{R}\int_{\xi_{R_1}}^{\xi_{R_2}} d\xi_R \tag{5.11}$$

where the limits of integration ξ_{R_1} and ξ_{R_2} depend on whether the point $\vec{x}$ is outside or inside the shell. If outside, the limits are $R - a$ to $R + a$ as the angle ϕ ranges from 0 to π. If inside, the limits are $a - R$ to $a + R$. Therefore,

$$\Phi^{\text{shell}}(R \geq a) = -\gamma\zeta_o\frac{a^2\Delta a}{R} = -\gamma\frac{V_f}{4\pi R} \tag{5.12}$$

$$\Phi^{\text{shell}}(R < a) = -\gamma\zeta_o a\Delta a = -\gamma\frac{V_f}{4\pi a} \tag{5.13}$$

where $V_f = 4\pi a^2(\Delta a)\zeta_o$ is the volume of fluid injected into the shell. V_f is analogous to mass, and ζ_o (fluid volume per unit volume) is analogous to density (mass per unit volume) in the gravitational analogy. The potential of a homogeneous, thin spherical shell is constant inside the shell. The constant value of the potential inside the shell means that the displacement and strain are zero inside the shell. The $1/R$ variation of the potential outside the shell is the same as if the fluid source were all concentrated at the center. The radial displacement outside the shell is

$$u_R^{\text{shell}}(R \geq a) = \frac{d\Phi}{dR} = \gamma\zeta_o\frac{a^2\Delta a}{R^2} \tag{5.14}$$

For a shell of finite thickness ($b \leq r \leq a$), Eqn. 5.12 is integrated with respect to radius a' of the shell, where a' is substituted in Eqn. 5.12 to distinguish it from a. For $R \geq a$,

$$\Phi^{\text{shell}}(R \geq a) = -\frac{\gamma\zeta_o}{R}\int_b^a a'^2 da' = -\frac{\gamma\zeta_o}{3R}(a^3 - b^3) = -\frac{\gamma\zeta_o V}{4\pi R} \tag{5.15}$$

where $V = (4\pi/3)(a^3 - b^3)$ is the volume of the spherical shell. Eqn. 5.15 is the same as Eqn. 5.12 for the thin shell if the potential is expressed in terms of injected fluid volume $V_f = \zeta_o V$. For a solid sphere, b is zero in Eqn. 5.15, and the potential is

$$\Phi^{\text{sphere}}(R \geq a) = -\frac{\gamma\zeta_o V}{4\pi R} \tag{5.16}$$

where $V = 4\pi a^3/3$ is the volume of the sphere. The displacement outside the solid sphere is

$$u_R^{\text{sphere}}(R \geq a) = \frac{\gamma\zeta_o V}{4\pi R^2} \tag{5.17}$$

That is, radial displacement varies inversely with the square of the distance from the center of the sphere. Inside ($R < b$) the spherical shell of finite thickness, integrating Eqn. 5.13 gives

$$\Phi^{\text{shell}}(R \leq b) = -\gamma\zeta_o \int_b^a a' da' = -\frac{\gamma\zeta_o}{2}(a^2 - b^2) \tag{5.18}$$

The potential within the finite shell itself ($b \leq R \leq a$) is obtained as the sum of the potential external to the shell $b \leq r \leq R$ and the potential internal to the shell $R \leq r \leq a$:

$$\Phi^{\text{shell}}(b \leq R \leq a) = -\frac{\gamma\zeta_o}{3}\frac{R^3 - b^3}{R} - \frac{\gamma\zeta_o}{2}(a^2 - R^2) \tag{5.19}$$

For a solid sphere ($b = 0$), the potential within the sphere is

$$\Phi^{\text{sphere}}(R \leq a) = -\frac{\gamma\zeta_o}{6}(3a^2 - R^2) \tag{5.20}$$

The volumetric strain within a spherical inclusion is given by the Laplacian of Φ:

$$\epsilon^{\text{sphere}}(R \leq a) = \frac{1}{R}\frac{d^2}{dR^2}(R\Phi) = \gamma\zeta_o \tag{5.21}$$

All the results in this section for spherical distributions of changes in increment of fluid content are easily applied to spherical distributions of pore-pressure changes simply by substituting p_o for ζ_o and c_m for γ. For example, the displacement outside a spherical region of radius a subjected to a uniform pore-pressure increment p_o is obtained by substituting $c_m p_o$ for $\gamma\zeta_o$ in Eqn. 5.17:

$$u_R^{\text{sphere}}(R \geq a) = \frac{c_m p_o V}{4\pi R^2} \tag{5.22}$$

Translating Eqn. 5.21 gives the volumetric strain within a spherical inclusion subjected to a uniform pressure change p_o:

$$\epsilon^{\text{sphere}}(R \leq a) = c_m p_o \tag{5.23}$$

Eqns. 5.23 and 5.23 were derived by Gambolati (1974, p. 1219). Because the volumetric strain within an unconstrained sphere would be $\alpha p_o/K$, Eqn. 5.23 shows that the volumetric strain within a spherical volume embedded in an infinite medium is smaller by the ratio K/K_v, since $c_m = \alpha/K_v$ and $K < K_v$. In other words, the elastic restraint of the surrounding porous medium reduces the volumetric strain from that of an unconstrained body to that of a uniaxially constrained body.

5.3 GREEN'S FUNCTIONS

In 1828, George Green developed a formal solution to Laplace's equation as the superposition of point-source solutions. The point-source solution is now known as the Green's function. This method is also known as the method of singularities because the displacement potential for a point source is infinite (a first-order pole) at the source. Green's function solutions have been used for problems of subsidence and radial flow in infinite and semi-infinite problem domains. When available, they have the advantage of being elegant, analytical solutions. Green's functions for a fluid volume source and a fluid pressure source are developed in this section.

> The preface to MacMillan's (1930) introduction to potential theory contains the following information on George Green:
>
> Green was almost entirely a self-taught mathematician. He did not receive his degree of Bachelor of Arts until 1837, at which time he was forty-four years of age. Notwithstanding these handicaps the brilliant originality of this paper marks it as one of the mathematical classics—a fact which should be stimulating to the more fortunate students of the present day.

In general, a Green's function is the response at a point $\vec{\mathbf{x}} = (x_1, x_2, x_3)$ due to a unit impulse at a point $\vec{\xi} = (\xi_1, \xi_2, \xi_3)$. In the theory of gravity or electrostatics, the Green's function is the potential due to a point mass or point charge. The increment of fluid content $\zeta(\xi_1, \xi_2, \xi_3)$ is a spatial distribution function of injected fluid volume per unit aquifer volume. Thus, the poroelastic equivalent of a point mass or point charge is a unit slug in the sense of well hydraulics. The addition of a unit volume of fluid at a point is represented by the three-dimensional Dirac delta function $\delta(x_1, x_2, x_3) = \delta(x_1)\delta(x_2)\delta(x_3)$. The Dirac delta function is loosely described as a unit impulse function that is zero everywhere except at $x = 0$, where it is infinite, and that has unit area under its graph (e.g., Carrier et al., 1966, p. 319). Thus, it is defined by the properties that $\delta(\vec{\mathbf{x}}-\vec{\xi}) = 0$ for $\vec{\mathbf{x}} \neq \vec{\xi}$ and $\delta(\vec{\mathbf{x}}-\vec{\xi}) \to \infty$ as $\vec{\mathbf{x}} \to \vec{\xi}$ such that

$$\int_{-\infty}^{\infty}\int_{-\infty}^{\infty}\int_{-\infty}^{\infty} \delta(x_1 - \xi_1, x_2 - \xi_2, x_3 - \xi_3) d\xi_1 d\xi_2 d\xi_3 = 1 \tag{5.24}$$

The delta function for a source of increment of fluid volume is the unit addition of fluid volume ($\zeta dV = 1$) at a point.

The Green's function $\Phi^*(x_1, x_2, x_3; \xi_1, \xi_2, \xi_3)$ is defined by the property that it satisfies Poisson's equation for a point source of unit fluid volume located at (ξ_1, ξ_2, ξ_3):

$$\nabla^2 \Phi^* = \gamma \delta(x_1 - \xi_1, x_2 - \xi_2, x_3 - \xi_3) \tag{5.25}$$

The potential due to a point source of fluid is given by Eqn. 5.16 for $\zeta_o V = 1$ in the limit that the radius of the sphere approaches zero. If the center of the sphere is located at $\vec{\xi}$, the Green's function is given by

$$\Phi^*(x_1, x_2, x_3; \xi_1, \xi_2, \xi_3) = -\frac{\gamma}{4\pi R} \tag{5.26}$$

The Green's function depends inversely on the distance $R = |\vec{x} - \vec{\xi}|$ between the points, which is the well-known result for the potential due to a point mass or point charge in the theories of gravity and electrostatics. In terms of Φ^*, Eqn. 5.5 can be written as

$$\Phi(x_1, x_2, x_3) = \int\int\int_D \Phi^*(x_1, x_2, x_3; \xi_1, \xi_2, \xi_3)\zeta(\xi_1, \xi_2, \xi_3) d\xi_1 d\xi_2 d\xi_3 \tag{5.27}$$

The physical interpretation of Eqn. 5.27 is that the potential is a superposition of potentials due to the injection of fluid volumes, $\zeta(\xi_1, \xi_2, \xi_3) d\xi_1 d\xi_2 d\xi_3$, into each volume element, $d\xi_1 d\xi_2 d\xi_3$.

Green's functions for displacement follow directly from Eqns. 5.26 and 5.1:

$$\begin{aligned} u_i^*(x_1, x_2, x_3; \xi_1, \xi_2, \xi_3) &= \frac{\partial \Phi^*}{\partial x_i} \\ &= \gamma \frac{x_i - \xi_i}{4\pi R^3} \end{aligned} \tag{5.28}$$

because

$$\frac{\partial R^{-1}}{\partial x_i} = -\frac{x_i - \xi_i}{R^3} \tag{5.29}$$

Green's functions for displacement u_i^* are directly proportional to the direction cosines $(x_i - \xi_i)/R$ and are inversely proportional to R^2. Because Green's functions u_i^* for displacement are the displacement components due to a source of unit volume injected at a point $\vec{\xi}$ (Fig. 5.2), the displacement components for a distribution of increment of fluid content are

$$u_i(x_1, x_2, x_3) = \int\int\int_D u_i^*(x_1, x_2, x_3; \xi_1, \xi_2, \xi_3)\zeta(\xi_1, \xi_2, \xi_3) d\xi_1 d\xi_2 d\xi_3 \tag{5.30}$$

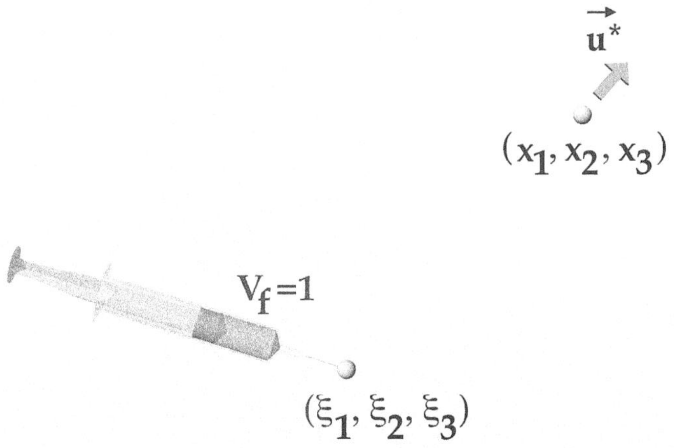

Figure 5.2: Physical interpretation of Green's function $\vec{u}^*$ for displacement. A unit fluid volume $V_f = 1$ is injected at the point (ξ_1, ξ_2, ξ_3).

As with the interpretation of the Poisson integral for potential, the displacement is the superposition of Green's functions for displacement weighted by the increment of fluid content at each point in the problem domain. Green's functions for stress can likewise be obtained from Φ^* and Eqn. 5.7.

The Green's function for a delta function pressure source $p = \delta(x_1 - \xi_1, x_2 - \xi_2, x_3 - \xi_3)$ is

$$\Phi^*(x_1, x_2, x_3; \xi_1, \xi_2, \xi_3) = -\frac{c_m}{4\pi R} \tag{5.31}$$

The physical meaning of a delta function pressure source is that it is a unit addition of energy ($pdV = 1$) at a point in the problem domain. That pressure is an energy density was first noted rigorously in the groundwater context by Hubbert (1940). The displacement components follow directly from Eqn. 5.31 or by substituting c_m for γ in Eqn. 5.28.

5.4 CENTER OF DILATATION

Combinations of point forces are known generally as *nuclei of strain* (Fig. 5.3) (Love, 1944, Chapter 8). A center of dilatation or tension center is built up as the superposition of three orthogonal double forces without moment (aligned opposing point forces of equal strength). In this section the Green's function (Eqn. 5.28) for the displacement field due to a point source

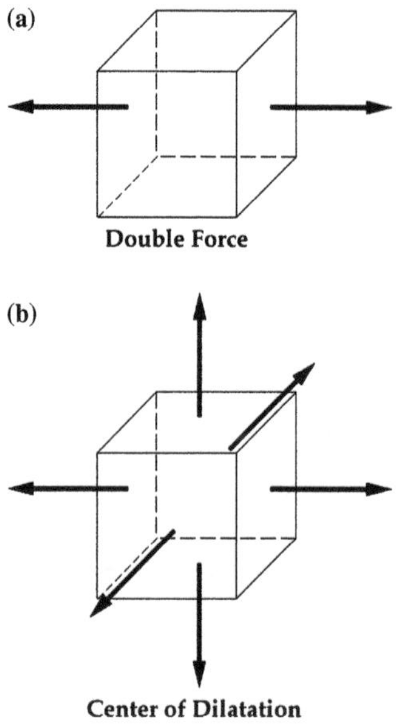

Figure 5.3: Nuclei of strain. (a) Double force. (b) Center of dilatation.

of fluid volume is shown to be identical to the displacement field due to a center of dilatation.

A double force without moment in the x-direction is formed from a point force $-P$ located at (0,0,0) and directed in the negative x-direction, and an opposing point force $+P$ located at $(d, 0, 0)$ and directed in the positive x-direction in the limit as d approaches zero. In this section, the coordinates and displacement components are written explicitly as (x, y, z) and (u, v, w), respectively. The two point forces can be represented by body-force distributions, $-P\delta(x, 0, 0)$ and $+P\delta(x - d, 0, 0)$, with units of force per unit volume. The point forces are recovered on performing the volume integration. The body-force distribution for the double force is

$$\begin{aligned} F_x^{\text{double}} &= \lim_{d\to 0}\left\{\frac{-(Pd)\delta(x,0,0) + (Pd)\delta(x-d,0,0)}{d}\right\} \\ &= -(Pd)\frac{\partial}{\partial x}\delta(x,0,0) \end{aligned} \tag{5.32}$$

A corresponding result can be written for the remaining two coordinate directions. The double-force strength m is defined to be $\lim_{d\to 0}(Pd)$.

A volume of fluid, V_f, added at the point $\vec{x} = \vec{\xi}$ is represented by $\zeta(\vec{x}) = V_f\delta(\vec{x} - \vec{\xi})$. The gradient of increment of fluid content is equivalent mathematically to a body force $F_i = -BK_u\partial\zeta/\partial x_i$ (cf. Eqn. 4.21), which for a point source of fluid at $\vec{x} = \vec{\xi}$ is

$$F_i = -BK_u V_f \frac{\partial}{\partial x_i}\delta(\vec{x} - \vec{\xi}) \tag{5.33}$$

Thus, comparing Eqn. 5.32 with Eqn. 5.33 shows that three orthogonal double forces of strength $Pd = BK_u V_f$ are equivalent to a point fluid source $V_f = Pd/BK_u$. Hence, the displacement field for a center of dilatation can be obtained from Eqn. 5.28 and Eqn. 3.79:

$$u_i^{\text{center}}(x, y, z) = \frac{Pd}{4\pi K_v^{(u)}} \frac{x_i - \xi_i}{R^3} \tag{5.34}$$

5.5 HALF SPACE WITH TRACTION-FREE BOUNDARY

Mindlin (1936) solved the elastostatic problem of a single force at a point in a half space with a traction-free boundary. His solution is the half-space equivalent of the Kelvin solution for a force at a point in an infinite medium (Love, 1944, Chapter 8). Mindlin and Cheng (1950a, 1950b) and Sen (1951) solved the problem of finding displacements for various nuclei of strain within a half space. The geometry for the half-space problem is shown in Figure 5.4. The result for a center of dilatation located at $(0, 0, z')$ is

$$\vec{\mathbf{u}}^{*\text{HS}} = \vec{\nabla}\Phi^*(x, y, z; 0, 0, z') + \vec{\nabla}_2\Phi^*(x, y, z; 0, 0, -z') \tag{5.35}$$

where the definition of the operator $\vec{\nabla}_2$ depends on whether the nucleus of strain is equivalent to a fluid volume source or to a fluid pressure source:

$$\vec{\nabla}_2 = (3-4\nu_u)\vec{\nabla}+2\vec{\nabla}z\frac{\partial}{\partial z}-4(1-\nu_u)\hat{\mathbf{k}}\nabla^2 z \quad \text{fluid volume source} \tag{5.36}$$

$$\vec{\nabla}_2 = (3-4\nu)\vec{\nabla}+2\vec{\nabla}z\frac{\partial}{\partial z}-4(1-\nu)\hat{\mathbf{k}}\nabla^2 z \quad \text{fluid pressure source} \tag{5.37}$$

In Eqn. 5.35 the operator $\vec{\nabla}_2$ is applied to Φ^* at the image point $(0, 0, -z')$. In Eqns. 5.36 and 5.37, $\hat{\mathbf{k}}$ is the unit vector in the z-direction. The undrained Poisson's ratio occurs in Eqn. 5.36 because a fluid volume source is equivalent to a center of dilatation in the form of the governing equation containing displacement and increment of fluid content as dependent variables (Eqn. 4.21). The drained Poisson's ratio occurs in Eqn. 5.37 because a fluid pressure source is equivalent to a center of dilatation in the form of the governing equation containing displacement and pressure as dependent variables

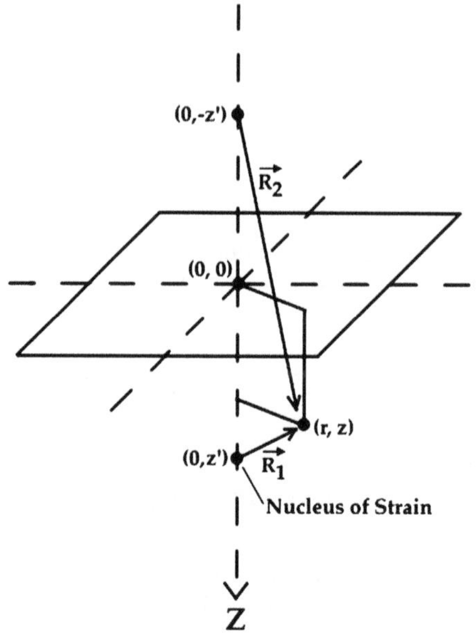

Figure 5.4: Geometry for the half-space solution for nucleus of strain at $(0, z')$. Image source is located at $(0, z')$ (after Geertsma, 1966).

(Eqn. 4.15). Green's function for displacement can be written explicitly from Eqn. 5.35 for a point source of fluid volume or fluid pressure using Eqn. 5.26 or Eqn. 5.31, respectively,

$$\vec{\mathbf{u}}^{*\mathrm{HS}} = \frac{\gamma}{4\pi}\left\{\frac{\vec{R}_1}{R_1^3} + \frac{(3-4\nu_u)\vec{R}_2}{R_2^3} - \frac{6z(z+z')\vec{R}_2}{R_2^5} - \frac{2\hat{\mathbf{k}}}{R_2^3}\right.$$
$$\left.\times[(3-4\nu_u)(z+z')-z]\right\} \quad \text{fluid volume source} \qquad (5.38)$$

$$\vec{\mathbf{u}}^{*\mathrm{HS}} = \frac{c_m}{4\pi}\left\{\frac{\vec{R}_1}{R_1^3} + \frac{(3-4\nu)\vec{R}_2}{R_2^3} - \frac{6z(z+z')\vec{R}_2}{R_2^5} - \frac{2\hat{\mathbf{k}}}{R_2^3}\right.$$
$$\left.\times[(3-4\nu)(z+z')-z]\right\} \quad \text{fluid pressure source} \qquad (5.39)$$

where $R_1^2 = r^2 + (z - z')^2$ and $R_2^2 = r^2 + (z + z')^2$. The half-space results are a weighted superposition of a center of dilatation at the source, a center of dilatation at the image point, a double force at the image point, and a doublet at the image point (Mindlin and Cheng, 1950b). The forces at the image point produce the traction-free boundary. The vertical component of the displacement for these two cases (Geertsma, 1966, p. 590) is

$$u_z^{*HS} = \frac{\gamma}{4\pi}\left\{\frac{z-z'}{R_1^3} + \frac{4\nu_u(z+z')-(z+3z')}{R_2^3} - \frac{6z(z+z')^2}{R_2^5}\right\}$$

fluid volume source (5.40)

$$u_z^{*HS} = \frac{c_m}{4\pi}\left\{\frac{z-z'}{R_1^3} + \frac{4\nu(z+z')-(z+3z')}{R_2^3} - \frac{6z(z+z')^2}{R_2^5}\right\}$$

fluid pressure source (5.41)

The displacement at the surface $z = 0$ for a point source at $(0, 0, z')$ is directly proportional to the infinite space solution u_z^*:

$$u_z^{*HS}(r, 0) = 4(1-\nu_u)u_z^*(r, 0) \qquad \zeta \text{ distribution} \qquad (5.42)$$

$$u_z^{*HS}(r, 0) = 4(1-\nu)u_z^*(r, 0) \qquad \text{pressure distribution} \qquad (5.43)$$

5.6 GRAVITY ANALOGY

The gravitational potential U satisfies Poisson's equation (e.g., Telford et al., 1990),

$$\nabla^2 U = -4\pi \mathcal{G} \rho(x, y, z) \qquad (5.44)$$

where $\mathcal{G}$ is the universal gravitational constant and $\rho(x, y, z)$ is the distribution of mass density. The components of the acceleration of gravity due to the anomaly are recovered as the gradient of U:

$$g_i = \frac{\partial U}{\partial x_i} \qquad (5.45)$$

Eqns. 5.44 and 5.45 establish the mathematical analogy between gravitational potential and displacement potential. Because the displacement components at the surface of a half space are proportional to the infinite space solution (Eqns. 5.42 and 5.43), formulas for the components of gravity at the earth's surface are proportional to the corresponding components of displacement. A point source, horizontal slab, and vertical cylinder are now discussed.

5.6.1 Point Source

In the case of gravity, a unit mass at depth z' below the origin is represented by $\rho = \delta(x, y, z - z')$. The vertical component of gravity on the plane $z = 0$

is readily obtained by resolving the force from Newton's inverse square law into the vertical direction (e.g., Telford et al., 1990, p. 35),

$$g_z^*(r, 0) = \mathcal{G}\frac{z'}{(r^2 + z'^2 s)^{3/2}} \tag{5.46}$$

where the asterisk indicates that Eqn. 5.46 is the Green's function for vertical acceleration of gravity per unit mass for a point source at $(0, z')$. The horizontal component of gravity per unit mass is

$$g_r^*(r, 0) = -\mathcal{G}\frac{r}{(r^2 + z'^2)^{3/2}} \tag{5.47}$$

Comparing the infinite space governing equations Eqns. 5.2 or 5.3 with Eqn. 5.44 and using the relationship between the full-space and half-space Green's functions for displacements on the surface $z = 0$ (Eqns. 5.42 and 5.43) give the equation

$$\frac{u_z^{*\mathrm{HS}}(r, 0)}{g_z^*(r, 0)} = -\frac{(1-\nu)c_m}{\pi \mathcal{G}} \qquad \text{pressure distribution} \tag{5.48}$$

for a pressure distribution and

$$\frac{u_z^{*\mathrm{HS}}(r, 0)}{g_z^*(r, 0)} = -\frac{(1-\nu_u)\gamma}{\pi \mathcal{G}} \qquad \zeta \text{ distribution} \tag{5.49}$$

for a distribution of increment of fluid content. By Eqn. 5.48, therefore, the vertical displacement on the free surface due to a unit energy point source $p = \delta(x, y, z - z')$ (cf. Geertsma, 1966, p. 590) is

$$u_z^{*\mathrm{HS}}(r, 0) = -\frac{c_m(1-\nu)}{\pi}\frac{z'}{(r^2 + z'^2)^{3/2}} \tag{5.50}$$

Eqn. 5.50 can also be obtained directly from Eqn. 5.41. Similarly, the horizontal displacement is

$$u_r^{*\mathrm{HS}}(r, 0) = \frac{c_m(1-\nu)}{\pi}\frac{r}{(r^2 + z'^2)^{3/2}} \tag{5.51}$$

5.6.2 Horizontal Slab

The vertical gravity anomaly for a thin horizontal slab of thickness h that extends between $-L \leq x \leq +L$ in width and infinitely in the y-direction (Telford et al., 1990, p. 40) is

$$g_z^{\text{slab}}(x, 0, 0) = 2\mathcal{G}\rho_o h \left\{ \tan^{-1} \frac{x - L}{h} - \tan^{-1} \frac{x + L}{h} \right\} \tag{5.52}$$

where ρ_o is the anomalous density.

The poroelastic result for vertical displacement is obtained from Eqn. 5.52 using the same ratio as for the respective point-source solutions (Eqns. 5.48 and 5.49) because the slab result can be obtained as an integration over point sources. Therefore, for a uniform decrease of increment of fluid content $-\zeta_o$, the vertical subsidence is

$$u_z^{\text{slab}}(x, 0, 0) = \frac{2(1 + \nu_u)B}{3\pi} \zeta_o h \left\{ \tan^{-1} \frac{x - L}{h} - \tan^{-1} \frac{x + L}{h} \right\} \tag{5.53}$$

Eqn. 5.53 was used by Segall (1989) to study subsidence at the Wilmington oil field. An equivalent solution holds for a uniform decrease of pressure $-p_o$.

5.6.3 Vertical Cylinder

Consider a vertical cylinder of density ρ_o, length h, and radius a, whose top is at a depth $z = z'$ below the surface. The vertical component of gravity at $z = 0$ on the axis of the cylinder (e.g., Telford et al., 1990, p. 37) is

$$g_z^{\text{cyl}}(0, 0) = 2\pi \mathcal{G}\rho_o \left\{ h + \sqrt{a^2 + z'^2} - \sqrt{a^2 + (z' + h)^2} \right\} \tag{5.54}$$

In the limit that the cylinder is very thin ($h \ll z'$), Eqn. 5.54 becomes

$$g_z^{\text{disk}}(0, 0) = 2\pi \mathcal{G}\rho_o h \left\{ 1 - \frac{z'}{\sqrt{a^2 + z'^2}} \right\} \tag{5.55}$$

By the argument as presented for the horizontal slab, the ratio of vertical displacement at $z = 0$ for a cylinder in a half space to the gravity anomaly is the same as that for a point source. Therefore, the vertical displacement at the origin for a uniform pressure decrease $-p_o$ in a vertical cylinder and disk, respectively, are as follows:

$$u_z^{\text{cyl}}(0,0) = 2(1-\nu)c_m p_o \left\{ h + \sqrt{a^2 + z'^2} - \sqrt{a^2 + (z'+h)^2} \right\} \quad (5.56)$$

$$u_z^{\text{disk}}(0,0) = 2(1-\nu)c_m p_o h \left\{ 1 - \frac{z'}{\sqrt{a^2 + z'^2}} \right\} \quad (5.57)$$

Eqn. 5.57 has been presented by Geertsma (1966) as an approximation to subsidence due to uniform pressure decline in a disk-shaped reservoir.

5.7 SUDDEN INJECTION OF FLUID AT A POINT IN INFINITE DOMAIN

The Green's functions for a point injection of fluid introduced thus far are for the elastostatic problem in an infinite problem domain. They allow for the calculation of displacements and stresses in the drained limit given the distribution of pressure sources, and in the undrained limit given the distribution of fluid sources. A Green's function for the time-dependent problem is obtained in this section for an instantaneous point source (*slug* in hydrogeology) in an infinite medium. The half-space problem, which is axially symmetric, is considered in Section 9.6.

Let the instantaneous injection of a fixed volume of fluid, V_f, at the origin be represented by

$$Q(\vec{x}, t) = V_f \delta(\vec{x}) \delta(t) \quad (5.58)$$

where Q is the source term in the diffusion equation (Eqn. 4.67) for ζ. The units of Q are fluid volume per unit bulk volume per unit time. The fluid volume V_f is recovered on integrating over a small region ΔV around the origin and over a small time Δt after time zero. In the absence of body forces, the well-known solution of Eqn. 4.67 for a point source (Carslaw and Jaeger, 1959, p. 256) is

$$\zeta(R, t) = \frac{V_f}{(4\pi c t)^{3/2}} \exp\left[-\frac{R^2}{4ct} \right] \quad (5.59)$$

where $R^2 = x^2 + y^2 + z^2$. The solution satisfies the requirement that integrating ζ over the infinite problem domain recovers the initial fluid volume V_f,

$$\int_{-\infty}^{\infty} dx \int_{-\infty}^{\infty} dy \int_{-\infty}^{\infty} dz \frac{V_f}{(4\pi c t)^{3/2}} \exp\left[-\frac{x^2 + y^2 + z^2}{4ct} \right] = V_f \quad (5.60)$$

This equation follows from the result that the value of the following one-dimensional integral is one:

$$\frac{1}{\sqrt{4\pi ct}}\int_{-\infty}^{\infty}\exp\left[-\frac{x^2}{4ct}\right]dx = \frac{2}{\sqrt{\pi}}\int_0^{\infty}\exp[-u^2]du = \mathrm{erf}(\infty) = 1 \quad (5.61)$$

The increment of fluid content is spherically symmetric as the fluid diffuses radially away from the origin. Therefore, the radial displacement at distance R from the origin is found by integrating the contributions from thin spherical shells of radius a and thickness Δa. Only shells whose radii are smaller than R contribute because the displacement inside a spherical shell source is zero (cf. Section 5.2). The variable ζ_o in Eqn. 5.14 is replaced by $\zeta(a,t)$ (Eqn. 5.59) to give

$$\begin{aligned} u_r(R,t) &= \gamma\int_0^R \zeta(a,t)\frac{a^2 da}{R^2} \\ &= \frac{\gamma}{R^2}\frac{V_f}{(4\pi ct)^{3/2}}\int_0^R \exp\left[-\frac{a^2}{4ct}\right]a^2 da \\ &= \frac{\gamma V_f}{4\pi R^2}g\left(\frac{R}{\sqrt{ct}}\right) \end{aligned} \quad (5.62)$$

where

$$g(\xi) \equiv \frac{1}{2\sqrt{\pi}}\int_0^{\xi} s^2\exp\left(-\frac{1}{4}s^2\right)ds = \mathrm{erf}\left(\frac{1}{2}\xi\right) - \frac{\xi}{\sqrt{\pi}}\exp\left(-\frac{1}{4}\xi^2\right) \quad (5.63)$$

and

$$\mathrm{erf}(z) \equiv \frac{2}{\sqrt{\pi}}\int_0^z \exp(-t^2)dt \quad (5.64)$$

Note that

$$\frac{dg}{d\xi} = \frac{1}{2\sqrt{\pi}}\xi^2\exp\left(-\frac{1}{4}\xi^2\right) \quad (5.65)$$

The Cartesian displacement components are obtained from the radial displacement by multiplying by the direction cosines (Rudnicki, 1986a) to yield

$$u_i(\vec{x},t) = \frac{\gamma V_f}{4\pi}\frac{x_i}{R^3}g\left(\frac{R}{\sqrt{ct}}\right) \quad (5.66)$$

As $t \to 0$ the displacement approaches the elastostatic solution for a point fluid source (Eqn. 5.28 with $\xi_i = 0$) because $g(R/\sqrt{ct}) \to 1$ as $t \to 0$. This result illustrates the connection between the elastostatic Green's function and the time-dependent Green's function for the instantaneous injection of a slug. As $t \to \infty$, $g(R/\sqrt{ct}) \to 0$, and the displacements approach zero.

The flow field for a point source in three dimensions is irrotational, as all flow vectors radiate from the point. Hence, $\zeta = Sp$ (cf. Section 4.10). Using this relationship in Eqn. 5.59 means that the pore pressure can be expressed variously as

$$\begin{aligned} p(\vec{x}, t) &= \frac{1}{S} \frac{V_f}{(4\pi ct)^{3/2}} \exp\left[-\frac{R^2}{4ct} \right] \\ &= \frac{\gamma}{c_m} \frac{V_f}{4\pi R^3} \xi \frac{dg}{d\xi} \\ &= \frac{BK_u(\lambda + 2G)}{\alpha(\lambda_u + 2G)} \frac{V_f}{4\pi R^3} \xi \frac{dg}{d\xi} \end{aligned} \tag{5.67}$$

where $\xi = R/\sqrt{ct}$. Finally, the stress components are computed from the stress-strain-pressure version of the constitutive equation (Eqn. 2.42):

$$\begin{aligned} \sigma_{ij}(\vec{x}, t) &= \frac{V_f}{2\pi R^3} \frac{GBK_u}{\lambda_u + 2G} \\ &\times \left[\delta_{ij} \left(g - \xi \frac{dg}{d\xi} \right) + \frac{x_i x_j}{R^2} \left(\xi \frac{dg}{d\xi} - 3g \right) \right] \end{aligned} \tag{5.68}$$

5.8 FLUID DIPOLE EQUIVALENCE TO BODY-FORCE DISTRIBUTIONS

In this section a body-force distribution is shown to be equivalent mathematically to a distribution of fluid dipoles in the fluid diffusion equation (Eqn. 4.67) (Cleary, 1977; Rudnicki, 1986a). The equivalence between a body force and a fluid dipole source is similar to the previously demonstrated equivalence between a fluid point source and double forces forming a center of dilatation in the force equilibrium equation (Eqn. 4.21) (cf. Section 5.4).

5.8.1 Single Dipole

The groundwater analogue of a charge dipole is a doublet of extraction and injection wells that are closely spaced and of equal strength.[2] Specifically, let $-V_f$ and $+V_f$ be the sink and source strengths located at $\vec{x}$ and $\vec{x} + \vec{d}$,

[2] This discussion parallels the treatment of the instantaneous and continuous doublet in heat conduction by Carslaw and Jaeger (1959, pp. 270–272).

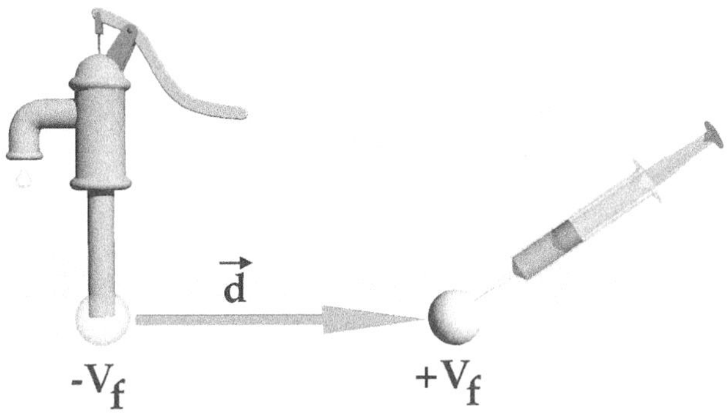

Figure 5.5: Dipole of increment of fluid content. The pole strength, V_f, is the fluid volume injected at a point in the aquifer. The dipole moment is $\vec{m} = V_f \vec{d}$.

respectively. The dipole moment $\vec{m}$ is the product of the pole strength, V_f, and the distance vector, $\vec{d}$, extending from the negative pole to the positive pole in the limit as $d \to 0$ (Fig. 5.5):

$$\vec{m} \equiv \lim_{d \to 0} V_f \vec{d} \tag{5.69}$$

5.8.2 Dipole Distribution

The concept of a single dipole can be extended to a dipole distribution consisting of a fluid-sink distribution and a replicated fluid-source distribution displaced by $\vec{d}$ in the limit as $\vec{d} \to 0$. This superposition means that the total fluid-source distribution function is the sum of the sink distribution $-Q(\vec{x}, t)$ and the displaced source distribution $+Q(\vec{x} - \vec{d}, t)$:

$$Q^{\text{dipole}}(\vec{x}, t) = -Q(\vec{x}, t) + Q(\vec{x} - \vec{d}, t) = -[Q(\vec{x}, t) - Q(\vec{x} - \vec{d}, t)] \tag{5.70}$$

In the limit $d \to 0$, the quantity in brackets in Eqn. 5.70 can be expressed in terms of the partial derivatives of Q,

$$Q^{\text{dipole}}(\vec{x}, t) = -\left[\frac{\partial Q}{\partial x_1} d_1 + \frac{\partial Q}{\partial x_2} d_2 + \frac{\partial Q}{\partial x_3} d_3\right] \tag{5.71}$$

where the components of the displacement vector $\vec{d}$ are (d_1, d_2, d_3). Alternative forms of Eqn. 5.71 include

$$Q^{\text{dipole}}(\vec{x}, t) = -\vec{\nabla} Q \cdot \vec{d} = -\frac{\partial Q}{\partial x_k} d_k \tag{5.72}$$

where $\vec{\nabla} Q \cdot \vec{d}$ is the directional derivative of Q in the direction $\vec{d}$.

The increment of fluid content, displacement, and fluid pressure due to the fluid dipole distribution, $Q^{\text{dipole}}(\vec{x}, t)$, given by Eqn. 5.72 can be obtained from the solutions ζ, u_i, and p due to the distribution for increment of fluid content, $Q(\vec{x}, t)$, according to the prescription

$$\zeta^{\text{dipole}} = -\vec{\nabla} \zeta \cdot \vec{d} \tag{5.73}$$

$$u_i^{\text{dipole}} = -\vec{\nabla} u_i \cdot \vec{d} \tag{5.74}$$

$$p^{\text{dipole}} = -\vec{\nabla} p \cdot \vec{d} \tag{5.75}$$

Eqns. 5.73–5.75 can be verified by inserting them into the governing Eqns. 4.67 and 4.21.

Now consider a body-force distribution, $F_i(\vec{x}, t) = P_i f(\vec{x}, t) = P_i f(x_1, x_2, x_3, t)$, where P_i is independent of position and time. The body-force term in Eqn. 4.67 is in the form of a dipole distribution (cf. Eqn. 5.72) with the following identifications:

$$d_i = -P_i \tag{5.76}$$

$$Q(\vec{x}, t) = \frac{k\gamma}{\mu} f(\vec{x}, t) \tag{5.77}$$

5.8.3 Single-Point Force

For a sudden single-point force, $f(\vec{x}, t)$ in Eqn. 5.77 is set equal to $\delta(\vec{x})\delta(t)$. Eqns. 5.58, 5.66, and 5.74 give

$$u_i^{(1)} = \frac{k\gamma^2 P_k}{4\pi\mu R^3} \left\{ g(\xi)\delta_{ik} + \frac{x_i x_k}{R^2} \left[\xi \frac{dg}{d\xi} - 3g(\xi) \right] \right\} \tag{5.78}$$

where $u_i^{(1)}$ is the solution of Eqn. 4.21 with $F_i = 0$. Then let $u_i^{(2)}$ be the solution of Eqn. 4.21 with $\zeta = 0$ (and hence $d\zeta/dx_i = 0$) for the specified force distribution $F_j(\vec{x}, t) = P_j \delta(\vec{x})\delta(t)$. That is, $u_i^{(2)}$ is the Kelvin solution

(Love, 1944, Chapter 8; Rudnicki, 1986a) for the standard elasticity problem with undrained constants:

$$u_i^{(2)} = \frac{P_j}{8\pi RG(\lambda_u + 2G)} \left\{ (\lambda_u + 3G)\delta_{ij} + (\lambda_u + G)\frac{x_i x_j}{R^2} \right\} \tag{5.79}$$

The displacement solution for the point force is the superposition of $u_i^{(1)}$ and $u_i^{(2)}$.

6

Uniaxial Strain

6.0 CHAPTER OVERVIEW

In this chapter the poroelastic constitutive relations and equations of mechanical equilibrium and fluid continuity are specialized to uniaxial strain. Only vertical displacements are allowed in problems of uniaxial strain because horizontal displacements are assumed to be zero. These assumptions have been used to model laboratory consolidation tests in which the walls of the sample container are considered to be rigid. The same assumptions are used to approximate a horizontal aquifer or reservoir of infinite lateral extent. The uniaxial strain problem is one dimensional because the stresses, displacements, strains, pore pressure, and increment of fluid content are functions only of the vertical coordinate. In the frequently encountered case of constant vertical stress, the pore-pressure field uncouples mathematically from the stress or strain field. Several example problems are considered in this chapter: consolidation of a half space and a finite layer, pore-pressure changes beneath a reservoir lake or in the seabed, sediment loading and natural hydraulic fracturing in a sedimentary basin, and water-level fluctuations in wells due to changes in barometric pressure.

6.1 CONSTITUTIVE EQUATIONS

The constitutive equations for uniaxial strain are obtained by inserting the constraint that $\epsilon_{xx} = \epsilon_{yy} = 0$ into the stress-strain-pore-pressure form of the constitutive equations (Eqns. 2.44–2.46) and noting that $\epsilon_{kk} = \epsilon_{zz}$:

$$\sigma_{xx}\Big|_{\epsilon_{xx}=\epsilon_{yy}=0} = \sigma_{yy}\Big|_{\epsilon_{xx}=\epsilon_{yy}=0} = \frac{2G\nu}{1-2\nu}\epsilon_{zz} - \alpha p \tag{6.1}$$

$$\sigma_{zz}\Big|_{\epsilon_{xx}=\epsilon_{yy}=0} = \frac{2G(1-\nu)}{1-2\nu}\epsilon_{zz} - \alpha p \tag{6.2}$$

Solving Eqn. 6.2 for ϵ_{zz} and substituting into Eqn. 6.1 yields

$$\begin{aligned}\sigma_{xx}\Big|_{\epsilon_{xx}=\epsilon_{yy}=0} = \sigma_{yy}\Big|_{\epsilon_{xx}=\epsilon_{yy}=0} &= \frac{\nu}{1-\nu}\sigma_{zz} - \frac{1-2\nu}{1-\nu}\alpha p \\ &= \frac{\nu}{1-\nu}\sigma_{zz} - 2\eta p \end{aligned} \tag{6.3}$$

For drained conditions ($p = 0$), Eqn. 6.3 gives the lateral stresses:

$$\sigma_{xx}\Big|_{p=0;\ \epsilon_{xx}=\epsilon_{yy}=0} = \sigma_{yy}\Big|_{p=0;\ \epsilon_{xx}=\epsilon_{yy}=0} = \frac{\nu}{1-\nu}\sigma_{zz}\Big|_{p=0;\ \epsilon_{xx}=\epsilon_{yy}=0} \tag{6.4}$$

Thus, the lateral stress increases proportionally to the vertical stress for the drained situation, and the expression is the usual result obtained from standard elasticity theory. For example, the horizontal stresses are one quarter the value of the lithostatic stress for a drained Poisson's ratio of 0.2. An expression for the drained vertical incompressibility (cf. Eqn. 3.55) is also obtained from Eqn. 6.2:

$$K_v \equiv \frac{1}{\beta_v} \equiv \frac{\delta\sigma_{zz}}{\delta\epsilon_{zz}}\Big|_{\epsilon_{xx}=\epsilon_{yy}=p=0} = \frac{2G(1-\nu)}{1-2\nu} \tag{6.5}$$

The constitutive equations for stress, strain, and increment of fluid content (Eqn. 2.55) give the following constitutive equations for conditions of uniaxial strain:

$$\sigma_{xx}\Big|_{\epsilon_{xx}=\epsilon_{yy}=0} = \sigma_{yy}\Big|_{\epsilon_{xx}=\epsilon_{yy}=0} = \frac{2G\nu_u}{1-2\nu_u}\epsilon_{zz} - \alpha M\zeta \tag{6.6}$$

$$\sigma_{zz}\Big|_{\epsilon_{xx}=\epsilon_{yy}=0} = \frac{2G(1-\nu_u)}{1-2\nu_u}\epsilon_{zz} - \alpha M\zeta \tag{6.7}$$

Solving Eqn. 6.7 for ϵ_{zz} and substituting into Eqn. 6.6 yields

$$\sigma_{xx}\Big|_{\epsilon_{xx}=\epsilon_{yy}=0} = \sigma_{yy}\Big|_{\epsilon_{xx}=\epsilon_{yy}=0} = \frac{\nu_u}{1-\nu_u}\sigma_{zz} - \frac{1-2\nu_u}{1-\nu_u}\alpha M\zeta \tag{6.8}$$

Eqns. 6.6–6.8 have the same form as Eqns. 6.1–6.3, but with ν_u in place of ν and $\alpha M\zeta$ in place of αp, where the Biot modulus, M, is the reciprocal of the storage coefficient for constant strain conditions (cf. Section 3.3). Thus, the increase of lateral stress as a function of vertical stress for the undrained situation is the same as the result obtained from standard elasticity theory but with the undrained Poisson's ratio in place of the drained value. Similarly, the expression for undrained vertical incompressibility is the same as for the drained vertical incompressibility (Eqn. 6.5) but with undrained Poisson's ratio:

$$K_v^{(u)} \equiv \frac{1}{\beta_v^{(u)}} \equiv \left.\frac{\delta\sigma_{zz}}{\delta\epsilon_{zz}}\right|_{\epsilon_{xx}=\epsilon_{yy}=\zeta=0} = \frac{2G(1-\nu_u)}{1-2\nu_u} \tag{6.9}$$

The induced pore pressure was shown previously to be proportional to σ_{zz} for uniaxial, undrained conditions (cf. Section 3.6.2),

$$p|_{\epsilon_{xx}=\epsilon_{yy}=\zeta=0} = -\frac{B(1+\nu_u)}{3(1-\nu_u)}\sigma_{zz} = -\gamma\sigma_{zz} \tag{6.10}$$

where γ is the loading efficiency.

Example: *Natural Hydraulic Fracturing*

The initiation of a vertical fracture in a sedimentary basin due to pore-pressure buildup can be approximated by the static constitutive equations for uniaxial strain (Engelder and Lacazette, 1990). The sedimentary layer is assumed to be in an initially drained state when a fluid source, such as oil or gas generation from kerogen maturation, increases the internal fluid pressure. The lateral stress given by Eqn. 6.3 decreases (i.e., becomes more compressive) linearly with pore pressure but at a rate only half that of the pore-pressure increase for $\nu = 0.25$ and $\alpha = 0.75$. If the tensile strength can be neglected, the critical pore pressure for initiating a natural hydraulic fracture is reached when it reduces the Terzaghi compressive horizontal stress $-(\sigma_{xx} + p)$ to zero. For example, if the lithostatic stress $\sigma_{zz} = -10$ MPa, the critical pore pressure is 6.7 MPa for the values of ν and α just given. The critical pore pressure depends strongly on the values of Poisson's ratio and the Biot-Willis parameter.

6.2 FORCE EQUILIBRIUM EQUATION

If the z-axis is vertical and the only body force is gravity, the mechanical equilibrium equations (Eqns. 4.7–4.9) reduce to the single equation

$$\frac{\partial\sigma_{zz}}{\partial z} = -F_z \tag{6.11}$$

If body forces are absent, σ_{zz} is independent of z, although it can still be a function of time.

The uniaxial strain condition means that the volumetric strain is equal to the vertical strain; hence, Eqn. 4.20 reduces to

$$\frac{\partial^2 w}{\partial z^2} = c_m \frac{\partial p}{\partial z} \tag{6.12}$$

Eqn. 6.12 can be integrated immediately to yield a first-order differential equation,

$$\frac{\partial w}{\partial z} = c_m p + g(t) \tag{6.13}$$

where the constant of integration $g(t)$ can be a function of time. Eqn. 6.13 is the uniaxial strain form of Eqn. 4.73 because the displacement field is irrotational.

6.3 FLUID DIFFUSION EQUATION

6.3.1 Stress Form

If body forces can be neglected and no fluid sources are present, using Eqn. 6.3 in Eqn. 4.75 means that, in the uniaxial strain case, Eqn. 4.77 becomes

$$\frac{\partial p}{\partial t} - c\frac{\partial^2 p}{\partial z^2} = -\gamma \frac{d\sigma_{zz}}{dt} \tag{6.14}$$

where c is the uniaxial hydraulic diffusivity and γ is the loading efficiency. The right-hand side of Eqn. 6.14 contains the total time derivative of σ_{zz} rather than the partial derivative, because σ_{zz} must be independent of z in the absence of body forces, as noted at the beginning of Section 6.2. Eqn. 6.14 can be expressed in several alternative forms based on different expressions for loading efficiency γ given in Section 3.6.2 (cf. Eqns. 3.78 and 3.83):

$$\frac{\partial p}{\partial t} - c\frac{\partial^2 p}{\partial z^2} = -\frac{B(1+\nu_u)}{3(1-\nu_u)}\frac{d\sigma_{zz}}{dt} \tag{6.15}$$

$$\frac{\partial p}{\partial t} - c\frac{\partial^2 p}{\partial z^2} = -\frac{\eta}{GS}\frac{d\sigma_{zz}}{dt} \tag{6.16}$$

If vertical stress is constant with time, the right-hand side of Eqns. 6.14–6.16 is zero, and pore pressure satisfies the standard transient flow equation used in hydrogeology:

$$\frac{\partial p}{\partial t} - c\frac{\partial^2 p}{\partial z^2} = 0 \tag{6.17}$$

In Eqn. 6.17 the diffusivity for the pore pressure is the same as that for the increment of fluid volume content in Eqn. 4.67. It must be emphasized that the reduction of the poroelastic problem to an uncoupled diffusion equation for pressure is not a general result but one specific to the conditions of one-dimensional, uniaxial strain with constant vertical stress.

6.3.2 Strain Form

A governing equation equivalent to Eqn. 6.14 for fluid diffusion can be written in terms of strain rate. For uniaxial strain ($\epsilon_{kk} = \epsilon_{zz}$), Eqn. 4.65 becomes

$$\frac{\partial p}{\partial t} - c_\epsilon\frac{\partial^2 p}{\partial z^2} = -\alpha M\frac{\partial \epsilon_{zz}}{\partial t} \tag{6.18}$$

where c_ϵ is the diffusivity based on the storage coefficient at constant strain. An important difference between Eqns. 6.18 and 6.14 is that the vertical strain ϵ_{zz} is a function of both z and t, unlike σ_{zz}, which is independent of z by force equilibrium. Hence, the time derivative of ϵ is a partial and not total derivative—that is, $\partial\epsilon/\partial t$ is a function of z.

6.4 STEP LOAD ON SEMI-INFINITE COLUMN

The sudden application of a uniform surface load by a permeable piston is the canonical consolidation problem. Also, a sedimentation or erosion event, which occurs over a short span of time relative to fluid diffusion, can be modeled as a step change in vertical load. The solutions for a semi-infinite column and a finite length column are treated in this and the following sections. *The z-axis is taken to be positive in the downward direction.*

6.4.1 Initial and Boundary Conditions

The problem is to find the pore pressure, $p(z,t)$, and vertical displacement, $w(z,t)$, after a uniform downward stress $-\sigma_o$ is suddenly applied at $t = 0$ on the surface ($z = 0$) of an infinite length vertical column that is constrained laterally (Fig. 6.1). The vertical load is held constant with time, and the top surface is assumed to be drained. The same boundary and initial conditions apply for a horizontally extensive blanket load, because the infinite lateral extent implies no lateral displacement in a vertical column. It is also implicitly assumed that no horizontal fluid flow occurs.

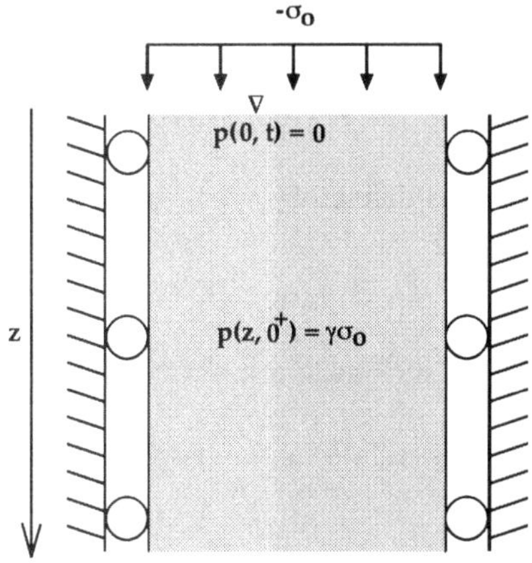

Figure 6.1: Initial and boundary conditions for step load on a half space (after Lewallen and Wang, 1998).

Mechanical equilibrium requires that the suddenly applied surface load $-\sigma_o$ be transmitted instantaneously throughout the column. This constant vertical stress at all depths induces an excess pore pressure equal to $p_o \equiv \gamma\sigma_o$ (Eqn. 6.10), likewise at all depths. Therefore, the induced pore pressure is a consequence of the instantaneous establishment of mechanical equilibrium following the sudden application of the surface load. Because the surface load is held constant with time, the uncoupled homogeneous pore-pressure diffusion equation (Eqn. 6.17) obtains. Therefore, the solutions apply for an initial uniform pore-pressure change p_o, whether it is due to a step surface load or to an internal source such as mineral dehydration or oil maturation.

The initial condition for pore pressure is $p(z, 0^+) \equiv p_o = \gamma\sigma_o$ for all z, where the notation 0^+ means "an infinitesimally short time" after the load is applied. The surface boundary condition is $p(0, t) = 0$ for $t > 0$, and the pore-pressure boundary condition at great depth is $dp/dz \to 0$ as $z \to \infty$.

6.4.2 Pore Pressure

The governing equation and initial and boundary conditions for the excess pore pressure are those of the classical Kelvin problem for the mathematically identical heat diffusion problem.[1] The Kelvin solution of Eqn. 6.17 for an

[1] Kelvin used the one-dimensional solution to estimate the age of the earth. The initial condition is that the entire earth be at the melting temperature, T_m, of rock. The surface boundary condition is that $T = 0$ at $z = 0$, and the boundary condition at great depth is that $dT/dz \to 0$ as $z \to \infty$. The time at which the calculated temperature gradient matches the measured values in deep mines is the age of the earth.

initial condition $p(z, 0) = p_o$ and a boundary condition $p(0, t) = p_s$ is

$$p(z, t) - p_o = (p_s - p_o)\mathrm{erfc}(z/\sqrt{4ct}) \tag{6.19}$$

where the error function is defined by

$$\mathrm{erf}(\xi) = \frac{2}{\sqrt{\pi}} \int_0^{\xi} \exp(-\xi'^2) d\xi' \tag{6.20}$$

and the complementary error function is defined by

$$\mathrm{erfc}(\xi) = 1 - \mathrm{erf}(\xi) \tag{6.21}$$

For $p_s = 0$, the solution is

$$p(z, t) = p_o \mathrm{erf}(z/\sqrt{4ct}) \tag{6.22}$$

The solution is shown in dimensionless form as a vertical profile of excess pore pressure (Fig. 6.2). Initially, the excess pressure decline is noticeable only at shallow depths, because the fluid at great depth does not "sense" the surface boundary condition. At a dimensionless depth of $z/\sqrt{4ct} \sim 1.0$, the

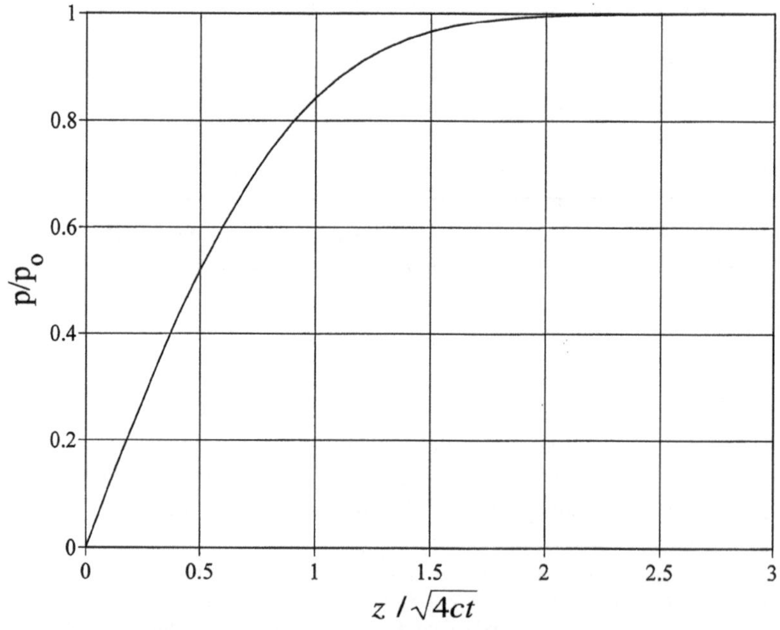

Figure 6.2: Pore pressure versus dimensionless depth in a laterally constrained half space under a suddenly imposed constant vertical load.

Pore-Pressure Step in Deeply Buried Layer. Eqn. 6.22 is an odd function when extended to infinite space. Therefore, an initial pore-pressure step p_o in a finite layer $-L \leq z \leq +L$ buried in infinite space can be represented as the superposition of a step load $p_o/2$ on a half space whose surface is at $z = -L$, and a step load $-p_o/2$ on a half space whose surface is at $z = +L$. This idealization of a deeply buried layer is limited to times less than that required for diffusive flow to reach the surface (cf. Eqn. 6.23). The solutions of the problem is the superposition of Kelvin solutions (Eqn. 6.22) of opposite sign and offset by $-L$ and $+L$:

$$p(z, t) = \frac{1}{2} p_o \left[\operatorname{erf} \frac{z+L}{\sqrt{4ct}} - \operatorname{erf} \frac{z-L}{\sqrt{4ct}} \right] \tag{6.24}$$

In the limit that the layer is infinitely thin ($L \to 0$), Eqn. 6.24 can be put in the form of the definition of the derivative $d[\operatorname{erf}(z/\sqrt{4ct})]/dz$ to yield

$$p(z, t) = \frac{L p_o}{\sqrt{\pi c t}} \exp(-z^2/4ct) \tag{6.25}$$

At the center of the layer ($z = 0$), the pressure decays inversely with the square root of time. For a hydraulic diffusivity of 10^{-4} m^2/s in the surrounding half spaces and a layer thickness $2L$ of 1000 m, the center pressure decays to half its initial value in about 100 years.

excess pressure has declined about 15% from its initial value. Based on this criterion, an estimate of the time it takes for the surface pressure to propagate diffusively to depth z is given by

$$t \sim \frac{z^2}{4c} \tag{6.23}$$

where Eqn. 6.23 defines a time lag that increases as the square of the distance traveled. This rule of thumb is useful for estimating the time constant for pore-pressure diffusion effects or for estimating when finite boundary effects preclude the use of the half-space solution. The only physical property governing the time lag is the hydraulic diffusivity, c. Values range from 1 m^2/s in some high-permeability rocks to 10^{-4} m^2/s in some low-permeability rocks (cf. Table 4.1). For example, the diffusive pressure wave takes about 7 hours to penetrate to a depth of 100 m for a diffusivity of 10^{-1} m^2/s, and about 29 days (a factor of 100 greater) to penetrate to a depth of 1000 m for the same diffusivity.

6.4.3 Displacement

An infinite column length means that the initial vertical displacement at the surface is infinite in response to the sudden application of the load. This artifact is removed in Section 6.5 in which a column of finite length or a layer of finite thickness is considered. However, the time-dependent settlement of the surface of the infinite column is finite. The change in pore pressure at every depth due to the flow of fluid to the surface drain is

$$\Delta p = p(z,t) - p_o = -p_o \text{erfc}\ (z/\sqrt{4ct}) \tag{6.26}$$

The decrease in fluid pressure at each depth leads to a change in vertical displacement, which is obtained by inserting Δp for p in Eqn. 6.13 and integrating from $z = \infty$ to the surface $z = 0$. The displacement at $z = \infty$ is assumed to be zero, which means that the constant of integration, $g(t)$, equals zero. Then

$$\Delta w(0,t) = -c_m p_o \int_{\infty}^{0} \text{erfc}\ (z/\sqrt{4ct}) dz \tag{6.27}$$

The definite integral from zero to infinity of the complementary error function has the value $\sqrt{1/\pi}$:

$$\int_0^\infty \text{erfc}(\xi) d\xi = \sqrt{\frac{1}{\pi}} \tag{6.28}$$

Making the change of variable $\xi = z/\sqrt{4ct}$ so that $d\xi = dz/\sqrt{4ct}$, and substituting $p_o = \gamma\sigma_o$ (cf. Eqn. 6.10) into Eqn. 6.27, yields

$$\Delta w(0,t) = 2c_m \gamma \sigma_o \sqrt{\frac{ct}{\pi}} \tag{6.29}$$

The vertical displacement due to the draining of the excess fluid pressure to the surface is proportional to the square root of time. The rate of settlement slows with time, but the time-dependent settlement grows without bound for an infinite column.

6.5 CONSOLIDATION OF A FINITE LAYER (TERZAGHI'S PROBLEM)

6.5.1 Initial and Boundary Conditions

In Terzaghi's classical consolidation test, a constant stress $-\sigma_o$ is applied suddenly on the surface $z = 0$ of a fluid-saturated sample of length L. The

z-axis is positive downward. The piston applying the load is permeable such that the top boundary is drained. The consolidation test (see Fig. 1.3) satisfies the uniaxial strain condition if the walls of the container are rigid. Following a step displacement, the sample consolidates gradually as fluid flows out the top drain. As with the semi-infinite column, the load produces an instantaneous undrained response, $p(z, 0^+) \equiv p_o = \gamma\sigma_o$, throughout the sample column. The experimental conditions translate to the following boundary conditions for stress, displacement, and pore pressure:

$$\sigma_{zz}(0,t) = -\sigma_o \tag{6.30}$$

$$w(L,t) = 0 \tag{6.31}$$

$$p(0,t) = 0 \tag{6.32}$$

$$\left.\frac{\partial p}{\partial z}\right|_{z=L} = 0 \tag{6.33}$$

These boundary conditions apply also to a suddenly applied, continuous blanket load over an areally extensive, finite-thickness layer. Also, the boundary conditions are equivalent to a problem of a finite layer of thickness, $2L$, drained at both $z = 0$ and $z = 2L$, because $z = L$ is a symmetry plane and satisfies a no-flow condition. A doubly drained layer can represent a confining layer sandwiched between two aquifers whose pressures are simultaneously lowered by an amount p_o. This model has been used to compute compaction of the confining layer by Domenico and Mifflin (1965) (see also Domenico and Schwartz, 1998, p. 227).

6.5.2 Pore Pressure

Because the vertical stress is independent of time, the pore pressure is uncoupled from the stress and satisfies the homogeneous diffusion equation. The initial condition and boundary conditions are the same as for a classical heat-conduction problem (Carslaw and Jaeger, 1959, Eqn. 6, p. 96). The initial pore pressure is the undrained response, $p_o = \gamma\sigma_o$. Hence, the solution is

$$p(z,t) = \frac{4\gamma\sigma_o}{\pi}\sum_{m=0}^{\infty}\frac{1}{2m+1}\exp\left[\frac{-(2m+1)^2\pi^2 ct}{4L^2}\right]\sin\left[\frac{(2m+1)\pi z}{2L}\right] \tag{6.34}$$

In the limit as $t \to 0^+$, the summation is the Fourier sine series representation of the constant $\pi/4$. Hence, $p(z, 0^+) = \gamma\sigma_o$. Thus, the series solution represents the initial undrained pore-pressure response. In the limit as $t \to \infty$, the summation in Eqn. 6.34 is zero, and $p(z, \infty) = 0$, the drained equilibrium pressure.

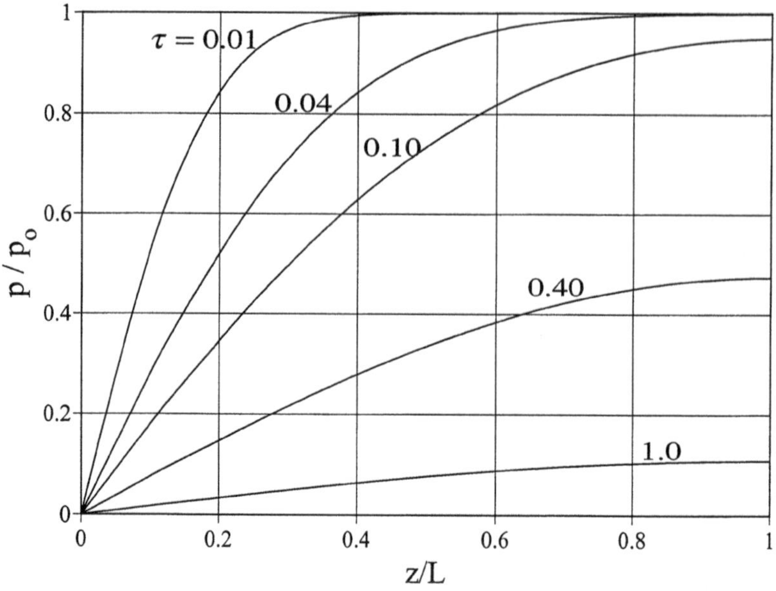

Figure 6.3: Evolution of pore pressure in a laterally constrained finite layer under a suddenly imposed constant vertical load. The successive contours are for dimensionless time $\tau = ct/L^2$.

Eqn. 6.34 is plotted for several values of dimensionless time in Fig. 6.3. Initially, the decline in pore pressure is limited to a region near the drain at $z = 0$. The effect of the finite thickness of the layer relative to the semi-infinite case becomes noticeable for dimensionless times greater than $\tau \equiv ct/L^2 = 0.04$. Conversely, the lack of a noticeable pressure change near the impermeable boundary at $z = L$ suggests that the semi-infinite solution accurately describes the pore-pressure evolution for dimensionless times less than $\tau = 0.04$. Eqn. 6.34 converges slowly for values of $\tau < 0.04$. A series expansion, which converges rapidly for *small* values of time (Carslaw and Jaeger, 1959, p. 97), is

$$p(z,t) = \gamma\sigma_o - \gamma\sigma_o \sum_{m=0}^{\infty}(-1)^m \left\{ \operatorname{erfc}\frac{(2m+1)L - (z-L)}{\sqrt{4ct}} + \operatorname{erfc}\frac{(2m+1)L + (z-L)}{\sqrt{4ct}} \right\} \tag{6.35}$$

6.5.3 Displacement

A finite column length means that the initial instantaneous displacement $w(z, 0^+)$, on application of the load, is finite. This instantaneous undrained response is followed by a time-dependent response, $\Delta w(z, t)$, during the pore-pressure diffusion phase. The total displacement is the sum of these two contributions.

Setting $\zeta = 0$ in Eqn. 6.7 gives the instantaneous undrained strain produced by the sudden application of the vertical stress $-\sigma_o$:

$$\frac{dw}{dz} = \epsilon_{zz}\Big|_{\zeta=0;\, \epsilon_{xx}=\epsilon_{yy}=0} = -\frac{1-2\nu_u}{2G(1-\nu_u)}\sigma_o = -\frac{1}{K_v^{(u)}}\sigma_o \tag{6.36}$$

Integrating Eqn. 6.36 from L to z with the boundary condition $w = 0$ at $z = L$ yields

$$w(z, 0^+) = \frac{\sigma_o(L-z)}{K_v^{(u)}} \tag{6.37}$$

The initial displacement is linear in z, and the displacement w_o at the top of the column, $z = 0$, is

$$w_o \equiv w(0, 0^+) = \frac{\sigma_o L}{K_v^{(u)}} \tag{6.38}$$

During the drainage phase, the pore pressure decreases at each depth. This change in pore pressure is

$$\Delta p(z, t) = p(z, t) - p_o = p(z, t) - \gamma\sigma_o \tag{6.39}$$

where $p(z, t)$ is given by Eqn. 6.34. The change in vertical displacement during drainage, $\Delta w(z, t)$, is then obtained by inserting Δp for p in Eqn. 6.13 and integrating the series solution term by term from $z = L$ (where $\Delta w = 0$ and hence $g(t) = 0$) to z. The result is

$$\Delta w(z, t) = c_m\gamma\sigma_o\left\{(L-z) - \frac{8L}{\pi^2}\sum_{m=0}^{\infty}\frac{1}{(2m+1)^2} \times \exp\left[\frac{-(2m+1)^2\pi^2 ct}{4L^2}\right]\cos\left[\frac{(2m+1)\pi z}{2L}\right]\right\} \tag{6.40}$$

Finally, the additional surface displacement during drainage is obtained by evaluating Eqn. 6.40 at $z = 0$:

$$\Delta w(0,t) = c_m\gamma\sigma_o L\left\{1 - \frac{8}{\pi^2}\sum_{m=0}^{\infty}\frac{1}{(2m+1)^2}\exp\left[\frac{-(2m+1)^2\pi^2 ct}{4L^2}\right]\right\} \tag{6.41}$$

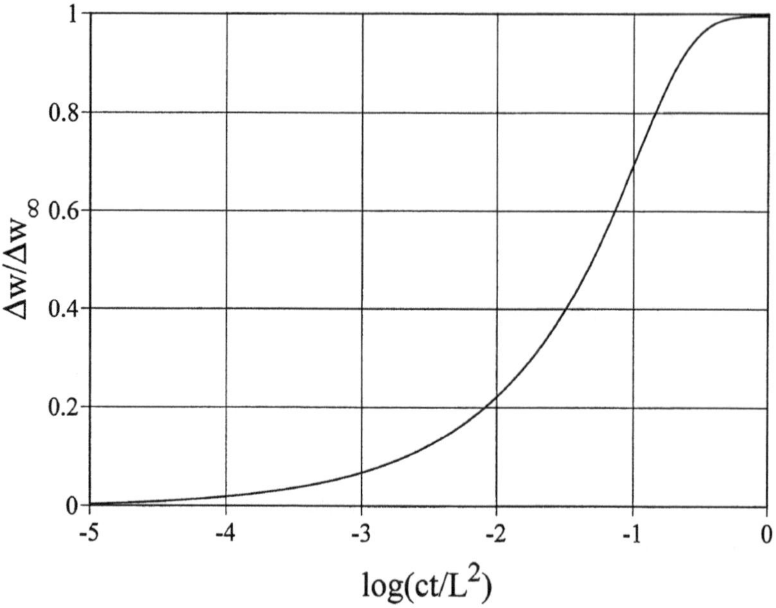

Figure 6.4: Normalized displacement change during the drainage phase following a step load on a finite layer. The horizontal axis is the log of dimensionless time $\tau = ct/L^2$.

The term in braces varies between 0 for small time and 1 for long time (Fig. 6.4). Therefore, the maximum settlement due to drainage is

$$\Delta w_\infty \equiv \Delta w(0, \infty) = c_m \gamma \sigma_o L \tag{6.42}$$

The total surface displacement is the sum of the initial undrained displacement (Eqn. 6.38) and the time-dependent displacement during drainage (Eqn. 6.42):

$$\begin{aligned} w(0, \infty) &= w_o + \Delta w(0, \infty) \\ &= \frac{\sigma_o L}{K_v^{(u)}} + c_m \gamma \sigma_o L \end{aligned} \tag{6.43}$$

In the long time limit, the total displacement must also equal the drained response $\sigma_o L / K_v$:

$$w(0, \infty) = \frac{\sigma_o L}{K_v} \tag{6.44}$$

Comparing these two expressions for the total drained displacement gives

$$c_m \gamma = \frac{1}{K_v} - \frac{1}{K_v^{(u)}} \tag{6.45}$$

Substituting Eqns. 6.5 and 6.9 into Eqn. 6.45 shows that the factor $c_m \gamma$ can also be expressed (cf. Detournay and Cheng, 1993, p. 146) as

$$c_m \gamma = \frac{\nu_u - \nu}{2G(1-\nu)(1-\nu_u)} \tag{6.46}$$

The consolidation example displays the physical reason for calling $1/K_v^{(u)}$ an *undrained* uniaxial compressibility and for calling $1/K_v$ a *drained* uniaxial compressibility. Biot (1941a) emphasized the temporal significance of the undrained and drained states by calling $\beta_v^{(u)} = 1/K_v^{(u)}$ the *initial compressibility* and $\beta_v = 1/K_v$ the *final compressibility*. The former governs the initial undrained vertical displacement, and the latter determines the final long-term displacement. Biot expressed the relationship between the drained and undrained uniaxial compressibilities as

$$\beta_v^{(u)} = \frac{\beta_v}{1 + \alpha^2 \beta_v M} \tag{6.47}$$

6.6 UNIFORMLY INCREASING LOAD ON A FINITE LAYER

The rapid accumulation of low-permeability sediments is one possible cause for the buildup of abnormally high pore pressures in a sedimentary basin. The reverse process of rapid erosion can be a cause for abnormally low pore pressures in the underlying rock layers.

A uniform loading rate is represented by $d\sigma_{zz}/dt = -\dot{\sigma}_o$, where positive values of $\dot{\sigma}_o$ correspond to an increasing surface load (sedimentation), and negative values of $\dot{\sigma}_o$ correspond to a decreasing surface load (erosion). Eqn. 6.14 then becomes

$$\frac{\partial^2 p}{\partial z^2} - \frac{1}{c}\frac{\partial p}{\partial t} = -\frac{A_o}{(k/\mu)} \tag{6.48}$$

where the constant

$$A_o \equiv \frac{k}{\mu c} \gamma \dot{\sigma}_o = c_m \dot{\sigma}_o \tag{6.49}$$

The physical situation considered in this section is a low permeability layer sandwiched between two very high permeability layers. The coordinate system is centered in the layer so that the upper ($z = -L$) and lower ($z = +L$) boundaries of the layer are drained. Eqn. 6.48 is mathematically equivalent to a heat-conduction problem in which heat is produced at a constant rate. The solution (Carslaw and Jaeger, 1959, p. 130; Palciauskas and Domenico, 1989) is

$$p(z,t) = \frac{A_o L^2}{2(k/\mu)} \left\{ 1 - \frac{z^2}{L^2} - \frac{32}{\pi^3} \sum_{m=0}^{\infty} \frac{(-1)^m}{(2m+1)^3} \right. \\ \left. \times \exp\left[\frac{-(2m+1)^2 \pi^2 ct}{4L^2} \right] \cos\left[\frac{(2m+1)\pi z}{2L} \right] \right\} \quad (6.50)$$

In Eqn. 6.50, the coefficient multiplying the quantity in braces is the steady-state pore pressure at the center of the layer ($z = 0$). It can be expressed variously as

$$\begin{aligned} p(0,\infty) &= \frac{A_o L^2}{2(k/\mu)} \\ &= \frac{\gamma L^2}{2c} \dot{\sigma}_o \\ &= \frac{1}{2} \tau (\gamma \dot{\sigma}_o) \end{aligned} \quad (6.51)$$

where $\tau \equiv L^2/c$ is the characteristic diffusion time. The solution is plotted in the dimensionless variables $p(z,t)/p(0,\infty)$ and z/L in Figure 6.5. As $t \to \infty$, the summation in Eqn. 6.50 goes to zero, and the pore pressure approaches its parabolic steady-state profile with the maximum pore pressure $p(0,\infty)$ at the center of the layer.

The two competing processes affecting the pore pressure in the loaded layer are dissipation due to fluid flow and buildup due to the Skempton effect. The maximum pore pressure $p(0,\infty)$ depends on three parameters (Eqn. 6.51): the characteristic fluid diffusion time, the loading efficiency, and the rate of stress buildup. The pore pressure reaches 90% of its maximum value, $p(0,\infty)$, after one time constant τ. If the layer were totally undrained, the pore pressure would be $\gamma \dot{\sigma}_o \tau$ after one time constant τ. Therefore, the ratio $p(0,\tau)/p^{(u)}(0,\tau)$ is approximately 0.45, that is, the pore-pressure buildup is about 45% of its undrained value at time $t = \tau$. Although additional loading with time increases the pore-pressure buildup by an additional 10%, it becomes a smaller and smaller fraction of the undrained value $\gamma \dot{\sigma}_o t$ for $t > \tau$ because the load grows without bound. For example, after a time duration of 3τ, the ratio $p(0,3\tau)/p^{(u)}(0,3\tau)$ is approximately 0.15.

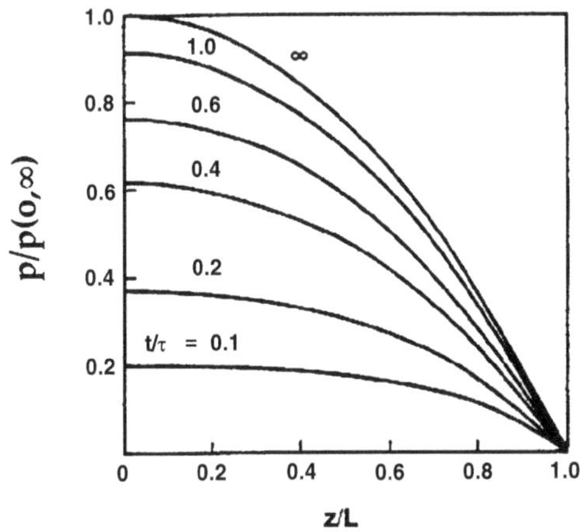

Figure 6.5: Pore-pressure profiles $p/p(0,\infty)$ from $z = 0$ to $z = L$ for several dimensionless times t/τ (after Palciauskas and Domenico, 1989, p. 210).

6.7 SEDIMENTATION ON IMPERMEABLE BASEMENT

The top surface of a sediment layer accumulating on impermeable basement rock is a moving, drained boundary. If the sediments were impermeable and Skempton's coefficient equal to one, the pore pressures would be lithostatic (i.e., equal to the weight of the overlying rock and water). In the other limit of highly permeable sediments, pore pressures in the sediment layer would be hydrostatic, as the induced pore pressures would drain off rapidly to the sediment-water interface.

Gibson (1958) presented analytical solutions of the problem of a moving boundary for a sediment layer whose thickness increases either linearly with time or as the square root of time. The z-axis is chosen to be positive in the upward direction (Fig. 6.6).

The applied vertical stress at a height z above the impermeable base, which induces changes in excess pore pressure, is the submerged weight of the overlying sediment. Therefore, the governing equation (Eqn. 6.14) for excess pore pressure becomes

$$\frac{\partial p}{\partial t} - c\frac{\partial^2 p}{\partial z^2} = \gamma(\rho - \rho_f)g\frac{d\ell}{dt} \tag{6.52}$$

where $d\ell/dt$ is the instantaneous sedimentation rate (e.g., 1 mm/yr).

6.7.1 Thickness Proportional to $\sqrt{t}$

Gibson showed that Eqn. 6.52 can be transformed into a solvable equation by separation of variables if $l(t) = a\sqrt{t}$, where a is a constant of proportionality.

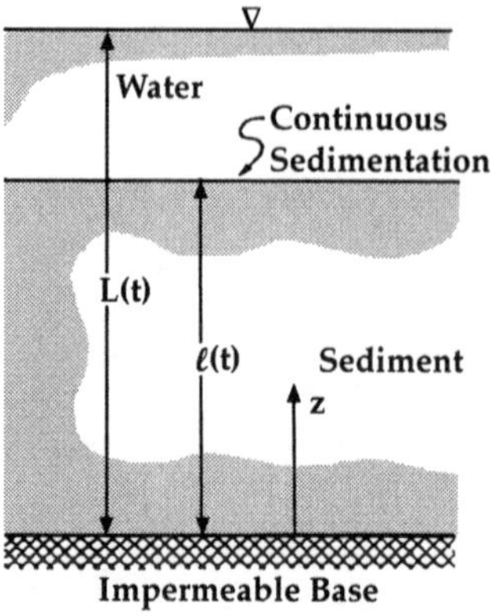

Figure 6.6: Problem of a moving boundary of uniformly growing sedimentary layer (after Gibson, 1958, and Bredehoeft and Hanshaw, 1968).

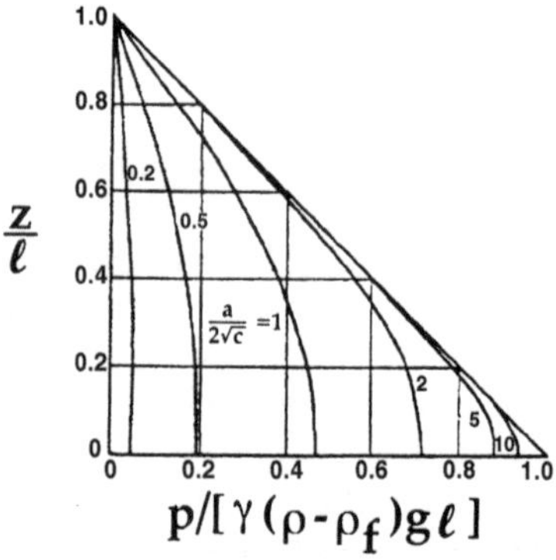

Figure 6.7: Excess pore pressure versus normalized depth for a sediment layer that increases in thickness proportionally to the square root of time (after Gibson, 1958, Fig. 3, p. 174).

The solution is

$$p(z,t) = \gamma(\rho - \rho_f) g a \sqrt{t} \left[1 - \left\{ \frac{\exp\left(-\frac{z^2}{4ct}\right) + \frac{z}{2}\left(\frac{\pi}{ct}\right)^{\frac{1}{2}} \operatorname{erf} \frac{z}{\sqrt{4ct}}}{\exp\left(-\frac{a^2}{4c}\right) + \frac{a}{2}\left(\frac{\pi}{c}\right)^{\frac{1}{2}} \operatorname{erf} \frac{a}{\sqrt{4c}}} \right\} \right] \tag{6.53}$$

Eqn. 6.53 is plotted in Figure 6.7 as profiles of the dimensionless variable $p/(\gamma(\rho - \rho_f)g\ell)$ versus z/ℓ for different values of the time-independent parameter $a/\sqrt{4c}$. Time is not an explicit variable labeling the profiles in Figure 6.7, because it appears within the braces on the right-hand side of Eqn. 6.53 only in the combination $z/\sqrt{4ct}$, which is equal to $(a/\sqrt{4c})(z/\ell)$.

The parameter $a/\sqrt{4c}$ is the ratio of a loading-rate factor relative to a dissipation-rate factor. Greater values of a or smaller values of c lead to greater excess pore pressures. Very large values of $a/\sqrt{4c}$ plot as a linear increase of excess pore pressure with a slope of one; that is, the excess pore pressure increases as γ times the lithostatic load minus hydrostatic pressure.

6.7.2 Thickness Proportional to *t*

For a constant rate of deposition, $\ell(t) = bt$, where b is the sedimentation rate, Gibson found the following integral expression for excess pore pressure:

$$p(z,t) = \gamma(\rho - \rho_f)gbt - \frac{\gamma(\rho - \rho_f)g}{\sqrt{\pi ct}} \exp\left(-\frac{z^2}{4ct}\right)$$
$$\times \int_0^\infty \xi \tanh\frac{b\xi}{2c} \cosh\frac{z\xi}{2ct} \exp\left(-\frac{\xi^2}{4ct}\right) d\xi \qquad (6.54)$$

Eqn. 6.54 is plotted in Figure 6.8 as profiles of the dimensionless variable $p/(\gamma(\rho - \rho_f)g\ell)$ versus z/ℓ for different values of b^2t/c. The dimensionless pressure profiles for the linear case are not invariant with time as with the case for square root of time. As the sediments thicken, the pore-pressure profile continually shifts to greater values of b^2t/c. Pore-pressure buildup is greater for longer times as well as for faster depositional rates or smaller hydraulic diffusivities. In the limit of very large times, the pore-pressure profile approaches the linear profile with a slope of one. A linear sedimentation rate adds sediment thickness sufficiently quickly that an undrained pore-pressure profile is created for long enough times.

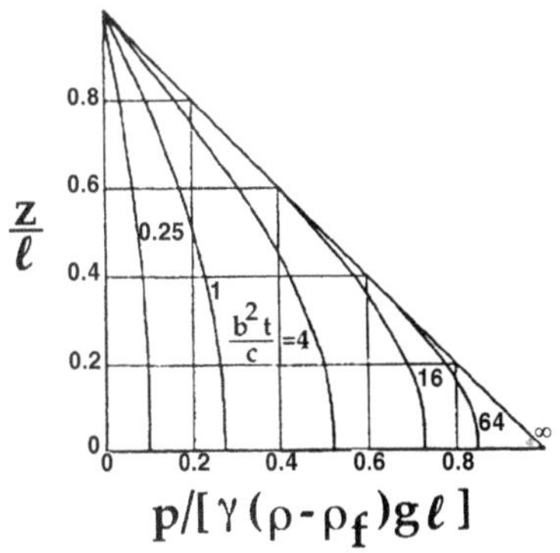

Figure 6.8: Excess pore pressure versus normalized depth for a sediment layer that increases in thickness linearly with time (after Gibson, 1958, Fig. 3, p. 174).

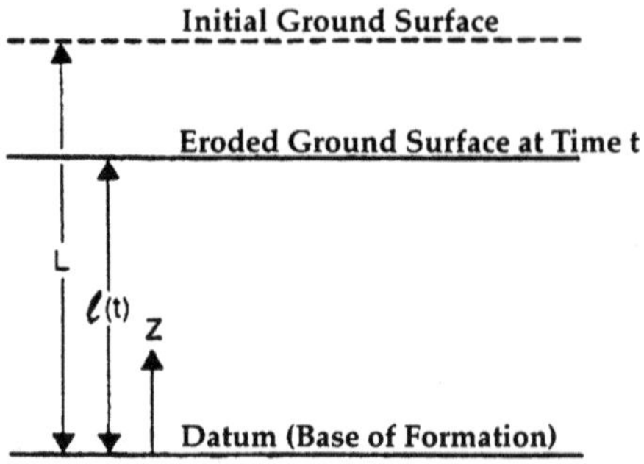

Figure 6.9: Coordinate system for the erosional unloading of a low-permeability layer (after Neuzil and Pollock, 1983, Fig. 1, p. 181).

6.8 EROSIONAL UNLOADING

Erosional or glacial unloading can produce transient underpressures in opposition to overpressures produced by sediment loading. Neuzil and Pollock (1983) obtained finite difference solutions for the pore pressure in a sedimentary layer thinned by erosion at a uniform rate, m, where the constant m is negative. The layer has an initial thickness L and rests on an impermeable base (Fig. 6.9). The moving surface is located at $\ell(t)$, which is the top of the hydrostatic column; that is, excess pore pressure is the amount greater than $\rho_f g(\ell(t) - z)$. The initial and boundary conditions are

$$\begin{aligned} \ell(0) &= L \\ p(z, 0) &= 0 \\ p(\ell, t) &= 0 \\ \left.\frac{\partial p}{\partial z}\right|_{z=0} &= 0 \end{aligned} \tag{6.55}$$

The initial condition corresponds to a hydrostatic state. The upper boundary condition of zero excess pore pressure at $z = \ell$ corresponds to a water table or hydrostatic state in an overlying permeable unit. The lower no-flow boundary condition corresponds to an underlying impermeable layer.[2]

[2] Neuzil and Pollock (1983) also presented results for the case of an underlying permeable unit, $p(0, t) = 0$.

Neuzil and Pollock posed the problem in the following dimensionless variables:

$$t^* = \frac{kt}{\mu S \ell^2}$$
$$p^* = \frac{p}{(\rho - \rho_f)gL}$$
$$z^* = z/\ell$$
$$L^* = \ell/L$$
$$m^* = m\frac{\ell^2}{Lc} \qquad (6.56)$$

They used a fully implicit finite difference scheme in which the time step was adjusted to remove one grid block at each time step. They presented their results in the form of the dimensionless underpressure, p^*, at the midpoint of the remaining section versus the dimensionless unloading rate, m^* (Fig. 6.10). The loading efficiency γ is assumed to be one. The three curves are for

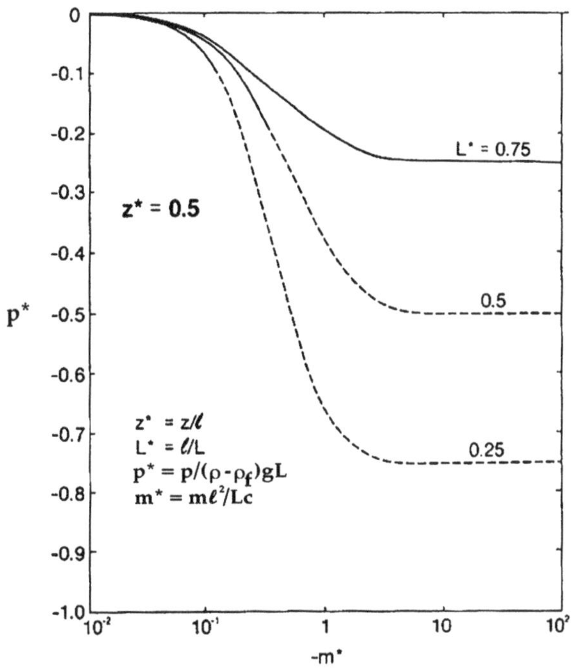

Figure 6.10: Dimensionless underpressure at the midpoint of the eroding layer versus dimensionless unloading rate for a low-permeability layer undergoing erosion at a constant rate (Neuzil and Pollock, 1983, Fig. 2*a*, p. 184).

residual layer thicknesses of 25, 50, and 75% of the original thickness. For example, a value of $L^* = 0.75$ means that 25% of the layer has been eroded. The time dependence is implicit in several of the dimensionless variables through the instantaneous thickness, $\ell(t)$, of the layer. The underpressure varies from 0 for small values of $-m^*$ to the undrained pore pressure ($-p^* = 1 - L^*$) for large values of $-m^*$. Large values of $-m^*$ correspond to large values of erosion rate or small values of hydraulic diffusivity.

6.9 PERIODIC LOAD ON A HALF SPACE

The problem treated in this and the next section is that of a harmonic stress applied at the surface of a half space and finite layer, respectively, while maintaining a drained pore-pressure condition at the surface. The finite layer and half-space solutions for a cyclic load have been applied to the problem of sea-wave loading of a structure overlying a poroelastic seabed (Mei and Foda, 1982). An arbitrary time-dependent load can be Fourier synthesized as the sum of harmonic loads of different amplitudes.

Mathematically, the time-dependent stress boundary condition is expressed as

$$\sigma_{zz}(0, t) = -\sigma_o \exp(\imath\omega t) \tag{6.57}$$

where σ_o is the amplitude of the (downward) compressive load, $\imath \equiv \sqrt{-1}$, and ω is the angular frequency of the time variation. The vertical load is *not* constant with time, and the pore-pressure diffusion equation (Eqn. 6.14) is therefore nonhomogeneous. The force equilibrium equation requires that the stress at all depths be the same as the applied surface load (i.e., $\sigma_{zz}(z, t) = \sigma_{zz}(0, t)$). The quasistatic assumption of instantaneous mechanical equilibrium requires that the loading period be long relative to the times for elastic wave propagation. Furthermore, the initial transient when the surface loading first starts is neglected. Under this pseudo steady-state approximation, the solution has the form

$$p(z, t) = \tilde{p}(z) \exp(\imath\omega t) \tag{6.58}$$

where $\tilde{p}(z)$ is the (complex) amplitude, which depends only on z. Substituting Eqn. 6.58 into Eqn. 6.14 leads to the following ordinary differential equation for $\tilde{p}(z)$:

$$\imath\omega\tilde{p} - c\frac{d^2\tilde{p}}{dz^2} = \imath\omega\gamma\sigma_o \tag{6.59}$$

The solution of Eqn. 6.59 for the boundary conditions $\tilde{p}(0) = 0$ and $d\tilde{p}/dz \to 0$ as $z \to \infty$ is

$$\begin{aligned} \tilde{p}(z) &= \gamma\sigma_o\{1 - \exp[-z\sqrt{\imath\omega/c}]\} \\ &= \gamma\sigma_o\{1 - \exp[-(1+\imath)z\sqrt{\omega/2c}]\} \\ &= \gamma\sigma_o\{1 - \exp[-z\sqrt{\omega/2c}]\exp[-\imath z\sqrt{\omega/2c}]\} \end{aligned} \tag{6.60}$$

The second form of Eqn. 6.60 follows from the formula $\sqrt{\imath} = \exp(\imath\pi/4) = (1+\imath)/\sqrt{2}$ because $\imath = \exp(\imath\pi/2)$. Let δ be the distance over which $\exp(-z\sqrt{\omega/2c})$ decays to $1/e$ times its surface value:

$$\delta \equiv \sqrt{\frac{2c}{\omega}} \tag{6.61}$$

For example, for a moderately low hydraulic diffusivity, $c = 10^{-2}$ m^2/s, and a period of one day, $\omega = 7.3 \times 10^{-5}$ s^{-1}, the decay distance δ is equal to 16.6 m. For high-frequency ($\omega \to \infty$) variations of the surface load, the half space responds in an undrained fashion because the boundary layer is negligible (very small δ). The pore pressure is in phase with the surface load, and its magnitude is $\gamma\sigma_o$. On the other hand, as the loading frequency $\omega \to 0$, δ becomes very large and the pore pressure approaches zero, its value for drained conditions. These limits show that undrained and drained conditions occur for very fast and very slow oscillations, respectively.

The amplitude of the pore pressure is given by

$$|\tilde{p}| = \gamma\sigma_o\left[1 - 2\exp\left(-\frac{z}{\delta}\right)\cos\frac{z}{\delta} + \exp\left(-\frac{2z}{\delta}\right)\right]^{1/2} \tag{6.62}$$

The amplitude is maximum when

$$\sin\frac{z}{\delta} + \cos\frac{z}{\delta} - \exp(-\frac{z}{\delta}) = 0 \tag{6.63}$$

The maximum value of the pore pressure occurs at a depth of $z = 2.284\delta$. The phase θ of the pore pressure is given by

$$\theta = \tan^{-1}\left\{\frac{\exp(-\frac{z}{\delta})\sin\frac{z}{\delta}}{1 - \exp(-\frac{z}{\delta})\cos\frac{z}{\delta}}\right\} \tag{6.64}$$

The magnitude and phase of $\tilde{p}/\gamma\sigma_o$ are plotted versus z/δ in Figure 6.11*a* and *b*, respectively. The maximum pore-pressure amplitude exceeds the undrained pore pressure $\gamma\sigma_o$ over a depth range $1.47\delta < z < 4.33\delta$. If the loading efficiency is one, the pore pressure can exceed the surface load by

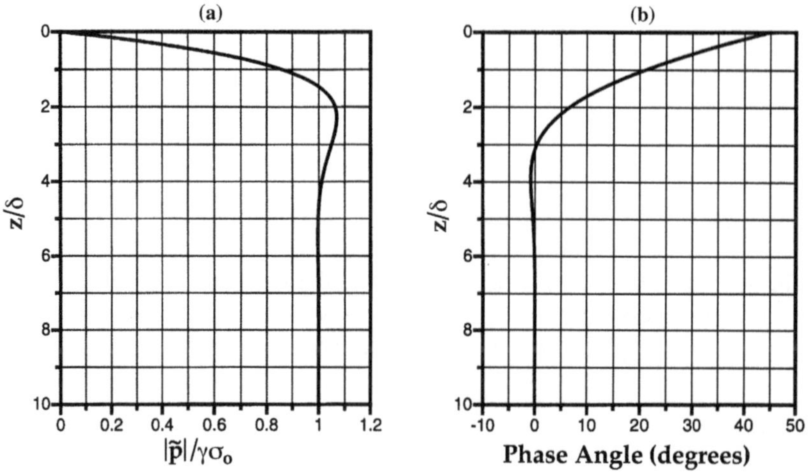

Figure 6.11: Periodic loading of a half space. (*a*) Normalized pore-pressure amplitude versus normalized depth. (*b*) Phase angle (in degrees) of pore pressure versus normalized depth.

up to 7%. This additional pore pressure is a result of the secondary coupling of pore pressure with stress. The phase angle versus depth contains a relative minimum between $\pi\delta$ and $2\pi\delta$. Diffusive pore-pressure effects are confined to a boundary layer of $z < 3\delta$. Below this layer, pore pressure is essentially in phase with the surface load, and its amplitude is the undrained value $\gamma\sigma_o$. The half-space solution provides a good approximation for a finite layer whose thickness is greater than 3δ.

6.10 PERIODIC LOAD ON A FINITE LAYER

6.10.1 Pore Pressure

Eqn. 6.59 must be solved for the boundary conditions $\tilde{p}(0) = 0$ and $d\tilde{p}/dz = 0$ at $z = L$. The homogeneous solution is assumed to have the form

$$\tilde{p}^{\text{homo}}(z) = A \sinh \frac{\lambda z}{L} + B \cosh \frac{\lambda z}{L} \tag{6.65}$$

where the variables A, B, and λ are parameters local to this section. Using Eqn. 6.65 in Eqn. 6.59 shows that

$$\lambda = L\sqrt{\frac{i\omega}{c}} \tag{6.66}$$

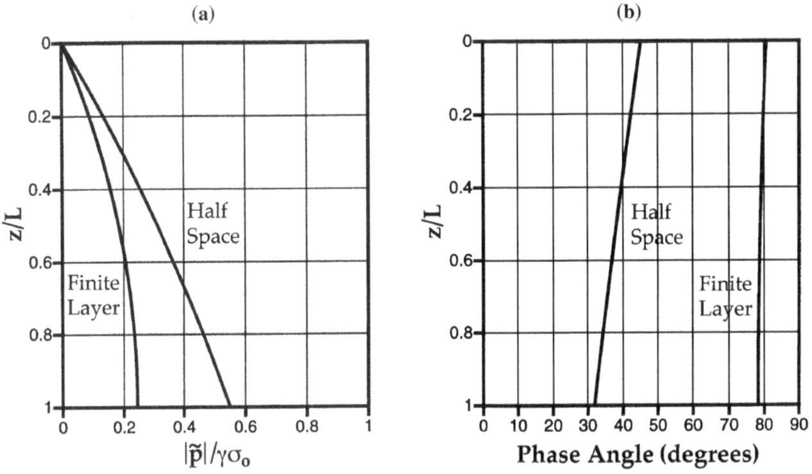

Figure 6.12: Comparison of periodic loading of a half space versus a finite layer whose thickness is half the decay distance ($L = \delta/2$). (*a*) Normalized pore-pressure amplitude versus normalized depth. (*b*) Phase angle (in degrees) of pore pressure versus normalized depth.

A particular solution of Eqn. 6.59 is $\gamma\sigma_o$. After adding the homogeneous and particular solutions, the constants A and B can be determined from the boundary conditions. The solution for the complex amplitude $\tilde{p}(z)$ (Detournay and Cheng, 1993, p. 148) is

$$\tilde{p}(z) = \gamma\sigma_o\left[1 + \tanh\lambda \sinh\left(\frac{\lambda z}{L}\right) - \cosh\left(\frac{\lambda z}{L}\right)\right] \tag{6.67}$$

The magnitude and phase of the complex amplitude are plotted in Figure 6.12*a* and *b*, respectively, for $L = \delta/2$. The finite-layer solution (Eqn. 6.67) differs significantly from the half-space solution (Eqn. 6.60) because the layer thickness L is one sixth the boundary layer thickness 3δ.

6.10.2 Displacement

Applying the boundary condition $p(0, t) = 0$ in Eqn. 6.13 gives the vertical strain at the surface:

$$\left.\frac{\partial w}{\partial z}\right|_{z=0} = \epsilon_{zz}(0, t) = g(t) \tag{6.68}$$

Substituting Eqns. 6.57 and 6.68 into Eqn. 6.2 and again using the boundary condition $p(0,t) = 0$ give

$$-\sigma_o \exp(\imath\omega t) = K_v g(t) \tag{6.69}$$

Solving Eqn. 6.69 for $g(t)$ and integrating Eqn. 6.68 from $z = L$ to $z = 0$ give

$$\tilde{w}(0) = \sigma_o L \left[\frac{1}{K_v^{(u)}} + c_m \gamma \frac{\tanh \lambda}{\lambda} \right] \tag{6.70}$$

where $w(0,t) = \tilde{w}(0) \exp(\imath\omega t)$ defines $\tilde{w}(0)$. In the drained limit ($\lambda \to 0$), $(\tanh \lambda)/\lambda$ approaches 1, and hence $\tilde{w}(0)$ approaches $\sigma_o L / K_v$ (cf. Eqn. 6.45). In the undrained limit ($\lambda \to \infty$), $(\tanh \lambda)/\lambda$ approaches 0, and hence $\tilde{w}(0)$ approaches $\sigma_o L / K_v^{(u)}$.

6.11 WATER LOAD ON A HALF SPACE

Variations in the level of a reservoir lake behind a dam or variations in sea level impose compressive stress and pore-pressure changes of equal magnitude at the water-rock interface. The pore-pressure response in a half space is obtained in this section for (1) a step change in water level and (2) a periodic fluctuation in water level. These solutions have been given by Roeloffs (1988) and Fang et al. (1993).

6.11.1 Step Change

If the water pressure at the base of a fluid layer ($z = 0$) is suddenly raised by an amount p_s and held constant, the vertical load is $\sigma_{zz}(0,t) = -\sigma_o = -p_s$ for $t > 0$, and the pore-pressure boundary condition is $p(0,t) = p_s$. Throughout the underlying half space ($z > 0$), the instantaneous effect of suddenly raising the water level is to produce a vertical stress, $-p_s$, and to induce an undrained pore pressure, $p_o = \gamma p_s$, at all depths. The instantaneously induced pore pressure profile then becomes the initial condition for subsequent fluid diffusion. The problem of fluid diffusion is uncoupled from the stress field due to the uniaxial strain and constant vertical stress conditions. The solution is obtained from Eqn. 6.19 with the following result:

$$p(z,t) = \gamma p_s + (1-\gamma) p_s \operatorname{erfc}(z/\sqrt{4ct}) \tag{6.71}$$

The pore pressure normalized by p_s is plotted as a function of dimensionless depth, $z/\sqrt{4ct}$, in Figure 6.13. The amount of possible increase in pore pressure during the diffusive phase depends on the value of γ. A value of

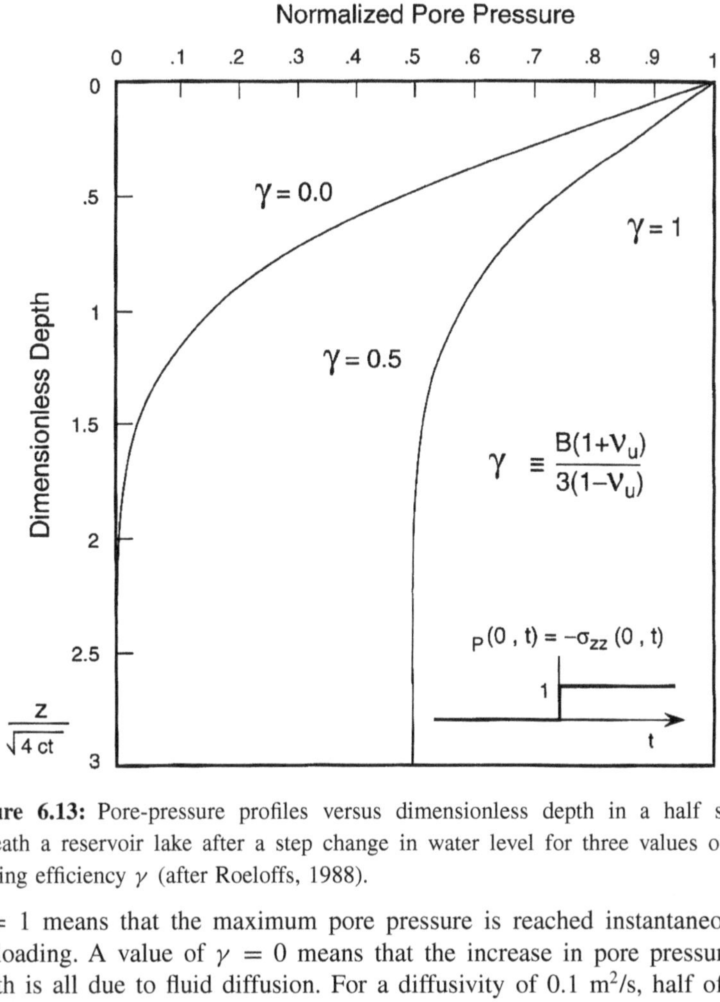

Figure 6.13: Pore-pressure profiles versus dimensionless depth in a half space beneath a reservoir lake after a step change in water level for three values of the loading efficiency γ (after Roeloffs, 1988).

$\gamma = 1$ means that the maximum pore pressure is reached instantaneously on loading. A value of $\gamma = 0$ means that the increase in pore pressure at depth is all due to fluid diffusion. For a diffusivity of 0.1 m^2/s, half of the possible increase in pore pressure due to fluid diffusion is reached in 35 years at a depth of 10 km. The evolution of pore pressure profiles from the initial condition, γp_s, toward the boundary value p_s near the surface can be seen more clearly with a specific example. In Figure 6.14, $c = 0.01$ m^2/s and $\gamma = 0.4$.

6.11.2 Periodic Variation

The solution for a periodic fluctuation of water level over a half space is obtained by superimposing the solutions for (1) a periodic stress variation (Eqn. 6.60) with a drained top surface and (2) a periodic pore-pressure variation, $p(0, t) = p_s \exp(i\omega t)$, with zero applied stress. For problem 2,

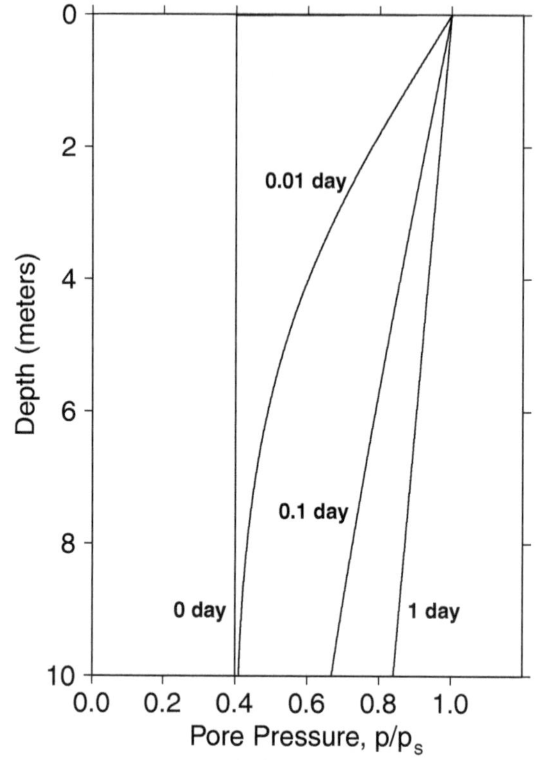

Figure 6.14: Normalized pore-pressure profiles in a half space for four durations lengths of time following a step change in water level. Depth is in meters for a value of c = 0.01 m^2/s and γ = 0.4.

the zero-stress condition means that Eqn. 6.14 reduces to the homogeneous diffusion equation for pressure. Assuming the solution to have the form $p(z,t) = \tilde{p}(z)\exp(\imath\omega t)$ leads to

$$\begin{aligned}\tilde{p}(z) &= p_s \exp\left(-z\sqrt{\imath\omega/c}\right)\\ &= p_s \exp\left(-(1+\imath)z\sqrt{\omega/2c}\right)\\ &= p_s \exp\left(-z\sqrt{\omega/2c}\right)\exp\left(-\imath z\sqrt{\omega/2c}\right)\end{aligned} \tag{6.72}$$

Eqn. 6.72 is well known from the identical thermal diffusion problem (e.g., Turcotte and Schubert, 1982, Section 4–14). The amplitude of the sinusoidally varying pore pressure attenuates exponentially with depth, and the phase is shifted relative to the pore pressure at the surface. The pore pressure is 180 degrees out of phase with the surface pressure at $z = \pi\delta$, where $\delta \equiv \sqrt{2c/\omega}$ (cf. Eqn. 6.61).

Summing Eqns. 6.60 with $\sigma_o = p_s$ and 6.72 leads to the solution for a periodic change in water level over a half space:

$$\tilde{p}(z) = \gamma p_s + (1-\gamma) p_s \exp\left(-z\sqrt{\omega/2c}\right) \exp\left(-\imath z\sqrt{\omega/2c}\right) \tag{6.73}$$

Thus, the solution is made up of an in-phase, undrained component plus $(1-\gamma)$ times the solution of the homogeneous diffusion equation for a periodic pore-pressure variation at the surface. The pore-pressure amplitude and phase as a function of depth depend on the frequency of loading and on two physical properties: hydraulic diffusivity and loading efficiency. The amplitude and phase versus dimensionless depth, z/δ, are plotted for several values of γ in Figure 6.15. The undrained limit is approached for high frequency (within the quasistatic approximation) or at great depth. The pore pressure is in phase with the sinusoidal surface pressure but is attenuated by the factor γ. If the loading efficiency $\gamma = 1$ (incompressible grains and fluid), the response in pore pressure is independent of time and is the undrained response at all depths. If the loading efficiency γ equals zero, the pore pressure is uncoupled from the stress variation, and the response reduces to the simple pore-pressure diffusion solution (Eqn. 6.72). The amplitude decays monotonically with depth, and the peak pore pressure lags the surface pore

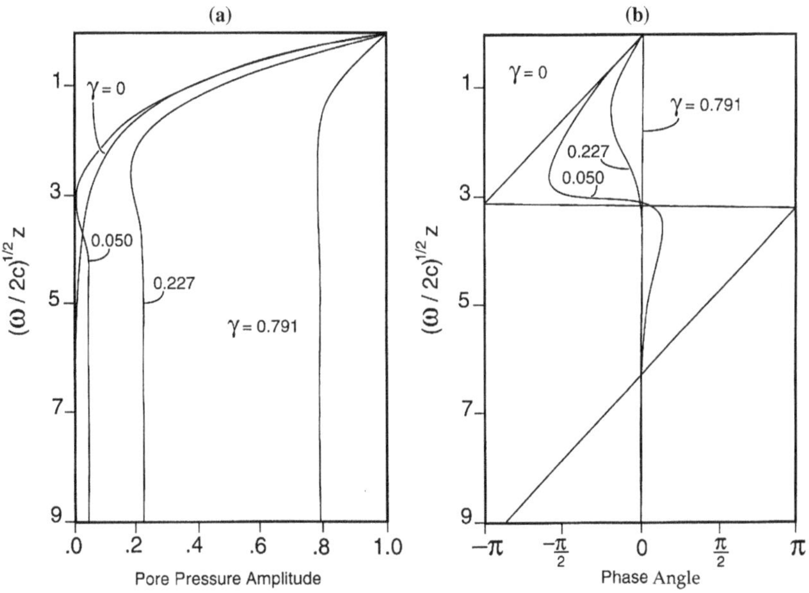

Figure 6.15: Periodic water loading (Roeloffs, 1988). (a) Normalized amplitude of pore pressure versus dimensionless depth. (b) Phase of pore pressure versus dimensionless depth.

pressure by an angle z/δ. For $\gamma > 0$, the pore-pressure amplitude essentially reaches the undrained value γp_s at a dimensionless depth $z/\delta \geq 2\pi$, and is within a few percent of this value at a dimensionless depth $z/\delta \approx \pi$. Below this depth, the pore pressures are essentially in phase with the surface load. For the special case that γ is identically zero, the phase continues to oscillate between $-\pi$ and π (cf. Fig. 6.15*b*), but the amplitude approaches zero. For an annual cycle, a dimensionless depth of π corresponds to an actual depth of 0.3 km for $c = 10^{-3}$ m^2/s, and 3 km for $c = 0.1$ m^2/s.

6.11.3 Induced Seismicity

The impoundment of a reservoir lake behind a large dam can produce earthquakes on an underlying fault (Roeloffs, 1988). A change in reservoir water level instantaneously changes the shear and normal stresses throughout the half space, and it induces a change in pore pressure by the Skempton effect. Subsequently the pore pressure continues to change as fluid diffuses from the lake bottom into the subsurface. The correlation between seismicity and variation of water level depends on fault type, orientation, and location as well as the poroelastic properties beneath the reservoir. The stability of a vertical strike-slip fault due to a cyclic water-level variation can be examined using a one-dimensional, half-space model. The critical shear stress according to Coulomb's friction law is

$$|\tau| = \mu_f(-S_n - p) = -\mu_f(S_n + p) \tag{6.74}$$

Thus, slip occurs when the magnitude of the shear stress, $|\tau|$, exceeds the frictional resistance, $\mu_f(-S_n - p)$, where μ_f is the coefficient of friction, $-S_n$ is the compressive normal stress, and p is the pore pressure. Changes in water level do not affect the preexisting shear stress on the strike-slip fault. Therefore, any change in fault stability results from a change in the effective normal stress, $S_n + p$, on the right-hand side of Eqn. 6.74. Positive values of $S_n + p$ reduce compressive normal stress and are destabilizing. The fault is chosen to lie in the yz-plane so that the normal stress S_n is σ_{xx}. The complex amplitude $\tilde{\sigma}_{xx}$ can be obtained from Eqn. 6.3, and the complex amplitude of the effective normal stress is then obtained by adding $\tilde{p}$ to $\tilde{\sigma}_{xx}$. For a cyclic water-load variation, $p_s \exp \imath\omega t$, the complex amplitude of the vertical stress $\tilde{\sigma}_{zz}$ is $-p_s$. Therefore,

$$\tilde{\sigma}_{xx} + \tilde{p} = \frac{\nu}{1-\nu}(-p_s) - \frac{1-2\nu}{1-\nu}\alpha\tilde{p} + \tilde{p} \tag{6.75}$$

where the complex amplitude of the pore pressure, $\tilde{p}$, is given by Eqn. 6.73. For values of $\nu = 0.25$ and $\alpha = 0.5$, the effective normal stress divided by

p_s for three values of γ is

$$\frac{\tilde{\sigma}_{xx}+\tilde{p}}{p_s} = -\frac{1}{3} + \frac{2}{3}\exp\left[-(1+i)z\sqrt{\omega/2c}\right] \qquad (\gamma = 0) \qquad (6.76)$$

$$\frac{\tilde{\sigma}_{xx}+\tilde{p}}{p_s} = \frac{1}{3}\exp\left[-(1+i)z\sqrt{\omega/2c}\right] \qquad (\gamma = 0.5) \qquad (6.77)$$

$$\frac{\tilde{\sigma}_{xx}+\tilde{p}}{p_s} = \frac{1}{3} \qquad (\gamma = 1) \qquad (6.78)$$

The amplitudes and phases of the effective normal stress are plotted for these three cases in Figure 6.16. For $\gamma = 0$, the effective normal stress approaches $-p_s/3$ below a depth $\pi\delta$ because there is no Skempton effect and diffusive flow effects are shallower. Thus, the effective normal stress is most compressive when the water level is highest; that is, the strike-slip fault

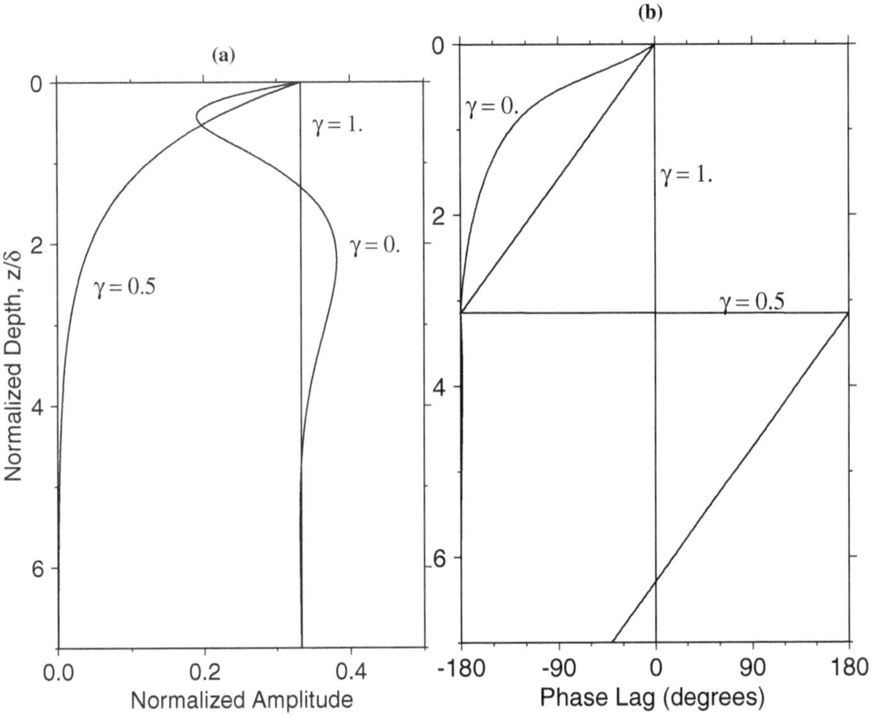

Figure 6.16: Amplitudes and phases of effective normal stress. (a) Normalized amplitude of effective normal stress, $(\tilde{\sigma}_{xx}+\tilde{p})/p_s$, versus normalized depth, z/δ, for periodic water loading of a semi-infinite vertical fault. (b) Phase (degrees) of effective normal stress versus normalized depth.

is most stable when the water level is highest and least stable when the water level is lowest. On the other hand, for $\gamma = 1$, the effective normal stress is least compressive at all depths when the water load is highest. Hence, the fault is destabilized at high water levels due to the pore-pressure buildup. For $\gamma = 0.5$, the (complex) effective normal stress is equal to one third the (complex) pore pressure for the pressure diffusion problem with a stress-free surface (cf. Eqn. 6.60). The amplitude of the effective normal stress goes to zero below a depth of $\pi\delta$, and fault stability is insensitive to surface water-level variations.

At shallower depths the situation is more complicated, because diffusive effects comingle with the Skempton effect. For $\gamma = 0$ and $z = \delta$, for example, the phase angle of the effective normal stress is -134.2 degrees, and the amplitude is $0.288\ p_s$.

The pressure variation with time at the surface and the effective normal stress at a depth $z = \delta$ are graphed in Figure 6.17 to illustrate the relationship between the two. For an annual cycle, the maximum (tensile) effective normal stress at $z = \delta$ on a vertical fault plane occurs about 135 days after the maximum water load.

6.11.4 Tidally Induced Pore Pressures in Seafloor Sediments

A probe for measuring in situ pore pressures in deep-ocean sediments has been described by Schultheiss (1990). Two pressure ports are located in the

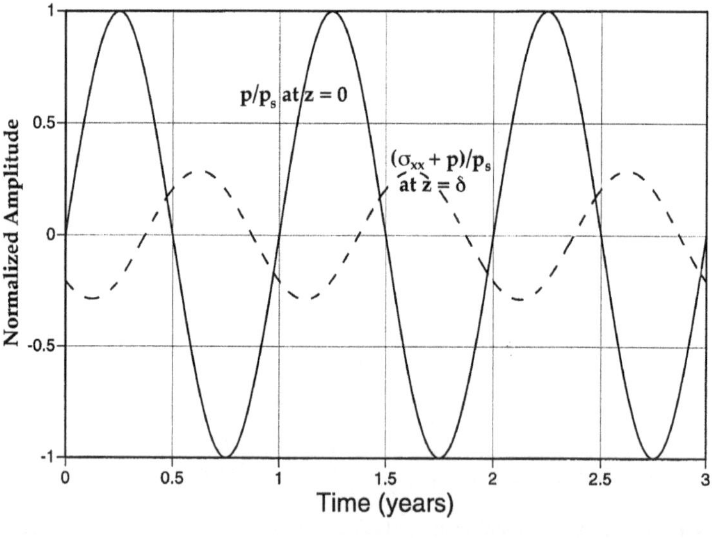

Figure 6.17: Normalized pore pressure at surface (solid) and normalized effective normal stress at $z = \delta$ (dashed) for $\gamma = 0$ and an annual water-level cycle of a reservoir lake. Negative values of $\sigma_{xx} + p$ are compressive and stabilizing.

portion of the probe embedded in the sediments. Each pressure port is connected via a water-filled tube to a differential pore-pressure transducer on the seafloor, which measures the difference in excess pressure, $\Delta p(z,t)$, between the pressure port and the seafloor,

$$\Delta p(z,t) = p(z,t) - p(0,t) \tag{6.79}$$

where depth z is measured downward from the seafloor.

Periodic tidal fluctuations in sea level are transmitted to the water-sediment interface as changes in surface load and fluid pressure. The differential pore pressure can be expressed as $\Delta p(z,t) = \widetilde{\Delta p}(z)\exp(\imath\omega t)$, where the complex amplitude, $\widetilde{\Delta p}(z)$, equals $\tilde{p}(z) - \tilde{p}(0)$. Substituting Eqn. 6.73 for $\tilde{p}(z)$ and the amplitude p_s for $\tilde{p}(0)$ gives

$$\widetilde{\Delta p}(z) = (1-\gamma)p_s\left\{\exp[-(1+i)z\sqrt{\omega/2c}] - 1\right\} \tag{6.80}$$

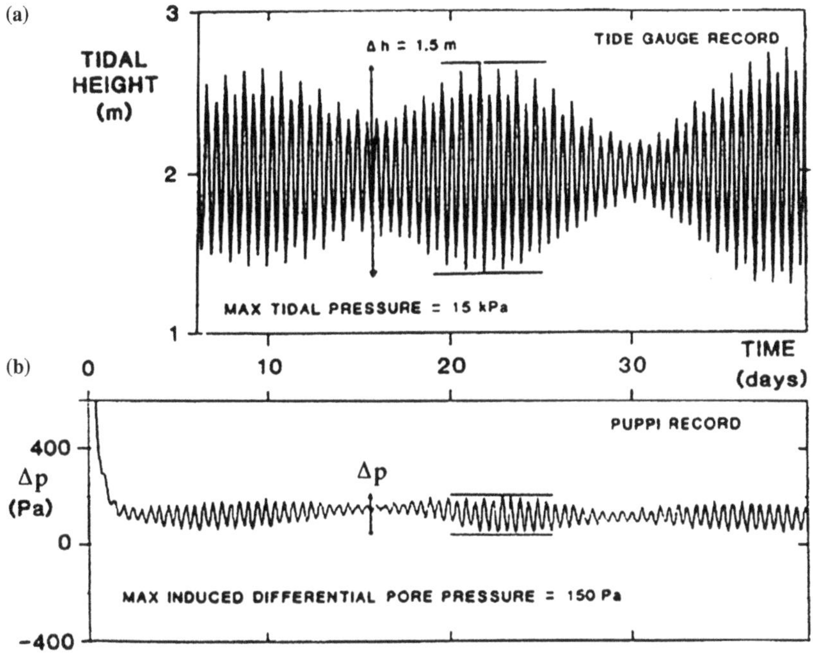

Figure 6.18: Tide gauge (a) and differential pore pressure (b) in seabed sediment (Schultheiss, 1990). The amplitude of the differential pore pressure at a depth 4 meters below the seabed is one percent of the amplitude at the seabed.

The real part of Eqn. 6.80 yields the amplitude of the differential pressure:

$$|\widetilde{\Delta p}| = (1-\gamma)p_s\left[\left(\exp\frac{-z}{\delta}\cos\frac{z}{\delta}-1\right)^2+\left(\exp\frac{-z}{\delta}\sin\frac{z}{\delta}\right)^2\right]^{1/2} \quad (6.81)$$

where $\delta = \sqrt{2c/\omega}$. The phase angle θ is given by

$$\theta = \tan^{-1}\left[\frac{-\exp\frac{-z}{\delta}\sin\frac{z}{\delta}}{\exp\frac{-z}{\delta}\cos\frac{z}{\delta}-1}\right] \quad (6.82)$$

For $z > 3\delta$, $|\widetilde{\Delta p}|$ is approximately $(1-\gamma)p_s$, and θ is approximately zero.

Typical records for tide gauge and differential pore pressure are shown in Figure 6.18. The principal lunar tide has a period of 12.42 hours. A one-meter change in sea level corresponds to about 10 kPa pressure. The amplitude

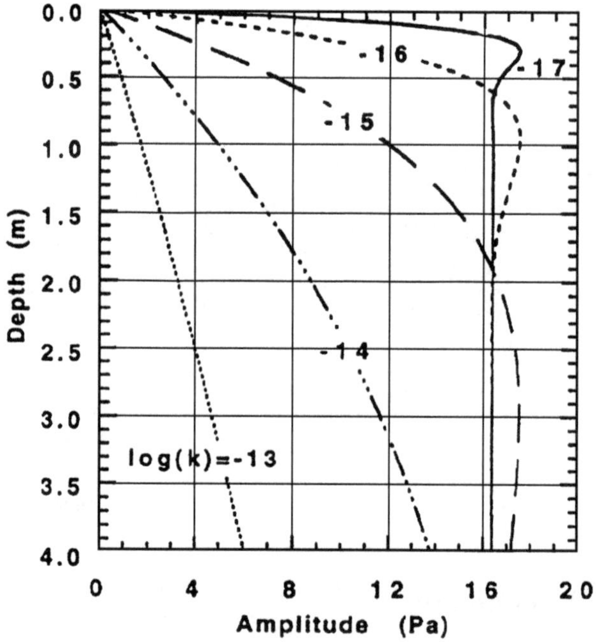

Figure 6.19: Differential pore pressure versus depth below the seafloor for different values of permeability k (Fang et al., 1993, Fig. 9*a*, p. 7298). The parameters are $p_s = 2000$ Pa, $1/K_v = 3.1\times10^{-8}$ Pa^{-1}, $1/K_f = 4.07\times10^{-10}$ Pa^{-1}, $\phi = 0.628$, and $\mu = 1.656\times10^{-3}$ Pa·s. The value of $1/K_v$ is 1.628 times that used by Fang et al., as their expressions for γ and S_s erroneously contained the factor $(1+\phi)$ multiplying $1/K_v$ (Wang and Davis, 1996).

of the differential pressure at a depth of 4 meters is about 1% of the tidal amplitude. Because the value of δ is approximately 0.2 m for a permeability of 10^{-16} m^2, the term in brackets in Eqn. 6.81 is essentially equal to one, and the loading efficiency γ is about 0.99. If $1/K_s \approx 0$ and $1/K_v \gg 1/K_f$, then $\gamma \approx 1 - \phi K_v/K_f$, and $S \approx 1/K_v + \phi/K_f$. With these assumptions, the variation of pore-pressure amplitude with depth below the seabed is plotted in Figure 6.19. For permeability values below 10^{-16} m^2, the amplitude of the differential pore pressure increases rapidly in the first few tenths of a meter, where pore-pressure diffusion contributes. It reaches a relative maximum within a meter of the seafloor before approaching a constant value, $p_s(1 - \gamma) \approx p_s \phi K_v/K_f$, as z exceeds the decay distance δ.

7

Plane Strain and Plane Stress in Rectangular Coordinates

7.0 CHAPTER OVERVIEW

Geomechanics problems, such as loading by a reservoir lake or seabed structure that is infinitely extensive in one direction on the earth's surface, can be represented by a perpendicular cross section in a state of plane strain. Geomechanics problems, such as water-level changes due to solid earth tides, can be interpreted assuming that a horizontal layer is in a state of plane stress.

Boundary-value problems for a half space in plane strain subjected to periodic or sudden loading have been solved using stress or displacement functions introduced by Biot (1956d). The advantage of stress or displacement functions is that their governing equations are uncoupled. The stress or displacement components are recovered as partial derivatives of the stress or displacement functions, respectively. Verruijt (1971) showed that McNamee and Gibson's (1960a, 1960b) displacement functions $\mathcal{E}$ and $\mathcal{S}$ used to solve the plane-strain problem of a suddenly applied load on the surface of a half space are a special case of Biot's.

Representative problems in plane strain considered in this chapter include (1) Mandel's (1953) problem, which is the canonical demonstration of how poroelastic coupling can produce nonmonotonic pore-pressure behavior following undrained loading[1]; (2) mechanical loading of a drained half space (Biot, 1941b; McNamee and Gibson, 1960b; Verruijt, 1971; Roeloffs, 1988); (3) fluid extraction from a long line of wells (Segall, 1985); and (4) shear slip on faults (Booker, 1974; Rice and Cleary, 1976; Roeloffs and Rudnicki, 1984/85; Rudnicki, 1986b, 1987; Rudnicki and Hsu, 1988; Rudnicki and Roeloffs, 1990).

[1] This phenomenon is called the Mandel-Cryer effect, because Cryer (1963) showed that the hydrostatic loading of a sphere induces similar nonmonotonic pore-pressure behavior at the center of the sphere.

7.1 CONSTITUTIVE EQUATIONS FOR PLANE STRAIN

If loading is infinitely long in the z-direction, the problem reduces to two dimensions in the (x, y) plane. The displacement components are independent of z; hence $\epsilon_{zz} = \epsilon_{xz} = \epsilon_{yz} = 0$. Substituting $\epsilon_{zz} = 0$ into Eqn. 2.32 gives

$$\sigma_{zz} = \nu(\sigma_{xx} + \sigma_{yy}) - (1 - 2\nu)\alpha p \tag{7.1}$$

The constitutive equations (Eqns. 2.30, 2.31, and 2.33), with stress and pore pressure as the independent variables, are then

$$\epsilon_{xx} = \frac{1 - \nu^2}{E}\sigma_{xx} - \frac{\nu(1 + \nu)}{E}\sigma_{yy} + \frac{\alpha(1 + \nu)}{3K}p \tag{7.2}$$

$$\epsilon_{yy} = -\frac{\nu(1 + \nu)}{E}\sigma_{xx} + \frac{1 - \nu^2}{E}\sigma_{yy} + \frac{\alpha(1 + \nu)}{3K}p \tag{7.3}$$

$$\epsilon_{xy} = \frac{1}{2G}\sigma_{xy} \tag{7.4}$$

Also, using Eqn. 7.1 in the constitutive equation for ζ (Eqn. 2.41) gives

$$\zeta = \frac{\alpha(1 + \nu)}{3K}\left[(\sigma_{xx} + \sigma_{yy}) + \frac{3}{B(1 + \nu_u)}p\right] \tag{7.5}$$

where Eqn. 3.33 is useful to arrive at the result.

7.2 GOVERNING EQUATIONS FOR PLANE STRAIN

The governing equations in plane strain are obtained in a manner that parallels the derivations of the three-dimensional equations (cf. Chapter 4).

7.2.1 Mechanical Equilibrium

The mechanical equilibrium equations in plane strain in the absence of body forces are

$$\begin{aligned} \frac{\partial \sigma_{xx}}{\partial x} + \frac{\partial \sigma_{xy}}{\partial y} &= 0 \\ \frac{\partial \sigma_{xy}}{\partial x} + \frac{\partial \sigma_{yy}}{\partial y} &= 0 \end{aligned} \tag{7.6}$$

Setting $\partial w/\partial z$ and body forces to zero in Eqn. 4.15 gives the displacement-pressure form of the mechanical equilibrium equations:

$$G\left[\frac{\partial^2 u}{\partial x^2}+\frac{\partial^2 u}{\partial y^2}\right]+\frac{G}{1-2\nu}\left[\frac{\partial^2 u}{\partial x^2}+\frac{\partial^2 v}{\partial x\partial y}\right]=\alpha\frac{\partial p}{\partial x} \tag{7.7}$$

$$G\left[\frac{\partial^2 v}{\partial x^2}+\frac{\partial^2 v}{\partial y^2}\right]+\frac{G}{1-2\nu}\left[\frac{\partial^2 u}{\partial y\partial x}+\frac{\partial^2 v}{\partial y^2}\right]=\alpha\frac{\partial p}{\partial y} \tag{7.8}$$

Similarly, setting $\partial w/\partial z$ and body forces to zero in Eqn. 4.22 gives the displacement-ζ form of the mechanical equilibrium equations:

$$G\left[\frac{\partial^2 u}{\partial x^2}+\frac{\partial^2 u}{\partial y^2}\right]+\frac{G}{1-2\nu_u}\left[\frac{\partial^2 u}{\partial x^2}+\frac{\partial^2 v}{\partial x\partial y}\right]=\alpha M\frac{\partial \zeta}{\partial x} \tag{7.9}$$

$$G\left[\frac{\partial^2 v}{\partial x^2}+\frac{\partial^2 v}{\partial y^2}\right]+\frac{G}{1-2\nu_u}\left[\frac{\partial^2 u}{\partial y\partial x}+\frac{\partial^2 v}{\partial y^2}\right]=\alpha M\frac{\partial \zeta}{\partial y} \tag{7.10}$$

7.2.2 Strain Compatibility

The strain-compatibility equation for plane strain is obtained by substituting Eqn. 7.1 into the three-dimensional strain-compatibility equation (Eqn. 4.26),

$$\nabla^2(\sigma_{xx}+\sigma_{yy}+2\eta p)=0 \tag{7.11}$$

where

$$\nabla^2=\frac{\partial^2}{\partial x^2}+\frac{\partial^2}{\partial y^2} \tag{7.12}$$

7.2.3 Fluid Diffusion

Inserting Eqn. 7.5 into the diffusion equation for ζ (Eqn. 4.67) in the absence of fluid sources gives

$$c\nabla^2\left[\sigma_{xx}+\sigma_{yy}+\frac{3}{B(1+\nu_u)}p\right]=\frac{\partial}{\partial t}\left[\sigma_{xx}+\sigma_{yy}+\frac{3}{B(1+\nu_u)}p\right] \tag{7.13}$$

The pore pressure–displacement form of the fluid diffusion equation is obtained from Eqn. 4.65:

$$\alpha\frac{\partial}{\partial t}\left[\frac{\partial u}{\partial x}+\frac{\partial v}{\partial y}\right]+S_\epsilon\frac{\partial p}{\partial t}=\frac{k}{\mu}\left[\frac{\partial^2 p}{\partial x^2}+\frac{\partial^2 p}{\partial y^2}\right] \tag{7.14}$$

7.3 MANDEL'S PROBLEM

Mandel (1953) presented the canonical example of nonmonotonic behavior in pore pressure following undrained loading. An infinitely long, rectangular plate is sandwiched between two rigid, impermeable plates (Fig. 7.1). A compressive force, $F = -2p_o a$, is suddenly applied at time zero and then held constant. The force F is a force per unit length in the z-direction, $-p_o$ is the normal stress $\sigma_{yy}(x, \pm b, t)$, and a is the half width of the bar. The left and right edges of the plate are stress free and drained. The rigid plate condition means that the total force is distributed across the plate such that

$$\int_{-a}^{a} \sigma_{yy}(x, b, t)dx = -2p_o a \tag{7.15}$$

The geometry and boundary conditions imply that horizontal planes remain parallel, and fluid flow takes place only in the horizontal direction. Thus, stress components and pore pressure are functions only of x and t. Because $\partial\sigma_{xy}/\partial y = 0$, force equilibrium implies that $\sigma_{xx}(x, t) = 0$. Integrating Eqn. 7.11 leads to

$$\sigma_{yy} + 2\eta p = g(t) \tag{7.16}$$

where $g(t)$ is a constant of integration that is a function of time only, because the flow field is irrotational. From the boundary condition $p(a, t) = 0$,

$$g(t) = \sigma_{yy}(a, t) \tag{7.17}$$

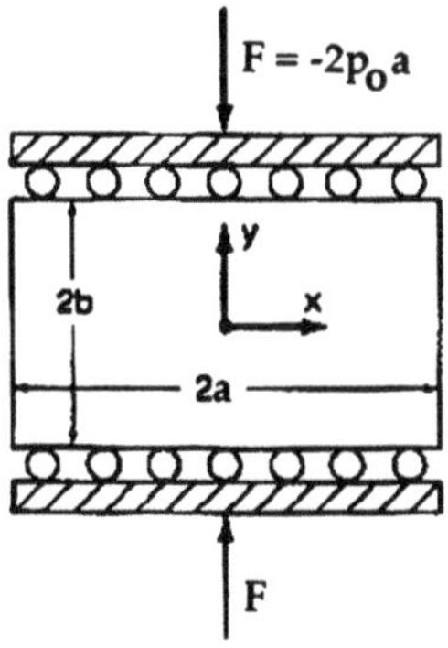

Figure 7.1: Geometry for Mandel's problem (after Abousleiman et al., 1996).

Also because $\sigma_{xx}(x, t) = 0$, Eqn. 7.13 simplifies to

$$c\frac{\partial^2}{\partial x^2}(\sigma_{yy} + A_1 p) = \frac{\partial}{\partial t}(\sigma_{yy} + A_1 p) \qquad (7.18)$$

where $A_1 = 3/B(1 + \nu_u)$.[2] Eqn. 7.18 is a diffusion equation for the effective stress $\sigma_{yy} + A_1 p$. Taking the Laplace transform of Eqn. 7.18 leads to the ordinary differential equation,

$$c\frac{d^2}{dx^2}\left(\tilde{\sigma}_{yy} + A_1\tilde{p}\right) = s\left(\tilde{\sigma}_{yy} + A_1\tilde{p}\right) \qquad (7.19)$$

where the tilde signifies the Laplace transform. The pore pressure $\tilde{p}$ is eliminated from Eqn. 7.19 by using Eqns. 7.16 and 7.17,

$$c\left(1 - \frac{A_1}{A_2}\right)\frac{d^2\tilde{\sigma}_{yy}}{dx^2} = s\left(1 - \frac{A_1}{A_2}\right)\tilde{\sigma}_{yy} + \frac{sA_1}{A_2}\tilde{\sigma}_{yy}(a, s) \qquad (7.20)$$

where $A_2 = 2\eta$. The general solution of Eqn. 7.20 is

$$\tilde{\sigma}_{yy}(x, s) = A(s)\cosh\sqrt{\frac{s}{c}}x + B(s)\sinh\sqrt{\frac{s}{c}}x - \frac{A_1}{A_2 - A_1}\tilde{\sigma}_{yy}(a, s) \qquad (7.21)$$

where $A(s)$ and $B(s)$ are coefficients to be determined from boundary conditions. The first two terms are the homogeneous solution, and the third term is the particular solution of Eqn. 7.20. Symmetry about $x = 0$ requires that $B(s) = 0$, so Eqn. 7.21 becomes

$$\tilde{\sigma}_{yy}(x, s) = A(s)\cosh\sqrt{\frac{s}{c}}x - \frac{A_1}{A_2 - A_1}\tilde{\sigma}_{yy}(a, s) \qquad (7.22)$$

The two unknowns at this point are $A(s)$ and $\tilde{\sigma}_{yy}(a, s)$. Substituting $x = a$ into Eqn. 7.22 gives

$$\tilde{\sigma}_{yy}(a, s) = \frac{A_2 - A_1}{A_2}A(s)\cosh\sqrt{\frac{s}{c}}a \qquad (7.23)$$

$$\tilde{\sigma}_{yy}(x, s) = A(s)\left\{\cosh\sqrt{\frac{s}{c}}x - \frac{A_1}{A_2}\cosh\sqrt{\frac{s}{c}}a\right\} \qquad (7.24)$$

[2] The two-dimensional loading efficiency γ_2 discussed later (Eqn. 7.62) is related to A_1 by $\gamma_2 = 2/A_1$.

Applying the Laplace transform of the force boundary condition (Eqn. 7.15) gives

$$A(s) = -\frac{\frac{p_o}{s} a \sqrt{\frac{s}{c}}}{\sinh \sqrt{\frac{s}{c}} a - \frac{a A_1}{A_2} \sqrt{\frac{s}{c}} \cosh \sqrt{\frac{s}{c}} a} \tag{7.25}$$

All the unknown parameters are thus established. Therefore, the Laplace transform of $p(x, t)$ is

$$\begin{aligned} \tilde{p}(x, s) &= \frac{1}{A_2} \{\tilde{\sigma}_{yy}(a, s) - \tilde{\sigma}_{yy}(x, s)\} \\ &= \frac{A(s)}{A_2} \left\{ \cosh \sqrt{\frac{s}{c}} a - \cosh \sqrt{\frac{s}{c}} x \right\} \end{aligned} \tag{7.26}$$

Returning to the time domain is achieved through use of the Laplace transform inversion theorem (Carslaw and Jaeger, 1959, p. 302),

$$p(x, t) = \frac{1}{2\pi \imath} \int_{\xi - \imath\infty}^{\xi + \imath\infty} \tilde{p}(x, s) \exp(st) ds \tag{7.27}$$

where ξ in Eqn. 7.27 is a constant greater than the real part of all singularities of $\tilde{p}(x, s)$ that lie on the negative real axis. The integral in Eqn. 7.27 is obtained by summing residues at all the poles of the integrand after completing the large semicircular contour (Fig. 7.2) because the contribution from the semicircular path tends to zero as the radius tends to infinity (Carslaw and Jaeger, 1959, p. 302). The poles are obtained by setting the denominator

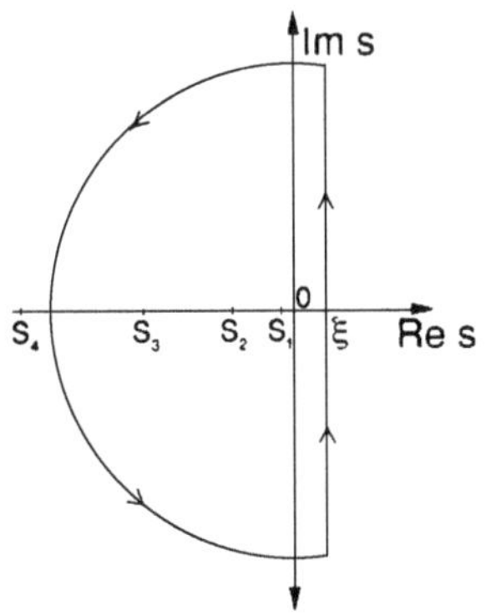

Figure 7.2: Contour integral for inverting the Laplace transform of an integrand containing simple poles along the negative real axis (after Mason et al., 1991).

of $\tilde{p}(x,s)$ to zero (cf. Eqn. 7.25),

$$\frac{\tanh\sqrt{\frac{s_n}{c}}a}{\sqrt{\frac{s_n}{c}}a} = \frac{A_1}{A_2} = \frac{3}{2\eta B(1+\nu_u)} \tag{7.28}$$

where the roots s_n of Eqn. 7.28 lie on the negative real axis. The residue of an integrand in the form $p(s)/q(s)$ at a pole s_n is given by $p(s_n)/q'(s_n)$ (LePage, 1961, p. 145), where the prime signifies the first derivative. Taking the derivative of the denominator of Eqn. 7.25, summing the residues over the roots s_n, and defining $(s_n/c)a^2 = -\lambda_n^2$ lead to

$$p(x,t) = \frac{2p_o}{A_1}\sum_{n=1}^{\infty}\frac{\sin\lambda_n}{\lambda_n - \sin\lambda_n\cos\lambda_n} \times\left(\cos\frac{\lambda_n x}{a} - \cos\lambda_n\right)\exp\left(-\frac{\lambda_n^2 ct}{a^2}\right) \tag{7.29}$$

The nonmonotonic pressure behavior is shown in Figure 7.3. The pore pressure rises above the undrained value p_o for small dimensionless times, $t^* = ct/a^2$. The behavior is explained by the contraction at the drained edges of the plate inducing additional pore-pressure buildup in the plate interior.

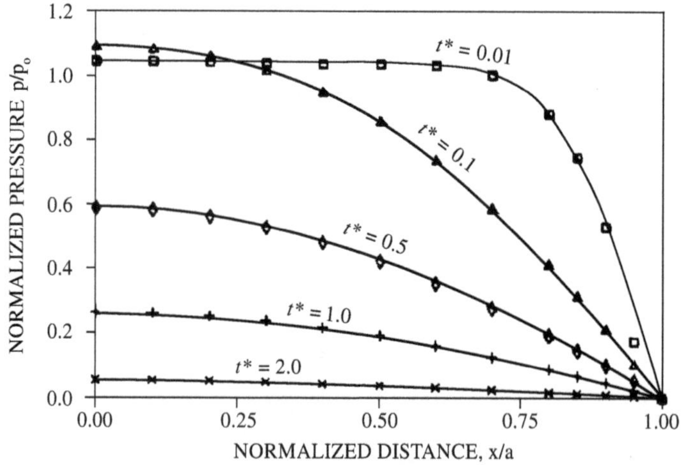

Figure 7.3: Pressure profiles across a plate in Mandel's problem. Note the nonmonotonic behavior of pore pressure in the central region of the sample between dimensionless times 0.01 and 0.1. Poroelastic constants used were $\nu = 0.2$, $\nu_u = 0.5$, and $B = 1$ (Cheng and Detournay, 1988).

Cheng and Detournay (1988) give equations for all the stress and displacement components. Abousleiman et al. (1996) give the results for an anisotropic medium whose principal axes are oriented either parallel or perpendicular to the load. Also, Cheng et al. (1993) present three-dimensional graphical results showing how pressure varies with x and t.

7.4 BIOT'S STRESS FUNCTION

Biot's stress function $F(x, y, t)$ (Biot, 1956d; Roeloffs, 1988) is defined by its relationship to stress components:

$$\sigma_{xx} = \frac{\partial^2 F}{\partial y^2} \tag{7.30}$$

$$\sigma_{yy} = \frac{\partial^2 F}{\partial x^2} \tag{7.31}$$

$$\sigma_{xy} = -\frac{\partial^2 F}{\partial x \partial y} \tag{7.32}$$

The force equilibrium equations (Eqns. 7.6) are automatically satisfied when the stress components are expressed in terms of the stress function F.

Governing equations involving the stress function are obtained by substituting Eqns. 7.30–7.32 into the compatibility equation (Eqn. 7.11) and the diffusion equation (Eqn. 7.13), respectively:

$$\nabla^2 \left(\nabla^2 F + 2\eta p\right) = 0 \tag{7.33}$$

$$c\nabla^2 \left(\nabla^2 F + \frac{3}{B(1+\nu_u)} p\right) = \frac{\partial}{\partial t} \left(\nabla^2 F + \frac{3}{B(1+\nu_u)} p\right) \tag{7.34}$$

These equations can be uncoupled by eliminating F or p between them. Eqn. 7.33 and its Laplacian are solved for $\nabla^4 F$ and $\nabla^6 F$, respectively, and substituted into the Laplacian of Eqn. 7.34 to yield

$$c\nabla^4 p = \frac{\partial}{\partial t} \nabla^2 p \tag{7.35}$$

A similar procedure gives

$$c\nabla^6 F = \frac{\partial}{\partial t} \nabla^4 F \tag{7.36}$$

The general solution of p is the sum of a solution that satisfies the diffusion equation and a solution that satisfies Laplace's equation (Biot, 1956d),

$$p = p_1 + p_2 \tag{7.37}$$

where

$$c\nabla^2 p_1 = \frac{\partial p_1}{\partial t} \tag{7.38}$$

and

$$\nabla^2 p_2 = 0 \tag{7.39}$$

Similarly, the general solution of F is the sum of a solution that satisfies the diffusion equation and a solution that satisfies the biharmonic equation,

$$F = F_1 + F_2 \tag{7.40}$$

where

$$c\nabla^2 F_1 = \frac{\partial F_1}{\partial t} \tag{7.41}$$

and

$$\nabla^4 F_2 = 0 \tag{7.42}$$

These governing equations are a useful formulation if boundary conditions are specified in terms of stress and pore pressure.

7.5 PERIODIC LOAD ON A HALF SPACE

Biot's stress function was used by Roeloffs (1988) to solve the plane-strain problem of a sinusoidally time-varying water load on a half space drained at its surface. This method is illustrated for water loading and for the similar problem of mechanical loading on a drained boundary. This same pair of problems was considered previously for uniaxial strain conditions (cf. Section 6.9). A combined treatment of these two problems was also made by Verruijt (1982).

7.5.1 Mechanical Load

A cyclic load with time variation $\exp(\imath\omega t)$ on the drained surface of a half space is represented by the surface boundary conditions

$$\sigma_{yy}(x,0,t) = -r(x)\exp(\imath\omega t) \tag{7.43}$$

$$\sigma_{xy}(x,0,t) = 0 \tag{7.44}$$

$$p(x,0,t) = 0 \tag{7.45}$$

where $r(x)$ is the spatial variation of the surface load. Separation of the space and time variables can be made for each function, p_1, p_2, F_1, and F_2, in Eqns. 7.38, 7.39, 7.41, and 7.42. For example,

$$p_1(x,y,t) = \tilde{p}_1(x,y)\exp(\imath\omega t) \tag{7.46}$$

where $\tilde{p}_1$ is a function of only x and y. The Fourier transforms with respect to x of Eqns. 7.38, 7.39, 7.41, and 7.42 (cf. Appendix A, Eqn. A.1) are

$$\left(-\lambda^2 + \frac{d^2}{dy^2} - \frac{\imath\omega}{c}\right)\bar{p}_1 = 0 \tag{7.47}$$

$$\left(-\lambda^2 + \frac{d^2}{dy^2}\right)\bar{p}_2 = 0 \tag{7.48}$$

$$\left(-\lambda^2 + \frac{d^2}{dy^2} - \frac{\imath\omega}{c}\right)\bar{F}_1 = 0 \tag{7.49}$$

$$\left(-\lambda^2 + \frac{d^2}{dy^2}\right)^2 \bar{F}_2 = 0 \tag{7.50}$$

where the bar denotes the Fourier transform with respect to x (e.g., $\bar{p}_1(\lambda,y) = \mathcal{F}\{\tilde{p}_1(x,y)\}$). The general solutions of Eqns. 7.47–7.50, bounded for $y \to \infty$, are

$$\bar{p}_1(\lambda,y) = A_1\exp(-ny) \tag{7.51}$$

$$\bar{p}_2(\lambda,y) = A_2\exp(-|\lambda|y) \tag{7.52}$$

$$\bar{F}_1(\lambda,y) = B_1\exp(-ny) \tag{7.53}$$

$$\bar{F}_2(\lambda,y) = B_2\exp(-|\lambda|y) + B_3 y\exp(-|\lambda|y) \tag{7.54}$$

where

$$n^2 = \lambda^2 + \frac{\imath\omega}{c} \tag{7.55}$$

The five coefficients in Eqns. 7.51–7.54 are obtained from the three boundary conditions (Eqns. 7.43–7.45) and the coupled governing equations (Eqns. 7.33 and 7.34),

$$A_1 = -\bar{r}(\lambda)\frac{\gamma_2(\imath\omega/2\eta c)}{D} \tag{7.56}$$

$$A_2 = \bar{r}(\lambda)\frac{\gamma_2(\imath\omega/2\eta c)}{D} \tag{7.57}$$

$$B_1 = \bar{r}(\lambda)\frac{\gamma_2}{D} \tag{7.58}$$

$$B_2 = \bar{r}(\lambda)\left(\frac{1}{\lambda^2} - \frac{\gamma_2}{D}\right) \tag{7.59}$$

$$B_3 = \bar{r}(\lambda)\frac{(\imath\omega/2\eta c)}{|\lambda| D} \tag{7.60}$$

where

$$\bar{r}(\lambda) = \mathcal{F}\{r(x)\} \tag{7.61}$$

is the Fourier transform of the spatial loading distribution. The two-dimensional, plane strain, loading coefficient is defined as (cf. Eqn. 3.78)

$$\gamma_2 \equiv \frac{2B}{3}(1+\nu_u) \tag{7.62}$$

The following definition is also made:

$$D \equiv \gamma_2\lambda^2 - \gamma_2|\lambda|n + (\imath\omega/2\eta c) \tag{7.63}$$

The inverse Fourier transform back to the spatial domain x must be performed numerically in most cases. Some special cases exist for which an analytical inversion is possible:

1. *Sinusoidal Spatial Load.* Let the sinusoidal surface load be represented by $r(x) = \cos Lx$ (Fig. 7.4). The Fourier transform of $r(x)$ is $\bar{r}(\lambda) = \pi[\delta(L+\lambda)+\delta(L-\lambda)]$. Because the forward transform involves δ-functions, the inverse transform is evaluated readily by substituting L and $-L$ in the integrand of Eqn. A.2 (see Appendix A). The result is a special case of the solution given by Verruijt (1982).
2. *Constant Spatial Load.* A uniform spatial load, $r(x) = 1$, on the surface of the half space has the Fourier transform $\bar{r}(\lambda) = 2\pi\delta(\lambda)$ (let $L \to 0$ in the transform of $\cos Lx$ given in No. 1). Using Eqns. 7.56 and 7.57

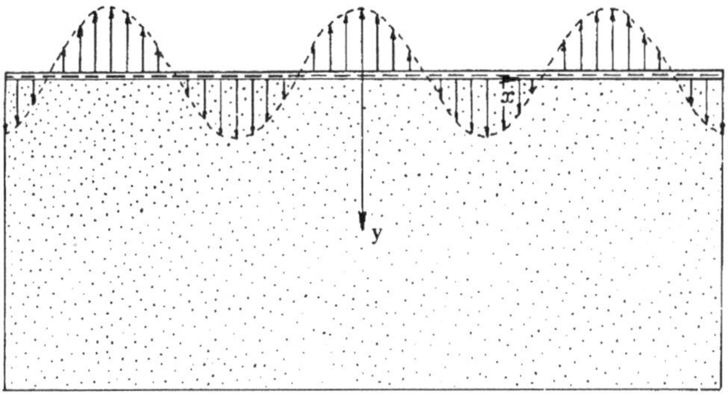

Figure 7.4: Sinusoidal, periodic load extending infinitely in the z-direction (Verruijt, 1982).

for A_1 and A_2, substituting into Eqns. 7.51 and 7.52, and taking the inverse Fourier transform give

$$\tilde{p}(x, y) = \gamma_2 \left[1 - \exp(-y\sqrt{\imath\omega/c})\right] \quad (7.64)$$

because $n = \sqrt{\imath\omega/c}$ and $D = \imath\omega/(2\eta c)$ for $\lambda = 0$. Eqn. 7.64 is independent of x, as expected for a uniform spatial load. The pore-pressure attenuation and phase shift are a function of depth only. If γ replaces γ_2 in Eqn. 7.64, the result is the same as the corresponding expression for pore pressure in a uniaxially constrained column (Eqn. 6.60) because γ_2 is the loading efficiency for the case of plane strain.

3. *Undrained Limit* ($\omega \to \infty$). Pore fluid diffusion has insufficient time to take place when frequencies ω are high. Therefore, the solution approaches the undrained response. The limit as $\omega \to \infty$ in Eqns. 7.56–7.60 can be obtained using l'Hôpital's rule and the relationship $dn/d\omega = i/(2cn) \to 0$:

$$A_1 = -\bar{r}(\lambda)\gamma_2 \quad (7.65)$$

$$A_2 = \bar{r}(\lambda)\gamma_2 \quad (7.66)$$

$$B_1 = 0 \quad (7.67)$$

$$B_2 = \bar{r}(\lambda)\frac{1}{\lambda^2} \quad (7.68)$$

$$B_3 = \bar{r}(\lambda)\frac{1}{|\lambda|} \quad (7.69)$$

Eqn. 7.67 means that $\bar{F}_1(\lambda, y) = 0$ (cf. Eqn. 7.53), and hence,

$$\lim_{\omega\to\infty} \bar{F}(\lambda, y) = B_2 \exp(-|\lambda|y) + B_3 y \exp(-|\lambda|y) \quad (7.70)$$

Inserting Eqn. 7.70 into Eqn. 7.30 and 7.31, and using Eqn. A.4, give

$$\lim_{\omega \to \infty} (\bar{\sigma}_{xx} + \bar{\sigma}_{yy}) = -2\bar{r}(\lambda) \exp(-|\lambda| y) \tag{7.71}$$

Because $\exp(-ny) \to 0$ as $\omega \to \infty$, $\bar{p}_1(\lambda, y) = 0$ (cf. Eqn. 7.51), and hence

$$\lim_{\omega \to \infty} \bar{p}(\lambda, y) = \bar{r}(\lambda)\gamma_2 \exp(-|\lambda| y) \tag{7.72}$$

Comparing Eqns. 7.71 and 7.72 shows that the pore pressure is γ_2 times the average compressive stress in the xy-plane, which again demonstrates the physical interpretation of γ_2 as the plane-strain loading efficiency. Furthermore, both compressive mean stress and pore pressure are in phase with the surface loading for $\omega \to \infty$.

4. *Undrained Limit for Line Load.* The inversion of Eqn. 7.71 for a line load, $r(x) = \delta(x)$, of unit strength at the origin can be obtained analytically, because $\bar{r}(\lambda) = 1$. Substituting into Eqn. 7.71 gives

$$\frac{\bar{\sigma}_{xx} + \bar{\sigma}_{yy}}{2} = -\exp(-|\lambda| y) \tag{7.73}$$

The inverse transform can be be found in any standard table (e.g., Erdélyi et al., 1954):

$$\frac{\sigma_{xx} + \sigma_{yy}}{2} = -\frac{1}{\pi} \frac{y}{x^2 + y^2} \tag{7.74}$$

The undrained solution is the same as that obtained in the classical theory of elasticity (e.g., Sneddon, 1951, p. 409).

5. *Undrained Limit for Strip Load.* For a compressive force per unit width, P, over the strip $-a \leq x \leq +a$, the convolution with the line-source solution (Eqn. 7.74) yields

$$\begin{aligned} \frac{\sigma_{xx} + \sigma_{yy}}{2} &= -\frac{P}{\pi} \int_{-a}^{a} \frac{y}{(x - x')^2 + y^2} dx' \\ &= -\frac{P}{\pi} \left[\tan^{-1} \frac{y}{x - a} - \tan^{-1} \frac{y}{x + a} \right] \end{aligned} \tag{7.75}$$

6. *Drained Limit ($\omega \to 0$).* Steady-state, drained conditions are approached as the frequency ω of the time variation approaches zero. The constants A_1 and A_2 equal zero in Eqns. 7.51 and 7.52, and hence $p(x, y)$ equals zero, as expected.

7.5.2 Water Load

The problem of a time-periodic water load on a half space is solved in the same manner as that for a purely mechanical load. The only change from the previous problem is that the pore-pressure boundary condition on the surface, Eqn. 7.45, is changed to

$$p(x,0,t) = r(x)\exp(\imath\omega t) \tag{7.76}$$

where $r(x)$ is now the spatial variation of the water load. The five unknowns in Eqns. 7.51–7.54 are given by (Roeloffs, 1988)[3]

$$A_1 = \bar{r}(\lambda)\frac{(1-\gamma_2)(\imath\omega/2\eta c)}{D} \tag{7.77}$$

$$A_2 = \bar{r}(\lambda)\frac{\gamma_2[\lambda^2 - n|\lambda| + (\imath\omega/2\eta c)]}{D} \tag{7.78}$$

$$B_1 = -\bar{r}(\lambda)\frac{1-\gamma_2}{D} \tag{7.79}$$

$$B_2 = \bar{r}(\lambda)\frac{\lambda^2 - \gamma_2 n|\lambda| + (\imath\omega/2\eta c)}{\lambda^2 D} \tag{7.80}$$

$$B_3 = \bar{r}(\lambda)\frac{\lambda^2 - n|\lambda| + (\imath\omega/2\eta c)}{|\lambda| D} \tag{7.81}$$

The analysis of different limiting cases can now be made as in Section 7.5.1 (see Roeloffs, 1988).

7.6 DISPLACEMENT FUNCTIONS

The displacement function method in which displacement components are obtained as various derivatives of displacement functions is similar to the stress-function method (Biot, 1956d). The governing equations for the displacement functions are uncoupled. The displacement function approach was used by McNamee and Gibson (1960a, 1960b), Gibson and McNamee (1963), and Gibson et al. (1970) to solve consolidation problems for a half space or a finite layer in plane strain or with axisymmetry. McNamee and Gibson assumed incompressible constituents ($K_s = K_f = \infty$). Verruijt (1971) generalized their formulation to the case that only $K_s = \infty$. Detournay and Cheng (1993) further generalized Verruijt's formulation by removing the restriction on K_s. Their formulation for plane strain is now given.

[3] The expression given for A_2 corrects a sign error in Roeloffs (1988).

The displacement components and increment of fluid content in plane strain are given in terms of two scalar displacement functions, $\mathcal{E}$ and $\mathcal{S}$:

$$u_x = -\frac{\partial \mathcal{E}}{\partial x} + y\frac{\partial \mathcal{S}}{\partial x} \tag{7.82}$$

$$u_y = -\frac{\partial \mathcal{E}}{\partial y} + y\frac{\partial \mathcal{S}}{\partial y} - (3-4\nu_u)\mathcal{S} \tag{7.83}$$

$$\zeta = -\frac{GS}{\eta}\nabla^2\mathcal{E} \tag{7.84}$$

Substitution of Eqns. 7.82–7.84 into Eqns. 7.9, 7.10, and 4.67 yields the uncoupled governing equations,

$$\nabla^2\mathcal{S} = 0 \tag{7.85}$$

$$c\nabla^4\mathcal{E} = \frac{\partial}{\partial t}\left(\nabla^2\mathcal{E}\right) \tag{7.86}$$

that is, $\mathcal{S}$ satisfies Laplace's equation, and $\mathcal{E}$ satisfies the biharmonic diffusion equation. The stresses and pore pressure are obtained from the displacement functions[4]:

$$\sigma_{xx} = 2G\left[\nabla^2\mathcal{E} - \frac{\partial^2\mathcal{E}}{\partial x^2} + y\frac{\partial^2\mathcal{S}}{\partial x^2} - 2\nu_u\frac{\partial \mathcal{S}}{\partial y}\right] \tag{7.87}$$

$$\sigma_{yy} = 2G\left[\nabla^2\mathcal{E} - \frac{\partial^2\mathcal{E}}{\partial y^2} + y\frac{\partial^2\mathcal{S}}{\partial y^2} - 2(1-\nu_u)\frac{\partial \mathcal{S}}{\partial y}\right] \tag{7.88}$$

$$\sigma_{xy} = 2G\left[-\frac{\partial^2\mathcal{E}}{\partial x\partial y} + y\frac{\partial^2\mathcal{S}}{\partial x\partial y} - (1-2\nu_u)\frac{\partial \mathcal{S}}{\partial x}\right] \tag{7.89}$$

$$p = -\frac{G}{\eta}\left[\nabla^2\mathcal{E} - \frac{2(\nu_u-\nu)}{1-\nu}\frac{\partial \mathcal{S}}{\partial y}\right] \tag{7.90}$$

The mean stress in the xy-plane is given by

$$\frac{\sigma_{xx}+\sigma_{yy}}{2} = G\left[\nabla^2\mathcal{E} + y\nabla^2\mathcal{S} - 2\frac{\partial \mathcal{S}}{\partial y}\right] \tag{7.91}$$

Verruijt (1971) showed that the McNamee and Gibson scalar displacement functions, $\mathcal{E}$ and $\mathcal{S}$, are a special case of a pair of scalar and vector

[4] Note that in Eqn. 116g in Detournay and Cheng (1993, pp. 137–138), the term $\partial^2\mathcal{S}/\partial z^2$ should be $\partial\mathcal{S}/\partial z$.

displacement functions, Φ and $\vec{\Psi}$, introduced earlier by Biot (1956d):

$$u_i = \frac{\partial}{\partial x_i}(\Phi + x_k \Psi_k) - 4(1 - \nu_u)\Psi_i \tag{7.92}$$

$$\zeta = \frac{GS}{\eta}\nabla^2\Phi \tag{7.93}$$

Biot noted that these functions are analogous to the Boussinesq-Papkovitch functions in ordinary linear elasticity. In the absence of body forces and sources, the components of $\vec{\Psi}$ must satisfy Laplace's equation,

$$\nabla^2\Psi_i = 0 \tag{7.94}$$

and Φ must satisfy the biharmonic diffusion equation:

$$\frac{\partial}{\partial t}(\nabla^2\Phi) - c\nabla^4\Phi = 0 \tag{7.95}$$

The relationship between the McNamee and Gibson displacement functions and the Biot functions is

$$\Phi = -\mathcal{E}(x, y, t) \tag{7.96}$$

$$\Psi_y = \mathcal{S}(x, y, t) \tag{7.97}$$

$$\Psi_x = \Psi_z = 0 \tag{7.98}$$

7.7 SUDDEN STEP LOAD ON A HALF SPACE

The problem of a suddenly applied surface load, $-r(x)H(t)$, can be handled in a manner similar to a periodic load. Using the Laplace transform for time and a subsequent Fourier transform for the horizontal spatial coordinate yields uncoupled ordinary differential equations. The unknown constants in the resulting general solutions are then found from the boundary conditions. The inversions back to the original coordinates must usually be performed numerically. The procedure is illustrated using displacement functions. An analytical solution is presented for a sinusoidally varying spatial load and for a strip load. The problem is solved for the general case in which both fluid and solid are compressible. The limiting case of incompressible constituents reduces to Biot's (1941b) solution.

Let the Laplace transform (Appendix A, Eqn. A.7) be denoted by a tilde, $\tilde{\mathcal{E}}(x, y, s) \equiv \mathcal{L}\{\mathcal{E}(x, y, t)\}$, and the subsequent Fourier transform with respect to x be denoted by an overbar, $\bar{\mathcal{E}}(\lambda, y, s) \equiv \mathcal{F}\{\tilde{\mathcal{E}}(x, y, s)\}$. The general solution of Eqn. 7.86 is

$$\mathcal{E} = \mathcal{E}_1 + \mathcal{E}_2 \tag{7.99}$$

where $\mathcal{E}_1$ and $\mathcal{E}_2$ satisfy $c\nabla^2\mathcal{E}_1 = \partial\mathcal{E}_1/\partial t$ and $\nabla^2\mathcal{E}_2 = 0$, respectively. The governing equations for $\bar{\mathcal{E}}_1$, $\bar{\mathcal{E}}_2$, and $\bar{\mathcal{S}}$ are then

$$\left(\frac{d^2}{dy^2} - \lambda^2 - \frac{s}{c}\right)\bar{\mathcal{E}}_1 = 0 \tag{7.100}$$

$$\left(\frac{d^2}{dy^2} - \lambda^2\right)\bar{\mathcal{E}}_2 = 0 \tag{7.101}$$

$$\left(\frac{d^2}{dy^2} - \lambda^2\right)\bar{\mathcal{S}} = 0 \tag{7.102}$$

where the initial conditions are assumed to be zero everywhere. These ordinary differential equations are the same as those obtained for periodic time loading $\exp(\imath\omega t)$ but with the Laplace transform variable s in place of $\imath\omega$. The general solutions that approach zero as $y \to \infty$ are[5]

$$\bar{\mathcal{E}}_1(\lambda, y, s) = A_1(\lambda, s)\exp(-ny) \tag{7.103}$$

$$\bar{\mathcal{E}}_2(\lambda, y, s) = A_2(\lambda, s)\exp(-|\lambda|y) \tag{7.104}$$

$$\bar{\mathcal{S}}(\lambda, y, s) = B_1(\lambda, s)\exp(-|\lambda|y) \tag{7.105}$$

where

$$n = \sqrt{\lambda^2 + \frac{s}{c}} \tag{7.106}$$

The boundary conditions are a specified normal load and zero shear stress on the drained surface of the half space:

$$\sigma_{yy}(x, 0, t) = -r(x) \tag{7.107}$$

$$\sigma_{xy}(x, 0, t) = 0 \tag{7.108}$$

$$p(x, 0, t) = 0 \tag{7.109}$$

The initial condition is that all these quantities are zero. The sequential Laplace and Fourier transform of $\sigma_{yy}(x, 0, t)$ in Eqn. 7.107 is $-\bar{r}(\lambda)/s$, where

[5] Note that the constants A_1, A_2, and B_1 have different meanings here than in Eqns. 7.56–7.58.

$\bar{r}(\lambda)$ is the Fourier transform of $r(x)$. The constants in Eqns. 7.103–7.105 are determined from the three boundary conditions (cf. Eqns. 7.88–7.90),

$$A_1 = \frac{\bar{r}(\lambda)}{s} \frac{\gamma_2}{2GD} \tag{7.110}$$

$$B_1 = -\frac{\bar{r}(\lambda)}{|\lambda|} \frac{1}{4G\eta c D} \tag{7.111}$$

$$A_2 = -\frac{n}{|\lambda|} A_1 + \frac{1-2\nu_u}{|\lambda|} B_1 \tag{7.112}$$

where

$$D = \gamma_2 |\lambda|^2 - \gamma_2 |\lambda| n + \frac{s}{2\eta c} \tag{7.113}$$

The relationship

$$\eta\gamma_2 = \frac{\nu_u - \nu}{1-\nu} \tag{7.114}$$

has also been used. Eqn. 7.113 is the same as Eqn. 7.63 but with s in place of $\imath\omega$. The solution for a periodic surface load (cf. Section 7.5) can be obtained from Eqns. 7.110–7.112 by first replacing $\bar{r}(\lambda)$ with $s\bar{r}(\lambda)$ and then replacing s with $\imath\omega$.

The transform of the vertical surface displacement turns out to be proportional to B_1:

$$\bar{u}_y(\lambda, 0, s) = -2(1-\nu_u) B_1 \tag{7.115}$$

$$= \frac{\bar{r}(|\lambda|)}{|\lambda|} \frac{1-\nu_u}{2G\eta c D} \tag{7.116}$$

7.7.1 Sinusoidal Spatial Load

If the surface load $r(x) = A \sin Lx$, the Fourier transform $\bar{r}(\lambda) = A\imath\pi[\delta(\lambda + L) - \delta(\lambda - L)]$, where δ is the Dirac delta function. Obtaining the inverse Fourier transform of Eqn. 7.116 directly from Eqn. A.2 (see Appendix A) yields

$$\tilde{u}_y(x, 0, s) = A \frac{1-\nu_u}{G} \frac{\sin Lx}{L} \frac{1}{2\eta\gamma_2 L^2 c - 2\eta\gamma_2 L c\sqrt{L^2 + s/c} + s} \tag{7.117}$$

The inverse Laplace transform can be obtained analytically. Only the last fraction appearing in Eqn. 7.117 contains the transform variable s. The denominator can be rationalized by completing the square, and the fraction put in the form

$$\frac{1}{s + 2\eta\gamma_2 L^2 c - 2\eta\gamma_2 L c\sqrt{L^2 + \frac{s}{c}}}$$
$$= \frac{1}{s + 4\eta\gamma_2 L^2 c(1-\eta\gamma_2)} + \frac{2\eta\gamma_2 L^2 c}{s[s + 4\eta\gamma_2 L^2 c(1-\eta\gamma_2)]}$$
$$+ \frac{2\eta\gamma_2 L\sqrt{c}\sqrt{L^2 c + s}}{s[s + 4\eta\gamma_2 L^2 c(1-\eta\gamma_2)]} \tag{7.118}$$

The transform pairs that are needed to invert Eqn. 7.118 are given in Table 7.1. The inversion of the last term in Eqn. 7.118 requires the third transform pair in Table 7.1, the convolution theorem (Appendix A, Eqn. A.15), and the integral (e.g., www.integrals.com):

$$\int_0^t \left\{ \frac{e^{-at}}{\sqrt{\pi t}} + \sqrt{a-b}e^{-bt}\mathrm{erf}[\sqrt{(a-b)t}] \right\} dt$$
$$= \frac{1}{\sqrt{a}}\mathrm{erf}\sqrt{at} + \frac{1}{b\sqrt{a}}(a-b)\mathrm{erf}\sqrt{at}$$
$$- \frac{\sqrt{a-b}}{b}e^{-bt}\mathrm{erf}(\sqrt{(a-b)t}) \tag{7.119}$$

The final result for the vertical displacement at the surface is

$$u_y(x, 0, t) = \frac{A(1-\nu)}{2G}\frac{\sin Lx}{L}F(\nu, \nu_u, L\sqrt{ct}) \tag{7.120}$$

TABLE 7.1.
Inverse Laplace Transform Pairs (Erdélyi et al., 1954)

$g(s) = \int_0^\infty e^{-st} f(t)dt$	$f(t)$
$\frac{1}{s+a}$	e^{-at}
$\frac{1}{s(s+a)}$	$\frac{1}{a}[1-e^{-at}]$
$\frac{\sqrt{s+a}}{s+b}$	$\frac{e^{-at}}{\sqrt{\pi t}} + \sqrt{a-b}e^{-bt}\mathrm{erf}\left[\sqrt{(a-b)t}\right]$

where

$$F(\nu, \nu_u, L\sqrt{ct}) \equiv 1 + \frac{1+\nu-2\nu_u}{1-\nu} e^{-\beta_2 L^2 ct} + \text{erf}(L\sqrt{ct})$$
$$-\sqrt{1-\beta_2} e^{-\beta_2 L^2 ct} \text{erf}(L\sqrt{(1-\beta_2)ct}) \quad (7.121)$$

and

$$\beta_2 \equiv 4\eta\gamma_2(1-\eta\gamma_2)$$
$$= \frac{4(\nu_u-\nu)(1-\nu_u)}{(1-\nu)^2} \quad (7.122)$$

1. *Incompressible Grain Limit* ($\nu_u = 1/2$). In the limit that $\nu_u = 1/2$, the parameter β_2 equals $(1-2\nu)/(1-\nu)^2$. The result for this case was originally given by Biot (1941b).
2. *Undrained Limit* ($t \to 0^+$). The limit as $t \to 0^+$ is obtained either by going back to the expression Eqn. 7.117 for the Laplace transform of u_y and using the initial value theorem for the Laplace transform (Eqn. A.16), or by inserting $t = 0$ into Eqn. 7.121:

$$u_y(x, 0, 0^+) = \lim_{s\to\infty} s\tilde{u}_y(x, 0, s)$$
$$= A\frac{1-\nu_u}{G}\frac{\sin Lx}{L} \quad (7.123)$$

   This displacement is an instantaneous initial settlement. The subsequent time-dependent settlement is obtained by subtracting Eqn. 7.123 from Eqn. 7.120.
3. *Drained Limit* ($t \to \infty$). The drained limit is obtained from Eqn. 7.117 and the final value theorem (Eqn. A.17), or by taking $t \to \infty$ into Eqn. 7.121:

$$u_y(x, 0, \infty) = \lim_{s\to 0} s\tilde{u}_y(x, 0, s)$$
$$= A\frac{1-\nu}{G}\frac{\sin Lx}{L} \quad (7.124)$$

   Therefore, the expected drained and undrained limits are confirmed.

7.7.2 Step Spatial Load

The step function $r(x) = -\sigma_o/2$ for $-\infty < x < 0$ and $r(x) = +\sigma_o/2$ for $0 \leq x < +\infty$, where σ_o is a constant vertical normal stress, can be represented in terms of the Fourier sine transform of $1/L$:

$$r(x) = \frac{\sigma_o}{\pi} \int_0^\infty \frac{\sin Lx}{L} dL \tag{7.125}$$

The displacement $u_y^{\text{step}}(x, 0, t)$ can then be obtained by superposition by integrating $u_y^{\sin}/L$ with respect to L from zero to infinity, where $u_y^{\sin}$ is given by Eqn. 7.120:

$$u_y^{\text{step}}(x, 0, t) = \frac{\sigma_o(1-\nu)}{2\pi G} \int_0^\infty \frac{F(\nu, \nu_u, L\sqrt{ct})}{L^2} (\sin Lx) dL \tag{7.126}$$

7.7.3 Rectangular Spatial Load

The solution for a rectangular strip load between $x = -a$ and $x = +a$ is obtained by superposition as $u_y^{\text{step}}(x+a) - u_y^{\text{step}}(x-a)$. An alternative method is to substitute the Fourier transform of the rectangular load, $\bar{r}(\lambda) = 2(\sin a\lambda)/\lambda$, into Eqn. 7.116 and then take the inverse Fourier and Laplace transforms.

7.8 UNDRAINED RESPONSE TO A LINE SOURCE

Fluid injection or withdrawal induces poroelastic displacements and stresses. The instantaneous or undrained displacement components for a unit line source (volume/length) located at $x = 0$ and $y = D$ in a half space in plane strain can be obtained by integrating the unit point-source solution for a half space (Eqn. 5.38) from $-\infty < z < +\infty$.[6] The Green's functions have also been derived by Segall (1985) based on a solution by Melan (1932) for point forces parallel and perpendicular to the free surface,[7]

$$u_x^{\text{line}}(x, y) = \frac{\gamma}{2\pi} \left\{ \frac{x}{r_1^2} + \frac{(3-4\nu_u)x}{r_2^2} - \frac{4y(y+D)x}{r_2^4} \right\} \tag{7.127}$$

$$u_y^{\text{line}}(x, y) = \frac{\gamma}{2\pi} \left\{ \frac{y-D}{r_1^2} + \frac{2y-(3-4\nu_u)(y+D)}{r_2^2} - \frac{4y(y+D)^2}{r_2^4} \right\} \tag{7.128}$$

[6] Note that the y and z axes are interchanged from their use in Eqn. 5.38.

[7] Note that the x and y axes in Eqns. 7.127 and 7.128 are interchanged from their use by Segall (1985).

where $\gamma = B(1+\nu_u)/(3(1-\nu_u))$ is the uniaxial strain-loading efficiency, $r_1^2 = x^2 + (y-D)^2$, and $r_2^2 = x^2 + (y+D)^2$. Inserting the strains computed by differentiating Eqns. 7.127 and 7.128 into the constitutive equations gives the stresses for a unit line source at $x = 0$ and $y = D$:

$$\sigma_{xx}^{\text{line}}(x,y) = \frac{\gamma G}{\pi}\left\{\frac{(y-D)^2 - x^2}{r_1^4} + \frac{(y+D)(3D-y) - 3x^2}{r_2^4} + \frac{16y(y+D)x^2}{r_2^6}\right\} \tag{7.129}$$

$$\sigma_{yy}^{\text{line}}(x,y) = \frac{\gamma G}{\pi}\left\{\frac{x^2 - (y-D)^2}{r_1^4} + \frac{(5y+D)(y+D) - x^2}{r_2^4} - \frac{16y(y+D)x^2}{r_2^6}\right\} \tag{7.130}$$

$$\sigma_{xy}^{\text{line}}(x,y) = \frac{\gamma G}{\pi}\left\{\frac{-2x(y-D)}{r_1^4} - \frac{2x(3y+D)}{r_2^4} + \frac{16y(y+D)^2 x}{r_2^6}\right\} \tag{7.131}$$

Eqns. 7.127–7.131 are readily generalized to a line source located at $x = \xi$ rather than at $x = 0$ by substituting $x - \xi$ for x.

Example: *Subsidence due to Oil Extraction*

Segall (1985) calculated poroelastic stress and displacement changes associated with the extraction of large volumes of fluid from two active oil fields near Coalinga, California, to explain nearby seismicity and subsidence. A state of plane strain was assumed for a cross section perpendicular to the line of wells extending for 10 km along the crest of the Coalinga anticline (Fig. 7.5). The 100-meter-thick reservoir is assumed to be a horizontal plane, embedded in impermeable rock above and below. Drainage to the line of

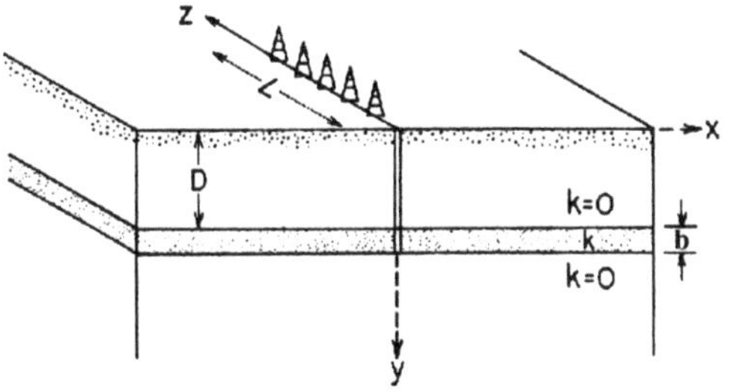

Figure 7.5: Poroelastic model for seismicity and subsidence near Coalinga, California (after Segall, 1985, Fig. 6).

wells is represented by a uniform volume flux, $-Q_o = -\dot{V}_o/Lb$, where Q_o is the volume rate of fluid extraction per unit area in the yz-plane, $\dot{V}_o$ is the average volumetric extraction rate, L is the length of the line of wells, and b is the thickness of the reservoir. Because the flow field is irrotational, pressure satisfies the same one-dimensional diffusion equation as does increment of fluid content (cf. Section 4.10.4). The solution has been given by Carslaw and Jaeger (1959, p. 75) for the analogous heat diffusion problem,

$$\zeta(x,t) = Sp(x,t) = -Q_o\sqrt{\frac{t}{c}}\,\mathrm{ierfc}(x^2/4ct)^{1/2} \tag{7.132}$$

where ierfc(x) is the first integral of the complementary error function

$$\begin{aligned}\mathrm{ierfc}(x) &\equiv \int_0^x \mathrm{erfc}(\xi)d\xi \\ &= \frac{e^{-x^2}}{\sqrt{\pi}} - x\,\mathrm{erfc}(x)\end{aligned} \tag{7.133}$$

The stress and displacement fields are then found by convolving the line-source solutions (Eqns. 7.127–7.131) with the increment of fluid content given by Eqn. 7.132. In particular, the vertical displacement at the surface $y = 0$ is obtained from Eqn. 7.128,

$$u_y(x,0,t) = \frac{2B(1+\nu_u)Q_o bD}{3\pi}\left(\frac{t}{c}\right)^{1/2}\int_{-\infty}^{+\infty}\frac{\mathrm{ierfc}(\xi^2/4ct)^{1/2}}{D^2+(x-\xi)^2}d\xi \tag{7.134}$$

where D is the depth to the reservoir, and b is the thickness of the reservoir. The average rate of fluid removal between 1905 and 1981 was $\dot{V}_o = 1.6\times 10^6$ m^3/year from a line of wells of length $L = 10$ km from a reservoir of thickness of $b = 100$ meters at a depth of $D = 1$ km. The subsidence at the free surface is shown at five different times for two different hydraulic diffusivities of the producing layer (Fig. 7.6).

7.9 SUDDEN FAULT SLIP

Time-dependent fault slip induces stress and pore-pressure changes in a poroelastic medium. For sudden faulting, the instantaneously induced pore pressures are given by the undrained response, and stresses and displacements are given by the elastic solution using undrained moduli. At long times, the pore pressures decay to zero, and stresses and displacements are given by the elastic solution using drained moduli. To a first approximation, the time-dependent pore pressure can be calculated from the diffusion equation

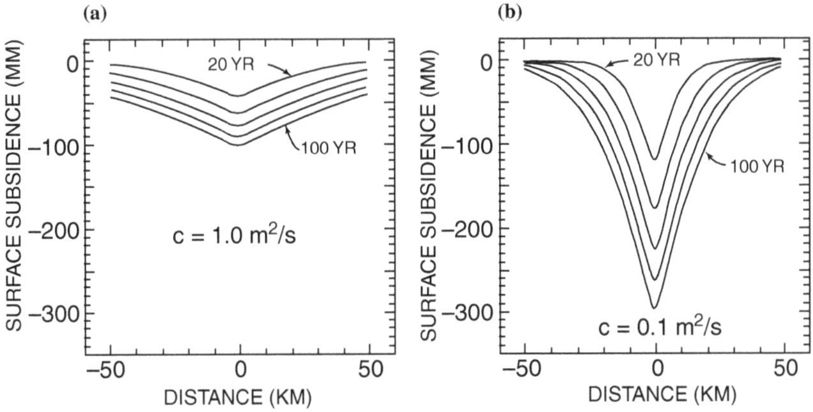

Figure 7.6: Surface subsidence for five times between 20 and 100 years due to a uniform rate of fluid extraction from a horizontal layer of thickness, 100 m, at a depth of 1 km (Segall, 1985, Fig. 8). The extraction rate $\dot{V}_o$ equals 2×10^6 m^3/yr over $L = 10$ km. The value of B is 0.6, the value of ν_u is 0.33, and the hydraulic diffusivity in (a) is 1.0 m^2/s and in (b) is 0.1 m^2/s.

using the undrained response as the initial condition. If, however, the rate of propagation of the fault is comparable to the rate of fluid diffusion, the full poroelastic solution is necessary.

7.9.1 Permeable Fault Plane

The fully coupled poroelastic model analyzed in this section is sudden slip across a semi-infinite plane extending from $-\infty < z < +\infty$ on the negative x-axis ($-\infty < x < 0$) (Fig. 7.7) (Booker, 1974; Rice and Cleary, 1976; Rudnicki, 1986b, 1987). The infinite extent of the plane in the z-direction allows the problem to be represented in plane strain. The slip is represented as a sudden displacement for $t > 0$ of the upper half plane $y > 0$ relative to the lower half plane $y < 0$ along the negative x-axis. Mathematically, the displacement boundary condition is

$$u_x(x, 0^+, t) - u_x(x, 0^-, t) = b_x(1 - H(x))H(t) = b_x H(-x)H(t) \quad (7.135)$$

where b_x is the displacement discontinuity, and $H(x)$ is the Heaviside step function defined by

$$H(x) = \begin{cases} 1 & \text{if } x \geq 0 \\ 0 & \text{if } x < 0 \end{cases} \quad (7.136)$$

The displacement components are antisymmetric with respect to the plane $y = 0$. The pore pressure is antisymmetric across the plane $y = 0$, as the

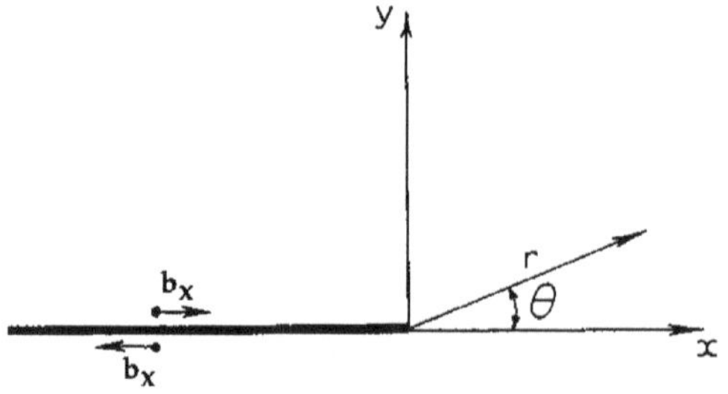

Figure 7.7: Sudden slip on a semi-infinite fault (Rudnicki, 1986b). The upper-half plane $y > 0$ is displaced a distance b_x in the $+x$-direction relative to the lower-half plane $y < 0$ along the negative x-axis ($-\infty < x < 0$). The displacement discontinuity is equivalent to an edge dislocation centered at the origin.

upper half plane is placed in compression and the lower half plane is placed in extension. Because the pore pressure is continuous across a permeable fault plane, the pore pressure must be zero on the plane $y = 0$ (i.e., $p(x, 0, t) = 0$). Similarly, the normal stress σ_{yy} is antisymmetric and continuous across the plane $y = 0$, and hence $\sigma_{yy}(x, 0, t) = 0$. Thus, the problem can be posed entirely in the domain $y \geq 0$, so that the pore pressure and normal stress σ_{yy} values on $y = 0$ serve as boundary conditions. The displacement boundary condition for $y = 0$ is one half the total displacement (i.e., $u_x(x, 0, t) = b_x H(-x) H(t)/2$). This boundary condition is more conveniently expressed in terms of the derivative $\partial u_x/\partial x$ because the Fourier transform of $\delta(x) = -dH(-x)/dx$ is identically equal to one.[8]

The solution can be obtained by using the three boundary conditions to identify the three constants, A_1, A_2, and B_1, in Eqns. 7.103–7.105:

$$A_1(\lambda, s) = \frac{b_x}{2} \frac{\nu_u - \nu}{(1-\nu)(1-\nu_u)} \frac{c}{s^2} \tag{7.137}$$

$$A_2(\lambda, s) = -\frac{b_x}{2} \left\{ \frac{\nu_u - \nu}{(1-\nu)(1-\nu_u)} \frac{c}{s^2} + \frac{1}{s\lambda^2} \right\} \tag{7.138}$$

$$B_1(\lambda, s) = -\frac{b_x}{2} \frac{1}{2(1-\nu_u)} \frac{1}{|\lambda| s} \tag{7.139}$$

The analytical inversions of the transforms are complicated and are illustrated only for pore pressure. Solutions for all the stress components obtained by Rice and Cleary (1976) and Rudnicki (1987) are then presented. The dou-

[8] Rudnicki (1987) used the constitutive equation for ϵ_{xx} to obtain $\partial u_x/\partial x = (1-\nu)\sigma_{xx}/2G$ at $y = 0$, because $\sigma_{yy}(x, 0, t) = p(x, 0, t) = 0$. His formulation of the displacement boundary condition leads to the same results for the coefficients A_1, A_2, and B_1 in Eqns. 7.137–7.139.

bly transformed pore pressure (cf. Eqn. 7.90) is

$$\bar{p}(\lambda, y, s) = \frac{b_x}{2}\frac{G}{\eta}\frac{\nu_u - \nu}{(1-\nu)(1-\nu_u)}\left[\frac{1}{s}e^{-|\lambda|y} - \frac{1}{s}e^{-ny}\right] \tag{7.140}$$

The inverse Laplace transform of $1/s$ in the first term of Eqn. 7.140 is simply one, and the inverse Fourier transform of $\exp(-|\lambda|y)$ is

$$\mathcal{F}^{-1}\{\exp(-|\lambda|y)\} = \frac{1}{\pi}\frac{y}{x^2+y^2} \tag{7.141}$$

The second term is more difficult. The Laplace transform of

$$s^{-1}\exp(-y\sqrt{\lambda^2+s/c})$$

is built up as follows (cf. Booker, 1974, Appendix 1):[9]

$$\mathcal{L}^{-1}\left\{e^{-y\sqrt{s}}\right\} = \frac{1}{2}\frac{y}{\sqrt{\pi t^3}}e^{-y^2/4t} \tag{7.142}$$

The shift theorem (Appendix A, Eqn. A.12) gives

$$\mathcal{L}^{-1}\left\{e^{-y\sqrt{\lambda^2+s}}\right\} = e^{-\lambda^2 t}\left\{\frac{1}{2}\frac{y}{\sqrt{\pi t^3}}e^{-y^2/4t}\right\} \tag{7.143}$$

The time-scaling theorem (Appendix A, Eqn. A.13) yields

$$\mathcal{L}^{-1}\left\{e^{-y\sqrt{\lambda^2+s/c}}\right\} = ce^{-\lambda^2 ct}\left\{\frac{1}{2}\frac{y}{\sqrt{\pi(ct)^3}}e^{-y^2/4ct}\right\} \tag{7.144}$$

Finally, the convolution theorem (Appendix A, Eqn. A.15) gives

$$\mathcal{L}^{-1}\left\{\frac{1}{s}e^{-y\sqrt{\lambda^2+s/c}}\right\} = \frac{y}{2\sqrt{\pi c}}\int_0^t \frac{1}{\tau^{3/2}}\exp[-y^2/(4c\tau) - \lambda^2 c\tau]d\tau \tag{7.145}$$

The integral in Eqn. 7.145 can be evaluated by making the substitution $\xi^2 = 1/\tau$, which puts the integrand into the form of Eqn. 7.4.33 of Abramowitz and Stegun (1964):

$$\mathcal{L}^{-1}\left\{\frac{1}{s}e^{-y\sqrt{\lambda^2+s/c}}\right\} = \frac{1}{2}\left\{e^{y|\lambda|}\operatorname{erfc}\left(\frac{y}{2\sqrt{ct}} + |\lambda|\sqrt{ct}\right)\right.$$
$$\left. + e^{-y|\lambda|}\operatorname{erfc}\left(\frac{y}{2\sqrt{ct}} - |\lambda|\sqrt{ct}\right)\right\} \tag{7.146}$$

[9] Just before Booker's Eqn. A1, the argument of $\mathcal{L}^{-1}$ should be $1/p^2$, not $1/p$, and the x in the final integral in Eqn. A1 should be x^2.

Because Eqn. 7.146 is an even function of the Fourier transform variable λ, the inverse of the exponential Fourier transform is one half the inverse of the cosine transform, which can be found as pair (15) in Erdélyi et al.'s (1954) tables. The final result is

$$\mathcal{F}^{-1}\left\{\mathcal{L}^{-1}\left\{\frac{1}{s}e^{-y\sqrt{\lambda^2+s/c}}\right\}\right\} = \frac{y}{\pi(x^2+y^2)}\exp\left[-(x^2+y^2)/4ct\right] \quad (7.147)$$

Combining Eqns. 7.140, 7.141, and 7.147 yields

$$\begin{aligned} p(x,y,t) &= \frac{G}{2\pi\eta}\frac{\nu_u-\nu}{(1-\nu)(1-\nu_u)}\frac{b_x y}{x^2+y^2}\left\{1-\exp[-(x^2+y^2)/4ct]\right\} \\ &= \frac{GB(1+\nu_u)}{3\pi(1-\nu_u)}\frac{b_x y}{x^2+y^2}\left\{1-\exp[-(x^2+y^2)/4ct]\right\} \end{aligned} \quad (7.148)$$

Rice and Cleary (1976) have derived the expressions in polar coordinates for all the stress components for the semi-infinite shear fault from a poroelastic generalization of Mushkelishvili's complex variables techniques:

$$\begin{aligned} \sigma_{\theta\theta} = \frac{G}{2\pi}\frac{\nu_u-\nu}{(1-\nu)(1-\nu_u)}\frac{b_x}{r}\sin\theta \Bigg\{ & 2\exp(-r^2/4ct) - \frac{1-\nu}{\nu_u-\nu} \\ & -\frac{4ct}{r^2}[1-\exp(-r^2/4ct)]\Bigg\} \end{aligned} \quad (7.149)$$

$$\begin{aligned} \sigma_{r\theta} &= \frac{G}{2\pi}\frac{\nu_u-\nu}{(1-\nu)(1-\nu_u)}\frac{b_x}{r} \\ &\quad\times\cos\theta\left\{\frac{1-\nu}{\nu_u-\nu} - \frac{4ct}{r^2}[1-\exp(-r^2/4ct)]\right\} \end{aligned} \quad (7.150)$$

$$\begin{aligned} \sigma_{rr} &= \frac{G}{2\pi}\frac{\nu_u-\nu}{(1-\nu)(1-\nu_u)}\frac{b_x}{r} \\ &\quad\times\sin\theta\left\{-\frac{1-\nu}{\nu_u-\nu} + \frac{4ct}{r^2}[1-\exp(-r^2/4ct)]\right\} \end{aligned} \quad (7.151)$$

$$p = \frac{G}{2\pi}\frac{\nu_u-\nu}{(1-\nu)(1-\nu_u)}\frac{b_x}{r}\frac{\sin\theta}{\eta}\left[1-\exp(-r^2/4ct)\right] \quad (7.152)$$

The short and long time limits can be verified to be the same as the undrained and drained elastic solutions, respectively.

7.9.2 Permeable versus Impermeable Fault Behavior

Rudnicki (1986b, 1987) presented solutions for shear stress and pore pressure on both a permeable and an impermeable fault plane:

$$\sigma_{xy}^{\text{perm}}(x, 0, t) = \frac{Gb_x}{2\pi(1-\nu_u)x} \times \left\{1 - \frac{\nu_u - \nu}{1-\nu}\frac{4ct}{x^2}\left[1 - e^{-x^2/(4ct)}\right]\right\} \tag{7.153}$$

$$\sigma_{xy}^{\text{imperm}}(x, 0, t) = \frac{Gb_x}{2\pi(1-\nu_u)x} \times \left\{1 - \frac{\nu_u - \nu}{1-\nu}\left[2e^{-x^2/(4ct)} - \frac{4ct}{x^2}\left(1 - e^{-x^2/(4ct)}\right)\right]\right\} \tag{7.154}$$

The time-dependent changes in shear stress are shown in Figure 7.8 in terms of the difference between its value at time t and its ultimate drained value normalized by the difference between its initial undrained value and its final drained value:

$$\frac{\sigma_{xy}(x, 0, t) - \sigma_{xy}(x, 0, \infty)}{\sigma_{xy}(x, 0, 0) - \sigma_{xy}(x, 0, \infty)} \tag{7.155}$$

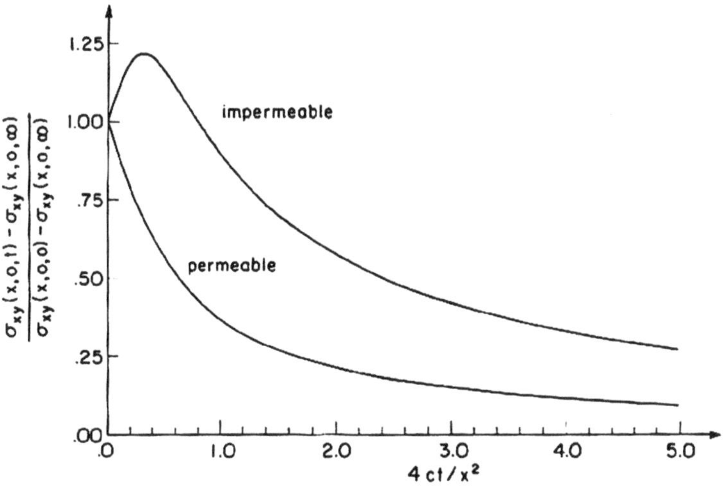

Figure 7.8: Shear stress on the fault plane versus the dimensionless quantity $4ct/x^2$ for an impermeable ($dp/dy = 0$), semi-infinite fault and a permeable ($p = 0$), semi-infinite fault (Rudnicki, 1986b).

The shear stress on the permeable fault decreases monotonically with time. On the other hand, the shear stress on the impermeable fault rises about 20% before decaying. This increase in shear stress due to poroelastic coupling is destabilizing and may lead to additional fault slip.

Pore-pressure contours are shown for $y > 0$ for the impermeable and permeable fault cases in Figure. 7.9. The pore pressure is antisymmetric in both cases. The no-flow condition across the plane $y = 0$ for the impermeable case means that flow must parallel this plane, and hence the pore pressure contours are perpendicular to it. The pore pressure near the terminus of the fault at the origin is significantly higher for the case of an impermeable fault. The pore pressures for the two cases converge for large distances from the fault.

7.9.3 Finite-Length Fault

Superposition of the solution for a fault of semi-infinite length can be used to obtain the solution for a finite-length fault. The superposition of a semi-infinite fault ending at $x = L$ with slip b_x and a semi-infinite fault ending at

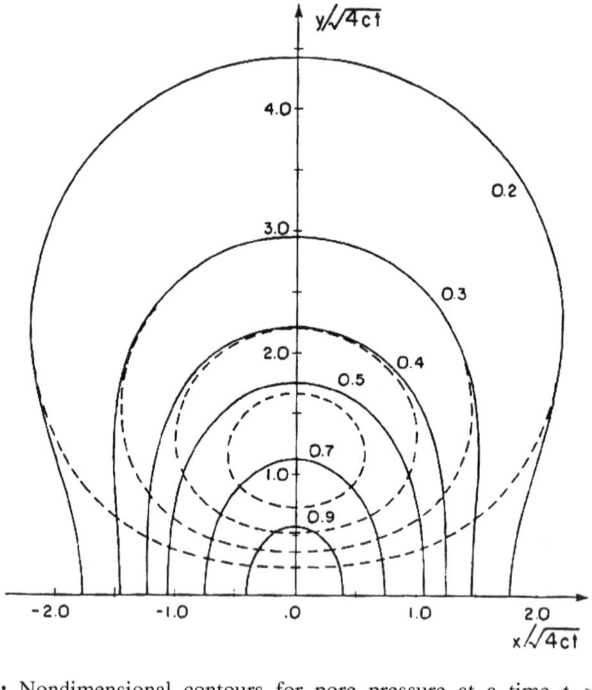

Figure 7.9: Nondimensional contours for pore pressure at a time $t > 0$ for an impermeable (solid curves), semi-infinite fault and a permeable (dashed curves) fault (Rudnicki, 1986b). The normalizing factor is $B(1 + \nu_u)Gb_x/[3\pi(1 - \nu_u)\sqrt{4ct}]$.

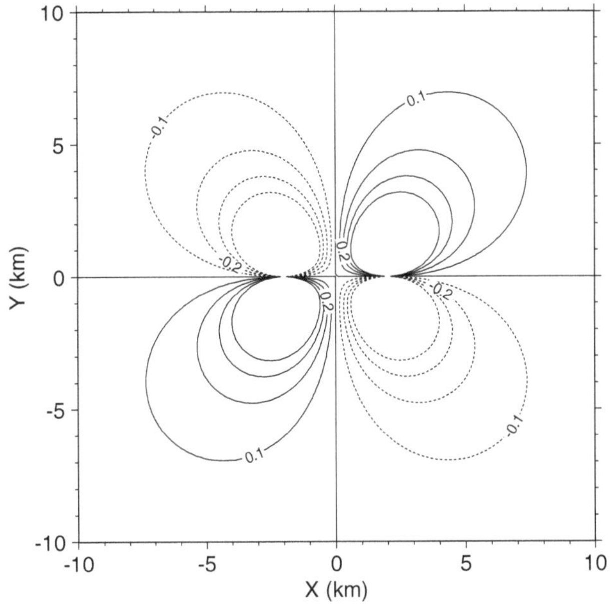

Figure 7.10: Pore-pressure (MPa) contours for a permeable, finite-length fault. Increases in pore pressure are solid curves, and decreases in pore pressure are dashed curves. The fault lies between $-2 < x < +2$ km, and its displacement is 1 m. Other parameters are $G = 10$ GPa, $B = 1$, and $\nu_u = 0.3$.

$x = -L$ with slip $-b_x$ yields slip of amount b_x over the interval $-L \leq x \leq +L$ and zero elsewhere. The different stress components and pore pressure can then be obtained by superposition:

$$p^{\text{finite}}(x, y, t) = p^{\text{inf}}(x - L, y, t) - p^{\text{inf}}(x + L, y, t) \tag{7.156}$$

The solution for pore pressure is shown in Figure 7.10 (Wang, 1997). Pore pressure is antisymmetric about the x and y axes. Water-level changes exhibit a four-quadrant pattern similar to the first motions due to an earthquake. The nodal planes are the fault plane and the auxiliary plane perpendicular to the fault and passing through the origin.

7.9.4 Steadily Propagating Slip

Roeloffs and Rudnicki (1985 and pers. comm.) solved the problem of a steadily propagating shear dislocation by superposition of the impulse solution.[10] Such a model can be used to interpret water-level changes associated with episodic creep on the San Andreas fault (Johnson et al., 1973). The terminus of a semi-infinite shear fault is assumed to move steadily in the $+x$-direction at uniform velocity V. The pore pressure, $p^*(x, y, t)$, for an impulse dislocation is obtained by differentiating the step function solution (Eqn. 7.148) with respect to t:

$$p^*(x, y, t) = \frac{GB(1+\nu_u)b_x}{3\pi(1-\nu_u)} \times \left\{ \frac{y}{x^2+y^2}\delta(t) - \frac{y}{4ct^2}\exp[-(x^2+y^2)/4ct] \right\} \quad (7.157)$$

The tip of the fracture is taken to be at $x = 0$ at $t = 0$, where x is a fixed reference frame. The dislocation is at $x = x' = Vt'$ when the time $t = t'$. Then by superposition,

$$p(x, y, t) = \frac{GB(1+\nu_u)b_x}{3\pi(1-\nu_u)} \times \left\{ \frac{y}{(x-Vt)^2+y^2} - \int_0^t \frac{y}{4c(t-t')^2} \times \exp\{-[(x-Vt')^2+y^2]/4c(t-t')\}dt' \right\} \quad (7.158)$$

The pore-pressure field for the moving dislocation is obtained for long times so that in a coordinate system, $X \equiv x - Vt$, moving with the dislocation, the pore pressure depends only on X and y. Substituting $x = X + Vt$ and making the change of variable $\tau = t - t'$ yields in the limit $t \to \infty$,

$$p(X, y) = \frac{GB(1+\nu_u)b_x}{3\pi(1-\nu_u)} \left\{ \frac{y}{R^2} - \frac{y}{4c}\exp\left(\frac{-VX}{2c}\right) \times \int_0^\infty \frac{1}{\tau^2}\exp\left[-\left(\frac{V^2\tau}{4c} + \frac{R^2}{4c\tau}\right)\right]d\tau \right\} \quad (7.159)$$

where $R^2 = X^2 + y^2$. Using the transformation $\xi = V^2\tau/4c$ in Eqn. 7.159 gives

$$p(X, y) = \frac{GB(1+\nu_u)b_x}{3\pi(1-\nu_u)} \left\{ \frac{y}{R^2} - \frac{y}{4c}\exp\left(\frac{-VX}{2c}\right)\frac{V^2}{4c} \times \int_0^\infty \frac{1}{\xi^2}\exp\left[-\left(\xi + \left(\frac{RV}{2c}\right)^2\frac{1}{4\xi}\right)\right]d\xi \right\} \quad (7.160)$$

[10] Rudnicki and Roeloffs (1990) presented an alternative derivation using Fourier transforms because the time derivative can be replaced by $-V\partial/\partial x$, where V is the speed of propagation.

The modified Bessel function, $K_o(z)$, can be expressed as an integral (cf. Carslaw and Jaeger, 1959, p. 490):

$$K_o(z) = \frac{1}{2}\int_0^\infty \exp\left[-\xi - \frac{z^2}{4\xi}\right]\frac{d\xi}{\xi} \tag{7.161}$$

Then, because $K_1(z) = -K_o'(z)$,

$$K_1(z) = \frac{z}{4}\int_0^\infty \exp\left[-\xi - \frac{z^2}{4\xi}\right]\frac{d\xi}{\xi^2} \tag{7.162}$$

Therefore,

$$p(X, y) = \frac{GB(1+\nu_u)}{3\pi(1-\nu_u)}\frac{b_x y}{R^2}\left\{1 - \frac{VR}{2c}K_1\left(\frac{VR}{2c}\right)\exp\left(\frac{-VX}{2c}\right)\right\} \tag{7.163}$$

In the limit $V \to 0$, $p(X, y)$ approaches zero, the drained response. In the limit $V \to \infty$, the term in braces approaches one, and $p(X, y)$ approaches the undrained solution. Figure 7.11 is a nondimensional contour plot of Eqn. 7.163.

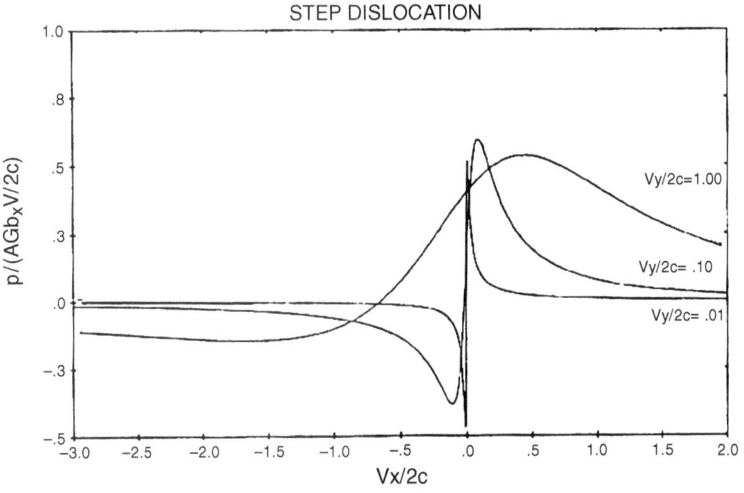

Figure 7.11: Nondimensional pore-pressure profiles along planes $y/(2c/V) = Vy/2c = 0.01$, 0.1, and 1.0 for a shear displacement propagating at velocity V along a permeable fault (after Roeloffs and Rudnicki, 1985, Fig. 3). The factor A in the normalizing factor is $B(1+\nu_u)/[3\pi(1-\nu_u)]$. The profiles can also be interpreted as the time dependence of pore pressure, with time increasing in the negative x-direction.

7.9.5 Time-Ramped Slip

Rudnicki and Hsu (1988) obtained the solution for both a permeable and an impermeable fault for a displacement, $b_x(t)$, that is uniform over the entire negative x-axis but increases linearly with time up to a value b_o before being constant with time:

$$b_x(t) = \begin{cases} b_o \frac{t}{t_o} & t < t_o \\ b_o & t \geq t_o \end{cases} \tag{7.164}$$

The solution can be obtained as a superposition of the impulse solution (Eqn. 7.157):

$$p(x, y, t) = \int_0^t b_x(t') p^*(x, y, t - t') dt' \tag{7.165}$$

The permeable fault result was expressed by Rudnicki and Hsu as

$$p(X, Y, T) = \frac{AGb_o}{\sqrt{4ct_o}} \{H(T)F(X, Y, T) - H(T-1)F(X, Y, T-1)\} \tag{7.166}$$

where

$$\begin{aligned} X &= x/\sqrt{4ct_o} \\ Y &= y/\sqrt{4ct_o} \\ R^2 &= X^2 + Y^2 \\ T &= t/t_o \\ A &= B(1 + \nu_u)/[3\pi(1 - \nu_u)] \\ F(X, Y, T) &= \frac{YT}{R^2}[1 - \exp(-R^2/T)] + YE_1(R^2/T) \end{aligned} \tag{7.167}$$

where $E_1(Z)$ is the exponential integral (Abramowitz and Stegun, 1964, Eqn. 5.1.1):

$$E_1(Z) = \int_Z^\infty s^{-1} \exp(-s) ds \tag{7.168}$$

The ramping of the displacement with time means that the peak pore pressure is smaller than for an instantaneous emplacement of a shear displacement of the same magnitude, because pore-pressure diffusion over the time

t_o dissipates some of the induced pore pressure. The full-fault displacement is attained at $T = 1$, at which time

$$p(X, Y, 1) = \frac{AGb_o}{\sqrt{4ct_o}} \left\{ \frac{Y}{R^2}(1 - \exp(-R^2)) + YE_1(R^2) \right\} \tag{7.169}$$

For $R = 1$, the ratio of the induced pore pressure for the ramp emplacement relative to that for instantaneous displacement (Eqn. 7.148 as $t \to 0$) is 0.851 at $\theta = 90^o$, 0.737 at $\theta = 60^o$, and 0.426 at $\theta = 30^o$.

7.10 CONSTITUTIVE EQUATIONS FOR PLANE STRESS

A state of plane stress in the xy-plane is defined by $\sigma_{zz} = \sigma_{zx} = \sigma_{zy} = 0$. The constitutive equations (Eqns. 2.30–2.32) with stress and pore pressure as the independent variables become

$$\epsilon_{xx} = \frac{1}{E}\sigma_{xx} - \frac{\nu}{E}\sigma_{yy} + \frac{\alpha}{3K}p \tag{7.170}$$

$$\epsilon_{yy} = -\frac{\nu}{E}\sigma_{xx} + \frac{1}{E}\sigma_{yy} + \frac{\alpha}{3K}p \tag{7.171}$$

$$\epsilon_{zz} = -\frac{\nu}{E}(\sigma_{xx} + \sigma_{yy}) + \frac{\alpha}{3K}p \tag{7.172}$$

where $E = 2G(1+\nu)$. The constitutive equations (Eqn. 2.51) with stress and increment of fluid content as the independent variables become

$$\epsilon_{xx} = \frac{1}{E_u}\sigma_{xx} - \frac{\nu_u}{E_u}\sigma_{yy} + \frac{B}{3}\zeta \tag{7.173}$$

$$\epsilon_{yy} = -\frac{\nu_u}{E_u}\sigma_{xx} + \frac{1}{E_u}\sigma_{yy} + \frac{B}{3}\zeta \tag{7.174}$$

$$\epsilon_{zz} = -\frac{\nu_u}{E_u}(\sigma_{xx} + \sigma_{yy}) + \frac{B}{3}\zeta \tag{7.175}$$

where $E_u = 2G(1 + \nu_u)$.

For undrained conditions ($\zeta = 0$), the vertical strain ϵ_{zz} is related to the areal strain $\epsilon_{xx} + \epsilon_{yy}$ by

$$\epsilon_{zz} = -\frac{\nu_u}{1 - \nu_u}(\epsilon_{xx} + \epsilon_{yy}) \tag{7.176}$$

The undrained volumetric strain is thus also proportional to the areal strain:

$$\epsilon_{kk} = \frac{1 - 2\nu_u}{1 - \nu_u}(\epsilon_{xx} + \epsilon_{yy}) \tag{7.177}$$

The undrained pore-pressure buildup is

$$p^{(u)} = -\frac{B}{3}(\sigma_{xx} + \sigma_{yy}) \tag{7.178}$$

Adding Eqns. 7.173 and 7.174 with $\zeta = 0$, solving for $\sigma_{xx} + \sigma_{yy}$, and substituting into Eqn. 7.178 give

$$p^{(u)} = -\frac{2GB}{3}\frac{1+\nu_u}{1-\nu_u}(\epsilon_{xx} + \epsilon_{yy}) \tag{7.179}$$

$$= -2G\gamma(\epsilon_{xx} + \epsilon_{yy}) \tag{7.180}$$

Example: *Solid Earth Tides*

Gravitational interactions between the earth and the moon and between the earth and the sun produce deformations of the earth's crust. In shallow, confined aquifers, these dilatations and expansions produce fluctuations in water-well levels that are on the order of several centimeters. Deeper wells generally show greater amplitudes (Melchior, 1983, Chapter 10). The two most frequently used tidal components in water-level analyses are the M_2 and O_1 lunar tides with periods of 12.42060 and 25.81930 hours, respectively (Roeloffs, 1995). The subscripts 1 and 2 indicate the approximate number of cycles per day.

The theory of earth tides is based on expanding the tidal potential (energy per unit mass) into harmonic components (Melchior, 1983, Chapter 1). The potential for the M_2 component, for example, is

$$W_2^{M_2} = 0.90812D\cos^2(\theta^*)\cos(2\tau) \tag{7.181}$$

The variable τ is mean lunar time equal to $t + h - s$, where t is mean solar time, h is mean tropic longitude of the sun, and s is mean tropic longitude of the moon. Astronomical observations have yielded expressions for h and s that are nearly linear with t. The variable θ^* is the latitude, and $D = 2.6277$ m^2/s^2 is "Doodson's constant" (Melchior, 1983, p. 18),

$$D = \frac{3}{4}\frac{M_m}{M_e}\frac{R_e^4}{d_m^3}g \tag{7.182}$$

where M_m is the mass of the moon, M_e is the mass of earth, R_e is the mean radius of earth, d_m is the half-major axis of the moon's orbit, and g is the acceleration of gravity.

The ratio W_2/g gives the height of the equilibrium tide, which is the theoretical deformation of a homogeneous incompressible fluid sphere of the same size and mass of the earth (Love, 1944, p. 261). However, the actual displacement of the solid earth will differ due to deformation. Based on the idea that the potential due to the deformation can be represented by the same spherical harmonic expansion as the tidal potential, the displacement in spherical coordinates at a point on the surface can be expressed in terms of coefficients multiplying the potential,

$$u_r = \bar{h}\frac{W_2}{g}$$

$$u_\theta = \frac{\bar{l}}{g}\frac{\partial W_2}{\partial \theta}$$

$$u_\phi = \frac{\bar{l}}{g \sin\theta}\frac{\partial W_2}{\partial \phi} \tag{7.183}$$

where θ is the colatitude of the point, and ϕ is east longitude of the point. The coefficients $\bar{h}$ and $\bar{l}$ are known as Love numbers. The components of strain in the horizontal plane are

$$\epsilon_{\theta\theta} = \frac{1}{r}\left(\frac{\partial u_\theta}{\partial \theta} + u_r\right) \tag{7.184}$$

$$\epsilon_{\phi\phi} = \frac{1}{r}\left(\frac{1}{\sin\theta}\frac{\partial u_\phi}{\partial \phi} + u_\theta \frac{\cos\theta}{\sin\theta} + u_r\right) \tag{7.185}$$

The areal strain in a horizontal plane is then

$$\epsilon_{\phi\phi} + \epsilon_{\theta\theta} = \frac{1}{rg}\left\{2\bar{h}W_2 + \bar{l}\left[\frac{1}{\sin\theta}\frac{\partial}{\partial\theta}\left(\sin\theta\frac{\partial W_2}{\partial\theta}\right) + \frac{1}{\sin^2\theta}\frac{\partial^2 W_2}{\partial\phi^2}\right]\right\} \tag{7.186}$$

The potential W_2 can be approximated as a spherical solid harmonic, $r^2 S_2(\theta, \phi)$, where S_2 is a spherical surface harmonic (i.e., W_2 is assumed to satisfy Laplace's equation). In Eqn. 7.186, the term in brackets following $\bar{l}$ is r^2 times the sum of the θ and ϕ derivative terms of the Laplacian in spherical coordinates. Therefore, the harmonic property implies that

$$r^2\left[\frac{1}{\sin\theta}\frac{\partial}{\partial\theta}\left(\sin\theta\frac{\partial S_2}{\partial\theta}\right) + \frac{1}{\sin^2\theta}\frac{\partial^2 S_2}{\partial\phi^2}\right] = -\left[\frac{\partial}{\partial r}\left(r^2\frac{\partial r^2}{\partial r}\right)\right]S_2$$
$$= -6r^2 S_2 \tag{7.187}$$

Therefore, the M_2 tide produces a surface areal strain in the horizontal plane,

$$\epsilon_{\theta\theta} + \epsilon_{\phi\phi} = (2\bar{h} - 6\bar{l})\frac{W_2^{M_2}}{R_e g} \tag{7.188}$$

where the Love numbers $\bar{h} = 0.638$ and $\bar{l} = 0.088$ (Melchior, 1983, p. 67), and $R_e = 6371$ km is the radius of the earth. In theory, measured amplitudes of the cyclic water levels and calculated amplitudes of tidal strains inserted into Eqn. 7.180 yield the coefficient $2G\gamma$. Phase lags occur due to nonideality of the undrained assumption (e.g., wellbore storage effects or drainage to the water table). The calibration of earth-strain measurements is problematic. Observed tidal strains, even at inland sites, are strongly affected by the effects of ocean loading (Beaumont and Berger, 1975; Berger and Beaumont, 1976).

8

Plane Strain in Polar Coordinates

8.0 CHAPTER OVERVIEW

A planar cross section of a long cylinder or borehole can be approximated to be in a state of plane strain. An example problem is hydraulic fracturing in an anisotropic horizontal stress field. Many cylindrical problems in plane strain simplify to be functions only of radius. Radially symmetric problems are irrotational, and hence the pore pressure and stress fields uncouple. Radially symmetric problems discussed in this chapter include (1) sudden pressurization or stress relief of a long core; (2) sudden stress release due to borehole excavation; (3) sudden pressurization of a borehole, which is a limiting case of the separately treated problem of a hollow cylinder; and (4) time-dependent radial flow to a well. The problem of a suddenly applied radial load on the surface of a cylinder produces a Mandel-Cryer effect. Axisymmetric poroelastic problems for which there is z dependence are discussed in the next chapter.

8.1 RADIAL SYMMETRY

The constitutive and governing equations are formulated in the next several subsections for radial symmetry, which is a special case of cylindrical coordinates (cf. Section 9.1). The displacements and stresses depend only on r and t. A general solution is developed for the Laplace transform of the radial displacement.

8.1.1 Strain

For radial symmetry, the only nonzero displacement is the radial displacement $u_r = u_r(r)$. The three strain components in polar coordinates for the case of plane strain (e.g., Boley and Weiner, 1985, p. 249) are

$$\epsilon_{rr} = \frac{\partial u_r}{\partial r} \tag{8.1}$$

$$\epsilon_{\theta\theta} = \frac{u_r}{r} \tag{8.2}$$

$$\epsilon_{r\theta} = 0 \tag{8.3}$$

The volumetric strain for the case of plane strain is, therefore, given by

$$\begin{aligned} \epsilon_{kk} &= \epsilon_{rr} + \epsilon_{\theta\theta} \\ &= \frac{\partial u_r}{\partial r} + \frac{u_r}{r} \end{aligned} \tag{8.4}$$

Note that a radial displacement u_r of an arc of a circle of radius r produces a tangential strain, u_r/r, of the arc (Jaeger and Cook, 1976, p. 51).

8.1.2 Constitutive Equations

The constitutive equations expressed in cylindrical coordinates for a state of plane strain are identical in form with those presented in Chapter 7. The r and θ coordinates are orthogonal and substitute for x and y, respectively. Thus, Eqn. 7.1 in polar coordinates is

$$\sigma_{zz} = \nu(\sigma_{rr} + \sigma_{\theta\theta}) - (1 - 2\nu)\alpha p \tag{8.5}$$

The constitutive equations for the in-plane normal stress components (cf. Eqns. 2.44 and 2.45) are

$$\sigma_{rr} = 2G\epsilon_{rr} + \frac{2G\nu}{1-2\nu}\epsilon_{kk} - \alpha p \tag{8.6}$$

$$\sigma_{\theta\theta} = 2G\epsilon_{\theta\theta} + \frac{2G\nu}{1-2\nu}\epsilon_{kk} - \alpha p \tag{8.7}$$

Alternatively, Eqns. 8.6 and 8.7 can be expressed using ζ in place of p (cf. Eqn. 2.53):

$$\sigma_{rr} = 2G\epsilon_{rr} + 2G\frac{\nu_u}{1-2\nu_u}\epsilon_{kk} - \alpha M\zeta \tag{8.8}$$

$$\sigma_{\theta\theta} = 2G\epsilon_{\theta\theta} + 2G\frac{\nu_u}{1-2\nu_u}\epsilon_{kk} - \alpha M\zeta \tag{8.9}$$

The shear stress, $\sigma_{r\theta}$, equals zero because $\epsilon_{r\theta}$ equals zero. The stress components σ_{rr}, $\sigma_{\theta\theta}$, and $\sigma_{r\theta}$ are shown in Figure 8.1.

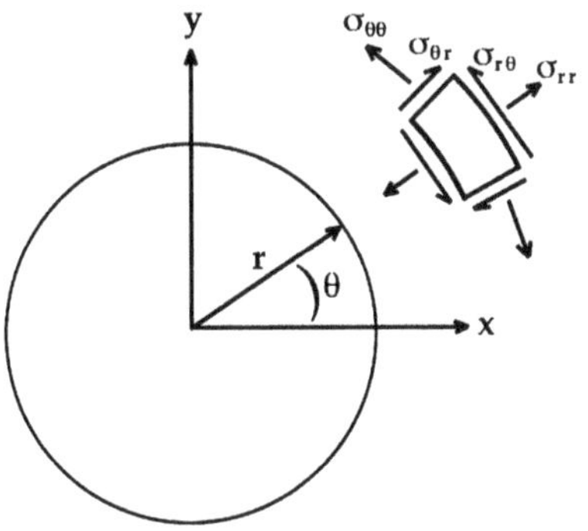

Figure 8.1: Stress components in polar coordinates (after Gould, 1983, p. 112).

Inverting Eqns. 8.6 and 8.7 gives the normal strains in the (r, θ) plane:

$$2G\epsilon_{rr} = \sigma_{rr} - \nu(\sigma_{rr} + \sigma_{\theta\theta}) + (1 - 2\nu)\alpha p \tag{8.10}$$

$$2G\epsilon_{\theta\theta} = \sigma_{\theta\theta} - \nu(\sigma_{rr} + \sigma_{\theta\theta}) + (1 - 2\nu)\alpha p \tag{8.11}$$

Eqns. 8.10 and 8.11 are equivalent to Eqns. 7.2 and 7.3.

The expression for the increment of fluid content in terms of stresses and fluid pressure is the same as Eqn. 7.5, because the sum of the normal stresses $\sigma_{rr} + \sigma_{\theta\theta}$ equals $\sigma_{xx} + \sigma_{yy}$,

$$\zeta = \frac{(1 - 2\nu)\alpha}{2G}\left[(\sigma_{rr} + \sigma_{\theta\theta}) + \frac{3p}{B(1 + \nu_u)}\right] \tag{8.12}$$

where the factor multiplying the square bracket term has been expressed differently (cf. Appendix B, Table B.1). The undrained pressure response is derived by setting $\zeta = 0$ in Eqn. 8.12 to yield

$$p^{(u)} = -\frac{B(1 + \nu_u)}{3}[\sigma_{rr} + \sigma_{\theta\theta}] \tag{8.13}$$

For irrotational flow, the plane-strain version of Eqn. 4.74 is obtained by using Eqn. 8.5,

$$\sigma_{rr} + \sigma_{\theta\theta} + 2\eta p = f(t) \tag{8.14}$$

where $f(t)$ is a function of time only.

8.1.3 Governing Equations

Because increment of fluid content, pressure, stress, and displacement are functions only of r, the governing equations are simplified considerably:

1. The increment of fluid content, ζ, satisfies a homogeneous diffusion equation if no fluid sources or body forces are present (cf. Eqn. 4.67),

$$c\left(\frac{1}{r}\frac{\partial}{\partial r}\left(r\frac{\partial \zeta}{\partial r}\right)\right) = \frac{\partial \zeta}{\partial t} \tag{8.15}$$

where the term in parentheses following c is $\nabla^2\zeta$ when $\zeta = \zeta(r, t)$.

2. Radial symmetry and plane strain imply that displacements are purely radial (i.e., the displacement field is irrotational). The inhomogeneous pressure diffusion equation (4.77) in the absence of fluid sources or body forces then becomes

$$\frac{\partial p}{\partial t} - c\left(\frac{1}{r}\frac{\partial}{\partial r}\left(r\frac{\partial p}{\partial r}\right)\right) = -(1-\nu)\gamma\frac{df}{dt} \tag{8.16}$$

3. A simple proportionality can be obtained between the radial derivatives of volumetric strain, ϵ_{kk} and ζ. The force-equilibrium equation for radial symmetry and plane strain (cf. Eqn. 9.6) is

$$\frac{\partial \sigma_{rr}}{\partial r} + \frac{\sigma_{rr} - \sigma_{\theta\theta}}{r} = 0 \tag{8.17}$$

Substituting the constitutive equations (Eqns. 8.8 and 8.9) into Eqn. 8.17 gives

$$\frac{\partial \epsilon_{kk}}{\partial r} = \frac{\alpha M(1-2\nu_u)}{2G(1-\nu_u)}\frac{\partial \zeta}{\partial r} \tag{8.18}$$

$$= \frac{\eta}{GS}\frac{\partial \zeta}{\partial r} \tag{8.19}$$

$$= \gamma\frac{\partial \zeta}{\partial r} \tag{8.20}$$

The loading efficiency for uniaxial strain (cf. Section 3.6.2) is the proportionality constant between $\partial\epsilon_{kk}/\partial r$ and $\partial\zeta/\partial r$.

4. The volumetric strain is related to the radial displacement by

$$\begin{aligned}\epsilon_{kk} &= \epsilon_{rr} + \epsilon_{\theta\theta} \\ &= \frac{\partial u_r}{\partial r} + \frac{u_r}{r} \\ &= \frac{1}{r}\frac{\partial (r u_r)}{\partial r}\end{aligned} \tag{8.21}$$

8.2 SUDDEN PRESSURIZATION OF A LONG CYLINDER

The problem of a sudden external change of fluid pressure on the surface of a long cylinder can be solved by superposition of the solutions for two problems: (1) a step change in stress while maintaining constant pore pressure, and (2) a step change in pore pressure while maintaining constant stress. This decomposition into two separate boundary-value problems is the same one used to solve the problem of a step change in water level over a half space (cf. Section 6.11.1). The stress-free boundary in the second problem means that poroelastic stresses are due only to the effects of a nonlinear pore-pressure distribution, and not to boundary stresses. The solution is obtained in Laplace transform space (Detournay and Cheng, 1993, pp. 149 ff.).

8.2.1 Laplace Transform Solution

The Laplace transform of Eqn. 8.15 is

$$\frac{d^2\tilde{\zeta}}{dr^2} + \frac{1}{r}\frac{d\tilde{\zeta}}{dr} - \frac{s}{c}\tilde{\zeta} = 0 \tag{8.22}$$

where

$$\mathcal{L}\{\zeta(r,t)\} \equiv \tilde{\zeta}(r,s) = \int_0^{\infty} e^{-st}\zeta(r,t)dt \tag{8.23}$$

In Eqn. 8.22 an initial condition, $\zeta(r,0) = 0$, of no fluid inflow or outflow to the cylinder is assumed. Eqn. 8.22 is the modified Bessel equation, which has the general solution (Carslaw and Jaeger, 1959, p. 327)

$$\tilde{\zeta}(r,s) = A_1(r,s)I_o(qr) + B_1(r,s)K_o(qr) \tag{8.24}$$

where $I_o(qr)$ and $K_o(qr)$ are hyperbolic or modified Bessel functions of zero order of the first and third kind, respectively (Fig. 8.2). The variable $q^2 \equiv s/c$ and $A_1(r,s)$ and $B_1(r,s)$ are constants determined from boundary conditions.

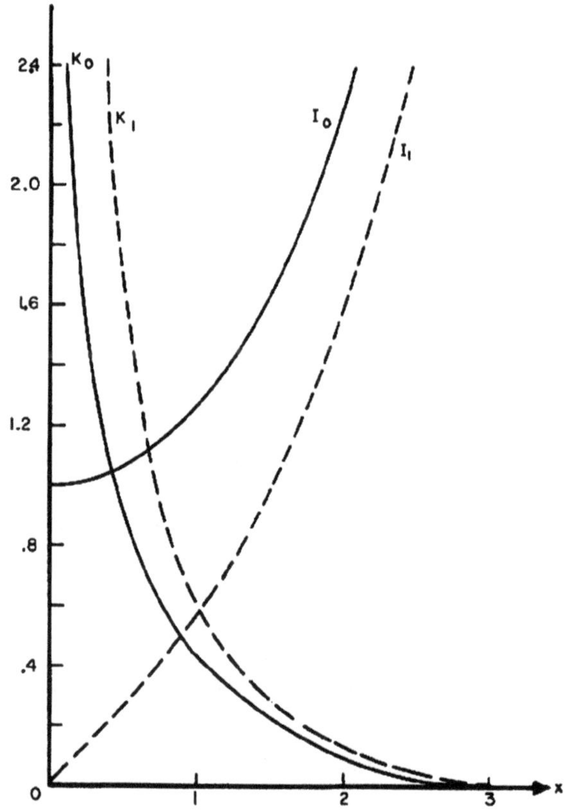

Figure 8.2: Hyperbolic Bessel functions $I_0(x)$, $I_1(x)$, $K_0(x)$, and $K_1(x)$. The hyperbolic Bessel functions do not oscillate like the ordinary Bessel functions. They bear a relationship analogous to that between a hyperbolic sine and cosine and an ordinary sine and cosine (Kraut, 1967, Fig. 6–6, p. 271).

The coefficient B_1 must be zero if $r = 0$ is in the problem domain, because $K_o(0)$ is unbounded. Substituting Eqn. 8.24 with $B_1 = 0$ into the Laplace transform of Eqn. 8.20 (cf. Eqn. A.11) and integrating give

$$\tilde{\epsilon}_{kk} = \gamma A_1 I_o(qr) + A_2 \tag{8.25}$$

where A_2 is a constant of integration. Eqn. 8.25 is then substituted into the Laplace transform of Eqn. 8.21. The required integration can be performed by using the following identity for hyperbolic Bessel functions (e.g., Carslaw and Jaeger, 1959, Appendix III, p. 489):

$$zI_1'(z) + I_1(z) = zI_o(z) \tag{8.26}$$

Or, equivalently,

$$\frac{1}{z}\frac{d(zI_1)}{dz} = I_o(z) \tag{8.27}$$

The final result is

$$\tilde{u}_r = \frac{\gamma}{q} A_1 I_1(qr) + \frac{A_2}{2} r \tag{8.28}$$

After applying boundary conditions, the radial displacement, stresses, and fluid pressure can be inverted numerically back to the time domain using the Stehfest algorithm (Appendix A, Section A.3).

8.2.2 Case 1: $\sigma_{rr}^{(1)}(b,t) = -P_b$ and $p^{(1)}(b,t) = 0$

The first boundary-value problem considered is the sudden application of a radial stress $\sigma_{rr}^{(1)}(b,t) = -P_b$, where b is the radius of the cylinder, for times greater than zero. The upper-case notation is used to distinguish the radial stress change from a fluid-pressure change of p_b used in the next subsection. The fluid pressure is maintained at zero on the cylinder surface (i.e., $p^{(1)}(b,t) = 0$ for $t > 0$). The initial condition is that both the fluid pressure and increment of fluid content are zero throughout the cylinder. This problem is the cylindrical equivalent of the Terzaghi problem for a finite layer. The two constants A_1 and A_2 in Eqn. 8.28 are determined from the stress and fluid pressure conditions at the boundary $r = b$.

1. *Pore-Pressure Boundary Condition.* Zero pressure at the cylinder surface means that $\tilde{p}^{(1)}(b,s) = 0$. Hence, $\tilde{\zeta}^{(1)}(b,s) = \alpha\tilde{\epsilon}_{kk}^{(1)}(b,s)$ (cf. Eqn. 2.25). Substituting Eqns. 8.24 and 8.25 for $\tilde{\zeta}^{(1)}$ and $\tilde{\epsilon}_{kk}^{(1)}$ gives $A_2^{(1)}/A_1^{(1)} = (1-\alpha\gamma)I_o(qb)/\alpha$.
2. *Radial-Stress Boundary Condition.* The Laplace transform of the radial stress boundary condition is

$$\tilde{\sigma}_{rr}^{(1)}(b,s) = -\frac{P_b}{s} \tag{8.29}$$

From Eqns. 8.1 and 8.28, the Laplace transform of the radial strain at $r = b$ is

$$\tilde{\epsilon}_{rr}^{(1)}(b,s) = \gamma A_1^{(1)}\left[I_o(qb) - \frac{1}{(qb)}I_1(qb)\right] + \frac{A_2^{(1)}}{2} \tag{8.30}$$

The Laplace transforms of all the terms in Eqn. 8.8 are now expressed in terms of $A_1^{(1)}$ and $A_2^{(1)}$ through Eqns. 8.24, 8.25, 8.29, and 8.30. After determining $A_1^{(1)}$ and $A_2^{(1)}$, the final result for the Laplace transform of the displacement (Detournay and Cheng, 1993, p. 150) is

$$\tilde{u}_r^{(1)}(r,s) = -\frac{P_b r}{2Gs} \frac{(1-2\nu_u)(1-\nu)I_o(qb) + 2(\nu_u - \nu)(qr)^{-1}I_1(qr)}{(1-\nu)I_o(qb) - 2(\nu_u - \nu)(qb)^{-1}I_1(qb)} \tag{8.31}$$

The undrained and drained limits can be obtained from these Laplace-transformed solutions.

1. *Undrained Limit.* The displacement in the undrained limit $t \to 0^+$ can be obtained from Eqn. 8.31 and the initial value theorem (Appendix A, Eqn. A.16) as the limit of $s\tilde{u}_r^{(1)}(r, s)$ as $s \to \infty$. This limit can be obtained using the following property of $I_n(x)$ (Kraut, 1967, p. 271):

$$\lim_{x \to \infty} I_n(x) = \frac{e^x}{\sqrt{2\pi x}} \tag{8.32}$$

The undrained radial displacement is then

$$u_r^{(1)}(r, 0^+) = -\frac{1 - 2\nu_u}{2G} P_b r \tag{8.33}$$

Alternatively, the undrained radial displacement can be obtained directly in the time domain. Combining Eqns. 8.20 and 8.21 shows that the radial displacement, u_r, satisfies

$$\frac{\partial}{\partial r}\left[\frac{1}{r}\frac{\partial}{\partial r}(ru_r)\right] = \gamma \frac{\partial \zeta}{\partial r} \tag{8.34}$$

The general solution is

$$u_r(r, t) = C_1(t)r + \frac{C_2(t)}{r} + \frac{\gamma}{r}\int_0^r \zeta(r', t)r'dr' \tag{8.35}$$

where $C_1(t)$ and $C_2(t)$ are constants of integration. The constant $C_2(t)$ must be zero for the displacement to be bounded at $r = 0$. The initial condition $\zeta(r, 0^+) = 0$ means that the integral on the right-hand side of Eqn. 8.35 is zero at $t = 0^+$, and hence $u_r^{(1)}(r, 0^+) = C_1(0^+)r$. Therefore, at $t = 0^+$, $\epsilon_{rr} = \epsilon_{\theta\theta} = C_1(0^+)$. From Eqns. 8.8 and 8.9, the two normal stresses are constant and hence equal to the boundary value for σ_{rr}:

$$\sigma_{rr}^{(1)}(b, 0^+) = \sigma_{\theta\theta}^{(1)}(b, 0^+) = \frac{2GC_1(0^+)}{1 - 2\nu_u} = -P_b \tag{8.36}$$

Eqn. 8.33 is then obtained after solving Eqn. 8.36 for $C_1(0^+)$. The undrained pore pressure is obtained from Eqns. 8.13 and 8.36:

$$p^{(u)} = \frac{2}{3}(1 + \nu_u) B P_b \tag{8.37}$$

2. *Drained Limit.* The displacement in the drained limit $t \to \infty$ can be obtained from Eqn. 8.31 and the final value theorem (Appendix A, Eqn. A.17) as the limit of $s\tilde{u}_r^{(1)}(r, s)$ as $s \to 0$. This limit can be taken using the result (Kraut, 1967, p. 271):

$$\lim_{x \to 0} I_n(x) = \frac{1}{n!}\left(\frac{x}{2}\right)^n \tag{8.38}$$

The drained radial displacement is then

$$u_r^{(1)}(r, \infty) = -\frac{1 - 2\nu}{2G} P_b r \tag{8.39}$$

Comparing Eqns. 8.33 and 8.39 shows that the drained radial displacement can be obtained from the undrained radial displacement by replacing ν_u with ν.

8.2.3 Mandel-Cryer Effect

The Laplace transform of the pore pressure is obtained by inserting Eqns. 8.24 and 8.25 into the Laplace transform of Eqn. 2.25:

$$\tilde{p}^{(1)}(r, s) = \frac{p^{(u)}}{s} \frac{(1 - \nu)[I_o(qb) - I_o(qr)]}{(1 - \nu) I_o(qb) - 2(\nu_u - \nu)(qb)^{-1} I_1(qb)} \tag{8.40}$$

A comparison of the pore pressure obtained from a complete poroelastic analysis with the uncoupled pore-pressure diffusion solution ($\nu_u = \nu$ in Eqn. 8.40) shows the Mandel-Cryer effect, whereby the pore pressure at the center of the cylinder rises above the initial undrained value before it decays to zero (Fig. 8.3). The physical explanation is that the cylinder contracts near the drained boundary due to dissipation of the induced pore pressure. Strain compatibility requires contraction and hence additional pore pressure buildup in the interior. Thus, the Mandel-Cryer effect can be viewed as a stress transfer toward the interior. The poroelastic pore pressure at the center is always greater than the uncoupled pore pressure obtained from the pore-pressure diffusion equation. The decay of the undrained buildup in pore pressure therefore takes longer than would be predicted from an uncoupled solution.

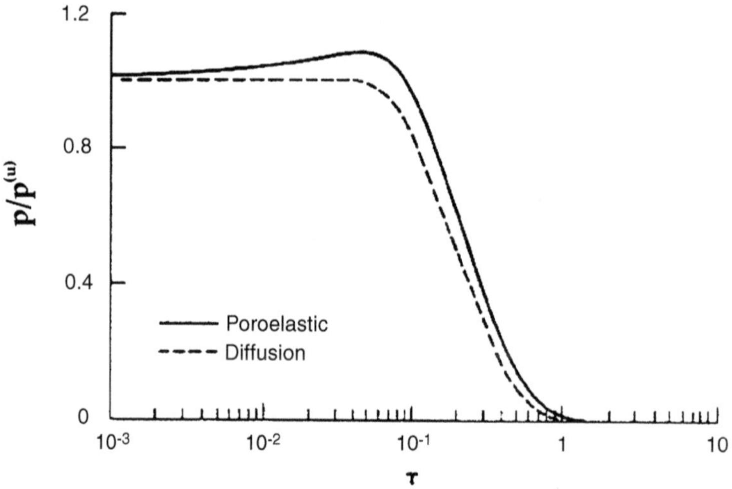

Figure 8.3: Pore pressure versus dimensionless time $\tau = ct/b^2$ at center of cylinder in plane strain subjected to suddenly applied radial load (Detournay and Cheng, 1993, Fig. 3, p. 151). Pore pressure is normalized by the undrained response $p^{(u)}$. The solid curve is the full poroelastic solution showing the Mandel-Cryer effect. The dashed curve is the solution for an uncoupled, diffusion equation. Elastic properties are $\nu = 0.15$ and $\nu_u = 0.31$.

8.2.4 Case 2: $\sigma_{rr}^{(2)}(b,t) = 0$ and $p^{(2)}(b,t) = p_b$

The sudden application of fluid pressure on the cylinder surface while maintaining constant stress means that $p^{(2)}(b,t) = p_b$ and $\sigma_{rr}^{(2)}(b,t) = 0$ for times greater than zero. This boundary-value problem is not associated with a practical physical situation, as fluid pressure generally cannot be applied independently of a compressive radial stress. The two constants in Eqn. 8.28 are determined by the same procedure as in Case 1 given earlier. The result for the Laplace transform of the displacement is

$$\tilde{u}_r^{(2)}(r,s) = \alpha(1-2\nu)\frac{p_b r}{2Gs}\frac{(qr)^{-1}I_1(qr) + (1-2\nu_u)(qb)^{-1}I_1(qb)}{(1-\nu)I_o(qb) - 2(\nu_u-\nu)(qb)^{-1}I_1(qb)} \quad (8.41)$$

The Laplace transform of the pore pressure is

$$\tilde{p}^{(2)} = \frac{p_b}{s}\frac{(1-\nu)I_o(qr) - 2(\nu_u-\nu)(qb)^{-1}I_1(qb)}{(1-\nu)I_o(qb) - 2(\nu_u-\nu)(qb)^{-1}I_1(qb)} \quad (8.42)$$

For long times, the boundary pore pressure, p_b, is diffused uniformly throughout the cylinder. The radial displacement is given by

$$u_r^{(2)}(r,\infty) = \alpha \frac{1-2\nu}{2G} p_b r \tag{8.43}$$

The steady-state radial displacement for a pore-pressure change, p_b, at the boundary is negative α times the displacement due to a radial compressive load at the boundary of equal magnitude (cf. Eqn. 8.39).

8.2.5 Circumferential Stresses at Short Times

A sudden discontinuity in pore pressure at the surface of a cylinder leads to an immediate change in fluid pressure in a thin boundary layer. If the interior pressure has the constant value p_i, and if a boundary pressure $p_b < p_i$ is suddenly imposed, the lower fluid pressure in the annulus at the rim actually leads to an extensional hoop stress. This occurs because the annulus circumference can be considered so thin that its circumferential dimension remains constrained. Conversely, if $p_b > p_i$, inward fluid diffusion produces a compressive hoop stress.

For Case 2, a discontinuity in pore pressure occurs on the surface of the cylinder because a sudden change in pore pressure is imposed on the boundary. For Case 1, a discontinuity in pore pressure occurs because a sudden stress change on the boundary induces an undrained pore-pressure response throughout the interior. The magnitude of the change in circumferential stress at short times is discussed first for Case 2, because changing the boundary pore pressure without changing the boundary stress displays the effect of the mismatch of the interior and boundary fluid pressures on the circumferential stress without the intermediary effect of the undrained pore-pressure change due to the applied stress change (Case 1).

Sudden Pore-Pressure Change: $\sigma_{rr}^{(2)}(b,t)=0$ *and* $p^{(2)}(b,t)=p_b$ *(Case* 2)

The integration constant in the plane-strain compatibility equation for radial symmetry (Eqn. 8.14) at time zero is $f(0) = 0$, because $\sigma_{rr}^{(2)}(r,0) = \sigma_{\theta\theta}^{(2)}(r,0) = p^{(2)}(r,0) = 0$. The notation $t = 0^{++}$ is used to indicate a very short time during which incipient fluid diffusion occurs. The time 0^{++} is longer than the undrained response 0^{+}. Integrating Eqn. 8.16 from $t = 0^{-}$ to $t = 0^{++}$ for any r within the interior of the cylinder that has not been penetrated by incipient fluid diffusion gives the result that $f(0^{++})$ is also zero, because the radial pressure distribution remains zero within the interior of the cylinder. Eqn. 8.14 then gives

$$\sigma_{\theta\theta}^{(2)}(b,0^{++}) = -2\eta p_b \tag{8.44}$$

because $\sigma_{rr}^{(2)}(b, 0^{++}) = 0$ is a boundary condition. Thus, an increase in the boundary pore pressure ($p_b > 0$) induces a compressive circumferential stress; and conversely, a decrease in the boundary pore pressure ($p_b < 0$) induces a tensile circumferential stress. Eqn. 8.44 can be obtained heuristically by the following argument (Detournay and Cheng, 1993, p. 156). Shortly after the application of the fluid-pressure boundary condition, the depth of penetration, δ, of the fluid diffusion is small compared with the radius b of the cylinder. Locally, the cylindrical surface is approximately planar and infinite in lateral extent, and hence a state of uniaxial strain obtains. The lateral stress given by Eqn. 6.3 for zero normal stress is then the same as the circumferential stress given by Eqn. 8.44. As stated by Detournay and Cheng (1993, p. 157): "the sudden variation of the tangential stress along the boundary takes place in order to preserve the tangential strain during the 'instantaneous' equilibration of the pore pressure to the imposed boundary value." Eqn. 8.44 can be generalized to account for an initial fluid pressure, p_i, before the application of the boundary pressure p_b:

$$\sigma_{\theta\theta}^{(2)}(b, 0^{++}) = -2\eta(p_b - p_i) \tag{8.45}$$

Sudden Stress Change: $\sigma_{rr}^{(1)}(b, t) = -P_b$ *and* $p^{(1)}(b, t) = 0$ (*Case* 1)

The sudden application of the boundary stress $\sigma_{rr}^{(1)}(b, t) = -P_b$ produces an undrained response in pore pressure given by Eqn. 8.37. The undrained response is instantaneous and occurs before incipient fluid flow due to the discontinuity between $p^{(u)}$ and p_b at the boundary. Notationally the undrained response is represented by superscript u, and the time at which the subsequent pressure change due to fluid flow occurs in a skin region near the boundary is represented by $t = 0^{++}$. The use of superscripts u and 0^{++} distinguishes these two nearly instantaneous but sequential states.

The change in the circumferential stress at the boundary due to incipient fluid flow following the undrained response is obtained from Eqn. 8.45,

$$\begin{aligned}\Delta\sigma_{\theta\theta}^{(1)}(b, 0^{++}) &= \sigma_{\theta\theta}^{(1)}(b, 0^{++}) - \sigma_{\theta\theta}^{(u)}(b)\\ &= -2\eta(0 - p^{(u)})\\ &= 2\eta p^{(u)}\end{aligned} \tag{8.46}$$

where $\sigma_{\theta\theta}^{(u)}(b) = \sigma_{\theta\theta}^{(1)}(b, 0^{+})$ is given by Eqn. 8.36. The zero pore-pressure boundary condition at $r = b$ is shown explicitly in the second line of Eqn. 8.46. A suddenly applied compressive radial stress at the boundary of a cylinder induces a positive undrained pore pressure. Hence, an initial outward fluid flow occurs for a boundary of zero pore pressure, and the change in circumferential stress is extensional.

8.2.6 Example: Sudden Removal of Core from a Borehole

Consider a rock mass with a uniform internal pore pressure, $p = p_o$, in a horizontally isotropic stress field, $\sigma_{rr} = -P_o$. Sudden removal of a cylindrical core from this environment is characterized by two boundary conditions (Fig. 8.4): (1) the stress relief means that the change in radial stress $\Delta\sigma_{rr}(b,t) = +P_o$ for $t > 0$; and (2) the pore-pressure relief means that the change in boundary pressure $\Delta p(b,t) = -p_o$ for $t > 0$. The stress relief induces a negative undrained change in pore pressure. On unloading, the discontinuity between the new internal pore pressure and atmospheric pressure produces a poroelastic circumferential stress. Because the core is stress free after removal, the circumferential stress at the cylinder surface at short time is obtained by summing Eqns. 8.44 and 8.46, with $P_b = -P_o$ and $p_b = -p_o$,

$$\begin{aligned}\sigma_{\theta\theta}^{\text{core}}(b, 0^{++}) &= \sigma_{\theta\theta}^{(2)}(b, 0^{++}) + \Delta\sigma_{\theta\theta}^{(1)}(b, 0^{++}) \\ &= 2\eta p_o + 2\eta p_{\text{relief}}^{(u)}\end{aligned} \tag{8.47}$$

where (cf. Eqn. 8.37)

$$p_{\text{relief}}^{(u)} = -\frac{2B(1+\nu_u)}{3}P_o = -\frac{2(\nu_u - \nu)}{\alpha(1-2\nu)}P_o \tag{8.48}$$

Therefore,

$$\sigma_{\theta\theta}^{\text{core}}(b, 0^{++}) = 2\eta p_o - \frac{2(\nu_u - \nu)}{1-\nu}P_o \tag{8.49}$$

The maximum undrained drop in pore pressure due to stress relief is $-P_o$ for incompressible constituents ($\alpha = 1$ and $\nu_u = 0.5$). For this case, the

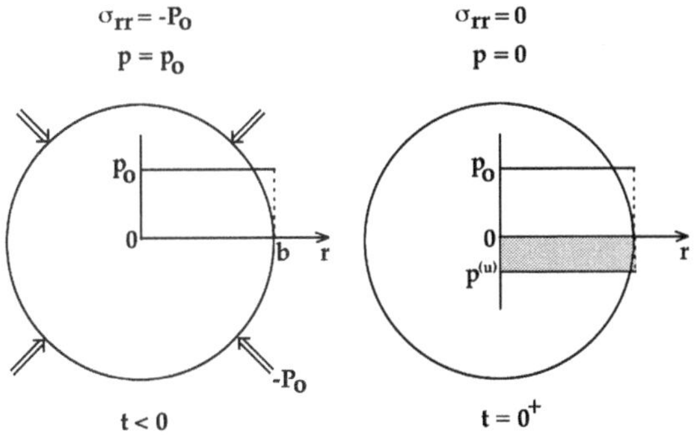

Figure 8.4: Initial and boundary conditions for stress-relieved core. The negative undrained response ($t = 0^+$) is shown as a shaded region.

circumferential stress is $2\eta(-P_o + p_o)$. If $P_o = p_o$, the circumferential stress is exactly zero. In general, $P_o > p_o$ so that the undrained decrease in pore pressure can be bigger than the initial pore pressure, and the undrained pressure response due to stress relief can produce unsaturated conditions within the core.

The complete time-domain solutions for displacement, strains, and stresses can be obtained by superimposing the Laplace transform of the solutions given in the previous two sections. They can be used to interpret strain relaxation of fluid-saturated core retrieved from drilling. The maximum tensile stress occurs at the cylinder surface immediately on unloading. Figure 8.5

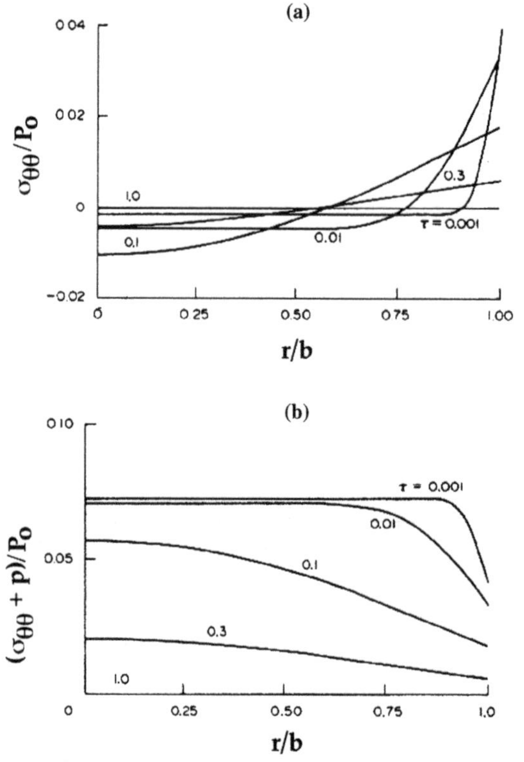

Figure 8.5: Radial profiles of circumferential stress in a suddenly recovered cylindrical core at various dimensionless times, $\tau = ct/b^2$, following stress relief to atmospheric conditions from in situ conditions $\sigma_{rr} = -P_o$ and $p = p_o$. Parameter values are $\nu = 0.18$, $\nu_u = 0.28$, $\alpha = 0.74$ (Ohio sandstone), and $p_o = 0.5\ P_o$ (Detournay and Cheng, 1993, p. 152). (a) Normalized circumferential stress $\sigma_{\theta\theta}/P_o$. (b) Normalized Terzaghi circumferential stress $(\sigma_{\theta\theta} + p)/P_o$. The Terzaghi circumferential stress decreases monotonically with radial distance.

shows radial profiles of circumferential stress for an instantly retrieved cylindrical core at various times after unloading.

8.3 SUDDEN PRESSURIZATION OF A BOREHOLE

The poroelastic response to sudden pressurization of a borehole can be treated under the assumptions of plane strain and radial symmetry. Initially, the fluid pressure throughout the areally infinite problem domain is taken to be zero (i.e., constant). The requirement that changes in stress and pore pressure approach zero in the far field ($r \to \infty$) means that $f(t) = 0$ in Eqn. 8.14. Hence, the time-dependent term in the nonhomogeneous pressure diffusion equation (Eqn. 8.16) is zero for the case of an areally extensive region surrounding the borehole, and the fluid-flow problem is uncoupled from the mechanical problem. Because a homogeneous pressure diffusion equation applies in a problem domain of infinite radius, the solution for these boundary-value problems can be drawn from the literature for heat flow and well hydraulics (e.g., Muskat, 1937; Carslaw and Jaeger, 1959; Streltsova, 1988).

The boundary conditions at the borehole wall ($r = a$) are

$$\sigma_{rr}(a, t) = -p_a \qquad t > 0 \tag{8.50}$$

$$p(a, t) = p_a \qquad t > 0 \tag{8.51}$$

The two boundary conditions will be broken up into two separate problems. For generality, $\sigma_{rr}^{(1)}(a, t)$ will be taken to be $-P_a$ to distinguish it from the boundary pore pressure, p_a.

8.3.1 Case 1: $\sigma_{rr}^{(1)}(a, t) = -P_a$ and $p^{(1)}(a, t) = 0$

The solution of the poroelastic problem of a uniform radial stress, $-P_a$, on the walls of a hole at which $p = 0$ is given by the classical Lamé solution:

$$u_r^{(1)}(r, t) = \frac{P_a}{2G} r \tag{8.52}$$

$$\sigma_{rr}^{(1)} = -P_a \frac{a^2}{r^2} \tag{8.53}$$

$$\sigma_{\theta\theta}^{(1)} = P_a \frac{a^2}{r^2} \tag{8.54}$$

The displacements and stresses are independent of time, and no pore pressures are induced because $\sigma_{rr}^{(1)} + \sigma_{\theta\theta}^{(1)} = 0$. Hence, no poroelastic effects occur because there is no undrained response in pore pressure.

8.3.2 Case 2: $\sigma_{rr}^{(2)}(a,t)=0$ and $p^{(2)}(a,t)=p_a$

Stress-free and drained boundary conditions at infinity imply that fluid pressure is uncoupled from stress and that it satisfies a homogeneous diffusion equation. The Laplace transform of the homogeneous pressure diffusion equation is identical to Eqn. 8.22 but with $\tilde{p}$ in place of $\tilde{\zeta}$,

$$\frac{d^2\tilde{p}}{dr^2} + \frac{1}{r}\frac{d\tilde{p}}{dr} - \frac{s}{c}\tilde{p} = 0 \tag{8.55}$$

where the initial condition $p(r,0) = 0$ has been used. The solution satisfying the boundary condition $p = p_a$ at $r = a$ (Carslaw and Jaeger, 1959, p. 335) is

$$\tilde{p}^{(2)}(r,s) = \frac{p_a}{s}\frac{K_o(qr)}{K_o(qa)} \tag{8.56}$$

where $q = \sqrt{s/c}$.

For an irrotational displacement field, ζ equals Sp (cf. Section 4.10.4), and Eqn. 8.19 therefore becomes $\partial\epsilon_{kk}/\partial r = (\eta/G)\partial p/\partial r$. The solutions for the Laplace transforms of the displacement and stresses can then be obtained in a manner similar to that used in Section 8.2 (Detournay and Cheng, 1993, p. 154):

$$\tilde{u}_r^{(2)}(r,s) = -\frac{a}{2G}\frac{2\eta p_a}{s}\left\{\frac{K_1(qr)}{(qa)K_o(qa)} - \frac{K_1(qa)}{(qr)K_o(qa)}\right\} \tag{8.57}$$

$$\tilde{\sigma}_{rr}^{(2)}(r,s) = \frac{2\eta p_a}{s}\left\{\frac{K_1(qr)}{(qr)K_o(qa)} - \frac{K_1(qa)}{(qr)(r/a)K_o(qa)}\right\} \tag{8.58}$$

$$\tilde{\sigma}_{\theta\theta}^{(2)}(r,s) = -\frac{2\eta p_a}{s}\left\{\frac{K_1(qr)}{(qr)K_o(qa)} - \frac{K_1(qa)}{(qr)(r/a)K_o(qa)} + \frac{K_o(qr)}{K_o(qa)}\right\} \tag{8.59}$$

Note that at $r = a$, $\tilde{\sigma}_{\theta\theta}^{(2)}(a,s) = -2\eta p_a/s$ and hence $\sigma_{\theta\theta}^{(2)}(a,t) = -2\eta p_a$.

8.3.3 Fluid Load: Summing Cases 1 and 2

The complete poroelastic solution for fluid pressurization of a borehole is obtained by inverting the Laplace transforms of the solutions for Case 2 back to the time domain, and adding it to the time-independent solution of Case 1. In general, the inverse Laplace transforms can be evaluated numerically using the Stehfest algorithm. The inversion of Eqn. 8.56, however, can also be expressed analytically (Carslaw and Jaeger, 1959, p. 335; see also Rice and Cleary, 1976, p. 237):

$$p(r,t) = p_a + \frac{2p_a}{\pi}\int_o^\infty \exp(-c\xi^2 t)\cdot\frac{J_o(\xi r)Y_o(\xi a) - J_o(\xi a)Y_o(\xi r)}{J_o^2(\xi a)+Y_o^2(\xi a)}\frac{d\xi}{\xi} \tag{8.60}$$

Eqn. 8.56 can be approximated for large values of s by using $\lim_{x\to\infty} K_o(x) = \sqrt{\pi/2x}\exp(-x)$ (Kraut, 1967, p. 271). Inverting (Carslaw and Jaeger, 1959, p. 336) then gives

$$p(r,t) \sim p_a(a/r)^{1/2}\text{erfc}[(r-a)/(4ct)^{1/2}] \tag{8.61}$$

where $t \ll r^2/4c$. For example, a pore pressure of $0.9p_a$ or greater will have penetrated up to a distance of about $0.2(ct)^{1/2}$ at time t (Rice and Cleary, 1976).

The circumferential stress is given by

$$\sigma_{\theta\theta}(r,t) = -\sigma_{rr}(r,t) - 2\eta p(r,t) \tag{8.62}$$

because $f(t) = 0$ in Eqn. 8.14. At the borehole wall, $r = a$, the pressure and radial stress must satisfy the boundary conditions, $p(a,t) = p_a$ and $\sigma_{rr}(a,t) = -p_a$, respectively. Therefore,

$$\sigma_{\theta\theta}(a,t) = p_a - 2\eta p_a = (1-2\eta)p_a \tag{8.63}$$

The fluid load solution differs from the Lamé solution by the factor $1 - 2\eta = 0.42$ for $\alpha = 0.74$ and $\nu = 0.18$. The smaller tensile circumferential stress is due to the effect of pressure diffusion. Eqn. 8.63 shows that the circumferential stress at the borehole wall is independent of time.

8.4 SUDDEN STRESS RELEASE DUE TO BOREHOLE EXCAVATION IN AN ANISOTROPIC STRESS FIELD

The circumferential stress at a borehole wall following its sudden excavation is examined in this section in the short and long time limits for the case of an anisotropic stress field. The complete solution in terms of Laplace transforms was given by Detournay and Cheng (1988; 1993, p. 155).

The preexcavation ($t < 0$), far-field stresses are, in general, anisotropic and may be represented in principal coordinates as

$$\begin{aligned}\sigma_{xx}^{\text{pre}} &= -(P_o - S_o)\\ \sigma_{yy}^{\text{pre}} &= -(P_o + S_o)\end{aligned} \tag{8.64}$$

where $-P_o$ is the mean compressive far-field stress, and S_o is the deviatoric component of the far-field stress. In Eqn. 8.64 the maximum compressive horizontal stress is in the y-direction, and the minimum compressive horizontal stress is in the x-direction. The sum of the predrilling horizontal normal stresses is invariant to the choice of coordinate system. In polar coordinates,

$$\sigma_{rr}^{\text{pre}} + \sigma_{\theta\theta}^{\text{pre}} = -2P_o \tag{8.65}$$

Immediately after drilling, the undrained stresses at the borehole wall are given by the elastic solution for a hole (Jaeger and Cook, 1976, p. 251). The radial normal stress is zero at the borehole wall, and the circumferential stress is

$$\sigma_{\theta\theta}^{\text{hole}}(a, \theta) = -2P_o - 4S_o \cos 2\theta \tag{8.66}$$

The change in mean horizontal normal stress at the borehole wall is

$$\Delta\sigma_{rr}(a, \theta) + \Delta\sigma_{\theta\theta}(a, \theta) = -4S_o \cos 2\theta \tag{8.67}$$

This change in mean horizontal normal stress induces an undrained fluid pressure change (cf. Eqn. 8.13):

$$p^{(u)}(a, \theta) = \frac{B(1 + \nu_u)}{3}\left[4S_o \cos 2\theta\right] \tag{8.68}$$

The maximum compressive circumferential stress is at the borehole wall in the direction of the minimum principal stress ($\theta = 0$ and $\theta = \pi$). Incipient pore pressure diffusion due to the induced fluid pressure modifies the circumferential borehole stress the same as for a suddenly pressurized cylinder (cf. Eqn. 8.46):

$$\sigma_{\theta\theta}^{(1)}(a, \theta = 0, 0^{++}) - \sigma_{\theta\theta}^{\text{hole}}(a, \theta) = 2\eta p^{(u)} \tag{8.69}$$

Also, at the instant the borehole is excavated, the pore pressure at the borehole wall changes from its in situ value of p_o to zero. The circumferential stress for the problem of a change in pore pressure at a stress-free borehole wall was obtained in the previous section. Setting $p_a = -p_o$ and $r = a$ in Eqn. 8.59 shows that the circumferential stress is constant with time. It is also independent of θ because the solution is for a radially symmetric problem:

$$\sigma_{\theta\theta}^{(2)}(a, \theta, t) = 2\eta p_o \tag{8.70}$$

Superposing Eqns. 8.69 and 8.70 and using Eqn. 8.66 give

$$\sigma_{\theta\theta}(a, \theta = 0, 0^{++}) = -2P_o + 2\eta p_o - 4\frac{1-\nu_u}{1-\nu}S_o \tag{8.71}$$

At long times, $p^{(u)}$ dissipates, and the maximum compressive circumferential stress at the borehole wall is the elastic solution for a hole (Eqn. 8.66) at $\theta = 0$ plus the time-independent circumferential stress $2\eta p_o$ (Eqn. 8.70):

$$\sigma_{\theta\theta}(a, \theta = 0, \infty) = -2P_o + 2\eta p_o - 4S_o \tag{8.72}$$

Thus, the maximum compressive circumferential stress at the borehole wall becomes increasingly compressive with time because $(1 - \nu_u)/(1 - \nu)$ in Eqn. 8.71 is less than one. If the far-field stress is isotropic ($S_o = 0$ in Eqns. 8.71 and 8.72), the circumferential stress at the borehole wall is independent of time. Combining Eqns. 8.66 and 8.70 for $\theta = \pi/2$ gives the minimum compressive circumferential stress at the borehole wall for long times:

$$\sigma_{\theta\theta}\left(a, \theta = \frac{\pi}{2}, \infty\right) = -2P_o + 2\eta p_o + 4S_o \tag{8.73}$$

8.5 HYDRAULIC FRACTURING

The poroelastic expression for the circumferential stress developed in the previous sections on borehole pressurization and borehole excavation can be used to estimate the breakdown pressure in hydraulic fracturing (Haimson and Fairhurst, 1969). The assumption is made that breakdown occurs when the Terzaghi effective stress equals the tensile strength, T:

$$\sigma_{\theta\theta} + p = T \tag{8.74}$$

Adding the drained minimum compressive circumferential stress (Eqn. 8.73) to the circumferential stress due to borehole pressurization (Eqn. 8.63) gives

$$\sigma_{\theta\theta}\left(a, \theta = \frac{\pi}{2}, \infty\right) = -2P_o + 2\eta p_o + 4S_o + (1 - 2\eta)p_a \tag{8.75}$$

where p_a is the pressure of the injected fluid. Substituting Eqn. 8.75 into the breakdown criterion (Eqn. 8.74) gives

$$p_a^{\text{breakdown}} = \frac{T + 2P_o - 4S_o - 2\eta p_o}{2(1-\eta)} \tag{8.76}$$

Eqn. 8.76 can also be expressed in terms of the maximum and minimum compressive horizontal stresses,

$$p_a^{\text{breakdown}} = \frac{T - (3\sigma^{\min} - \sigma^{\max}) - 2\eta p_o}{2(1-\eta)} \tag{8.77}$$

where $\sigma^{\min} = -(P_o - S_o)$ and $\sigma^{\max} = -(P_o + S_o)$ (cf. Eqn. 8.64). Eqn. 8.77 is usually referred to as the Haimson-Fairhurst criterion. Detournay and Cheng (1993) showed that it represents a lower bound due to slow pressurization. The nonporoelastic breakdown pressure, known as the Hubbert-Willis (1957) criterion, represents the fast pressurization limit. It is obtained from Eqn. 8.76 by setting $\eta = 0.5$, not $\eta = 0$. This paradox has been discussed by Schmitt and Zoback (1992).

8.6 SUDDEN INTERNAL PRESSURIZATION OF A HOLLOW CYLINDER

An internally pressurized hollow cylinder can be used in laboratory testing to determine the tensile strength of rocks and to simulate hydraulic fracturing. The pressurization of a borehole (cf. Section 8.3) in a horizontally infinite region is the limiting case of a hollow cylinder with an infinite outer radius. The poroelastic problem for a finite outer radius is more complicated, because $f(t)$ in Eqn. 8.14 is nonzero and the pore-pressure diffusion equation is nonhomogeneous. In this section, the short and long time limits are discussed in detail (Rice and Cleary, 1976, pp. 236–237), and radial profiles of pressure and circumferential stress are presented graphically (Detournay and Carvalho, 1989; Schmitt et al., 1993).

The geometry and boundary conditions are illustrated in Figure 8.6. The stress boundary conditions for the hollow cylinder for $t > 0$ are that the radial stress on the inner wall is $\sigma_{rr}(a, t) = -p_a$, and the radial stress on the outer wall is $\sigma_{rr}(b, t) = 0$, where a and b are the inner and outer radii, respectively. The fluid pressure on the inner wall is $p(a, t) = p_a$, and the fluid boundary condition on the outer wall is $p(b, t) = 0$.

8.6.1 Short Time Limit

The circumferential stress in the short time limit at the interior surface, $r = a$, of a hollow cylinder due to sudden pressurization can be understood in the same manner as that for the short time behavior at the exterior surface of a solid cylinder (Section 8.2.5). The poroelastic boundary conditions at the inner cylinder wall are decomposed into two separate boundary value problems: (1) applied radial stress and zero pore pressure, and (2) boundary pore pressure and no applied stress. Poroelastic stresses in the latter case result

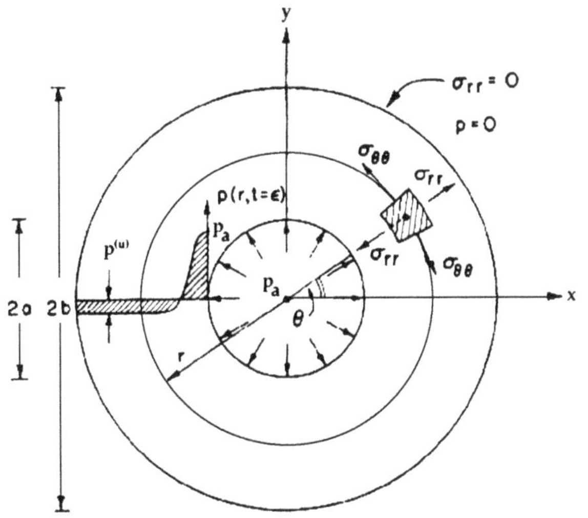

Figure 8.6: Fluid loading of a hollow cylinder (after Rice and Cleary, 1976, p. 236). A steep pore-pressure gradient exists near the inner surface shortly after application of the fluid pressure. The pore pressure has its boundary value p_a at $r = a$ and decreases to a (negative) undrained value $p^{(u)}$ in the interior of the sample due to the radial load on the inner boundary.

from the variation of pore pressure with radius, even in the absence of an applied boundary stress. Each problem leads to a discontinuity in pore pressure and subsequent incipient fluid infiltration, which changes the circumferential stress. In the first case, the discontinuity arises from the undrained pore pressure induced by the sudden change in stress on the borehole wall. In the second case, the discontinuity arises from the boundary condition itself.

Case 1: $\sigma_{rr}^{(1)}(a,t) = -p_a$ *and* $p^{(1)}(a,t) = 0$

The stresses for Case 1 in a hollow cylinder with finite outer radius and $p_b = 0$ are not independent of time, as they are in the infinite radius case (cf. Section 8.3). For very short times, the applied stress on the internal cylindrical surface produces displacements associated with undrained conditions; that is, the solution is obtained by setting $\zeta(r,t) = 0$ in the general displacement solution (Eqn. 8.35). Substituting the displacement solution back into the constitutive equations and applying the stress boundary conditions then yield the undrained stresses at the instant of loading:

$$\sigma_{rr}^{(1)}(r, 0^+) = \frac{p_a a^2}{b^2 - a^2}\left[1 - \frac{b^2}{r^2}\right] \tag{8.78}$$

$$\sigma_{\theta\theta}^{(1)}(r, 0^+) = \frac{p_a a^2}{b^2 - a^2}\left[1 + \frac{b^2}{r^2}\right] \tag{8.79}$$

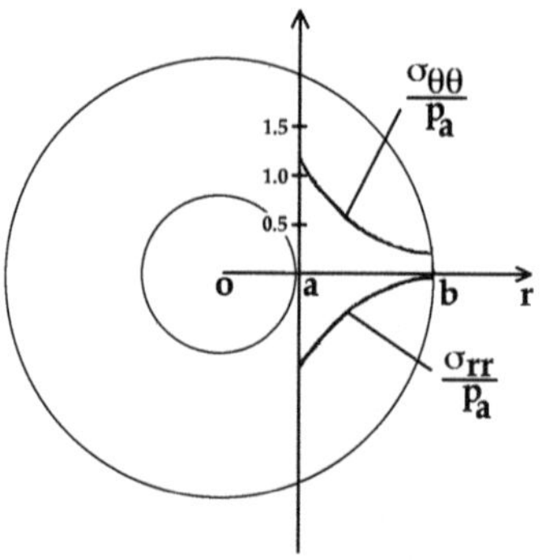

Figure 8.7: Stresses in a hollow cylinder for $b/a = 3$ (after Gould, 1983, p. 118).

This classical solution for stresses in a pressure vessel is independent of the elastic constants and applies in both the undrained and drained limits. The two components of stress are shown in Figure 8.7 for $b/a = 3$. The internal pressure puts the cylinder in tension in the circumferential direction but produces compressional radial stress throughout the cylinder. From Eqn. 8.79 the circumferential stress at the inner boundary is

$$\sigma_{\theta\theta}^{(1)}(a, 0^+) = p_a \frac{b^2 + a^2}{b^2 - a^2} \tag{8.80}$$

The undrained response in pore pressure is obtained using Eqn. 8.13:

$$p^{(u)}(r, 0^+) = -\frac{B(1+\nu_u)}{3}\{\sigma_{rr}^{(1)}(r, 0^+) + \sigma_{\theta\theta}^{(1)}(r, 0^+)\} \tag{8.81}$$

$$= -\frac{B(1+\nu_u)}{3}\frac{2a^2 p_a}{(b^2 - a^2)} \tag{8.82}$$

$$= -\frac{\nu_u - \nu}{\eta(1-\nu)}\frac{a^2 p_a}{(b^2 - a^2)} \tag{8.83}$$

The initial undrained response, $p^{(u)}$, is a negative pressure change because the loading on the interior cylinder surface produces a tensile mean stress. (Note that $p^{(u)} = 0$ as $b \to \infty$.) The pressure $p^{(u)}$ has a constant value throughout the cylinder, except at the boundary where it is p_a. For example, $p^{(u)} = -p_a/8$ for $b/a = 3$ and for incompressible constituents ($B = 1$ and $\nu_u = 0.5$). The poroelastic circumferential stress at the interior cylinder wall due to the discontinuity in pore pressure is given by Eqn. 8.46:

$$\sigma_{\theta\theta}^{(1)}(a, 0^{++}) - \sigma_{\theta\theta}^{(1)}(a, 0^+) = 2\eta p^{(u)} \tag{8.84}$$

Because $p^{(u)} < 0$, the tangential stress following incipient fluid diffusion is less tensile than the undrained tangential stress, $\sigma_{\theta\theta}^{(1)}(a, 0^+)$, given by Eqn. 8.80.

Case 2: $\sigma_{rr}^{(2)}(a,t) = 0$ *and* $p^{(2)}(a,t) = p_a$

No undrained pore pressure is generated in Case 2 because the applied loads are zero. However, a tangential stress due to incipient fluid flow occurs because of the pressure discontinuity between the initial condition of zero pore pressure and the boundary pore pressure, p_a. This tangential stress (cf. Eqn. 8.44) is

$$\sigma_{\theta\theta}^{(2)}(a, 0^{++}) = -2\eta p_a \tag{8.85}$$

Case 1 + Case 2

Summing Eqns. 8.84 and 8.85 gives

$$\sigma_{\theta\theta}(a, 0^{++}) = \left[\frac{2(1-\nu_u)}{(1-\nu)}\frac{a^2}{(b^2-a^2)} + (1-2\eta)\right] p_a \tag{8.86}$$

Taking the limit $b \to \infty$ recovers the result for borehole pressurization in an infinite domain (cf. Eqn. 8.63).

8.6.2 Long Time Limit

The development in this section follows that for thermoelasticity given by Boley and Weiner (1985, pp. 289–290). The constitutive equations are written in terms of the displacement, u_r, and its derivatives and are substituted into the radial equilibrium equation (Eqn. 8.17) to yield

$$\frac{\partial^2 u_r}{\partial r^2} + \frac{1}{r}\frac{\partial u_r}{\partial r} - \frac{u_r}{r^2} = \frac{\alpha}{3K}\left(\frac{1+\nu}{1-\nu}\right)\frac{\partial p}{\partial r} \tag{8.87}$$

Or, equivalently,

$$\frac{\partial}{\partial r}\left[\frac{1}{r}\frac{\partial}{\partial r}(r u_r)\right] = c_m \frac{\partial p}{\partial r} \tag{8.88}$$

The general solution is

$$u_r = C_1 r + \frac{C_2}{r} + \frac{c_m}{r}\int_a^r p(r')r'dr' \tag{8.89}$$

where C_1 and C_2 are constants fixed by the two stress boundary conditions.

Again, the problem is decomposed into Cases 1 and 2. For Case 2, the zero-stress condition, $\sigma_{rr} = 0$, on the cylindrical surfaces $r = a$ and $r = b$ yields

$$u_r^{(2)}(r,t) = \frac{c_m}{r}\left\{\int_a^r p(r',t)r'dr' + \frac{(1-2\nu)r^2 + a^2}{b^2 - a^2}\int_a^b p(r',t)r'dr'\right\} \tag{8.90}$$

$$\sigma_{rr}^{(2)}(r,t) = \frac{2\eta}{r^2}\left\{\left[\frac{r^2 - a^2}{b^2 - a^2}\right]\int_a^b p(r',t)r'dr' - \int_a^r p(r',t)r'dr'\right\} \tag{8.91}$$

$$\sigma_{\theta\theta}^{(2)}(r,t) = \frac{2\eta}{r^2}\left\{\left[\frac{r^2 + a^2}{b^2 - a^2}\right]\int_a^b p(r',t)r'dr' + \int_a^r p(r',t)r'dr' - p(r)r^2\right\} \tag{8.92}$$

The superscript 2 indicates that these three equations are the poroelastic stresses and displacements resulting from a nonconstant pore pressure as a function of radius, while zero stresses exist on the boundaries.

In the long time limit, the pore-pressure profile approaches the steady-state solution of the homogeneous diffusion equation:

$$p(r,\infty) = p_a \frac{\log_e(b/r)}{\log_e(b/a)} \tag{8.93}$$

Inserting Eqn. 8.93 into Eqns. 8.91 and 8.92 gives

$$\sigma_{rr}^{(2)}(r,\infty) = \eta p_a \left\{\frac{\log_e \frac{r}{a}}{\log_e \frac{b}{a}} - \left(\frac{b}{r}\right)^2\left(\frac{r^2 - a^2}{b^2 - a^2}\right)\right\} \tag{8.94}$$

$$\sigma_{\theta\theta}^{(2)}(r,\infty) = \eta p_a \left\{\frac{1}{\log_e \frac{b}{a}} + \frac{\log_e \frac{r}{a}}{\log_e \frac{b}{a}} - \left(\frac{b}{r}\right)^2\left(\frac{r^2 + a^2}{b^2 - a^2}\right)\right\} \tag{8.95}$$

Because the radial and tangential stresses for Case 1 are the same for the undrained and drained limits, Eqns. 8.78 and 8.79 must be added to those for Case 2 to obtain the total solution. At the boundary $r = a$, the tangential stress is

$$\sigma_{\theta\theta}(a,\infty) = p_a\left\{\frac{2a^2(1-\eta)}{b^2 - a^2} + \frac{\eta}{\log_e(b/a)} + (1 - 2\eta)\right\} \tag{8.96}$$

Eqn. 8.96 is also given by Rice and Cleary (1976, Eqn. 71, p. 237).

8.6.3 Complete Time Solution

The complete solution must incorporate the nonhomogeneous term in the radial governing equation (8.16). The function $f(t)$ is found by first solving Eqn. 8.14 for $\sigma_{\theta\theta}$ and then substituting it into the stress equilibrium equation 8.17:

$$\frac{\partial}{\partial r}(r^2\sigma_{rr}) = r[f(t) - 2\eta p] \tag{8.97}$$

Integrating from $r = a$ to $r = b$ and applying the boundary conditions yield

$$f(t) = \frac{1}{b^2 - a^2}\left\{2a^2 p_a + 4\eta \int_a^b r' p(r', t) dr'\right\} \tag{8.98}$$

Hence, Eqn. 8.16 is an integro-differential equation as $f(t)$ in Eqn. 8.98 contains $p(r, t)$. Profiles of pore pressure, total circumferential stress, and effective circumferential stress are plotted in Figure 8.8. Negative pore pressures due to the undrained response persist through much of the cylinder until $t = 1$ s. A steady-state pressure profile is approached by about $t = 10$ s (Fig. 8.8*a*). The total circumferential stress near the borehole wall increases from its initial value of $0.57p_a$, due to incipient fluid flow from the undrained response (Eqn. 8.86), to a long-term value of $0.78p_a$ (Eqn. 8.96). The total circumferential stress change does not decrease monotonically with radius at short times (Fig. 8.8*b*). However, the effective circumferential stress, $\sigma_{\theta\theta} + p$, does decrease monotonically with radius (Fig. 8.8*c*).

8.7 LINE SOURCE IN PLANE STRAIN

The poroelastic problem of specified radial flow due to a line source in an areally extensive region in plane strain (Rudnicki, 1986a; Helm, 1994) is considered in this section. The flow field is irrotational, with the consequence that pore pressure is proportional to mean stress, and hence to increment of fluid content. The increment of fluid content satisfies the diffusion equation (Eqn. 4.67), with the boundary condition $\lim_{r\to\infty} d\zeta/dr = 0$ and the initial condition $\zeta = 0$. A sudden injection at time zero of a specified volume of fluid per unit length is represented by V_o', and a constant rate of fluid volume injection per unit length is represented by Q_o'. The general solution for the radial component of displacement is given by Eqn. 8.35. The constants C_1 and C_2 must be zero because the displacement is bounded at the origin and as $r \to \infty$. Therefore, Eqn. 8.35 for an infinite problem domain becomes

$$u_r(r, t) = \frac{\gamma}{r}\int_0^r \zeta(r', t) r' dr' \tag{8.99}$$

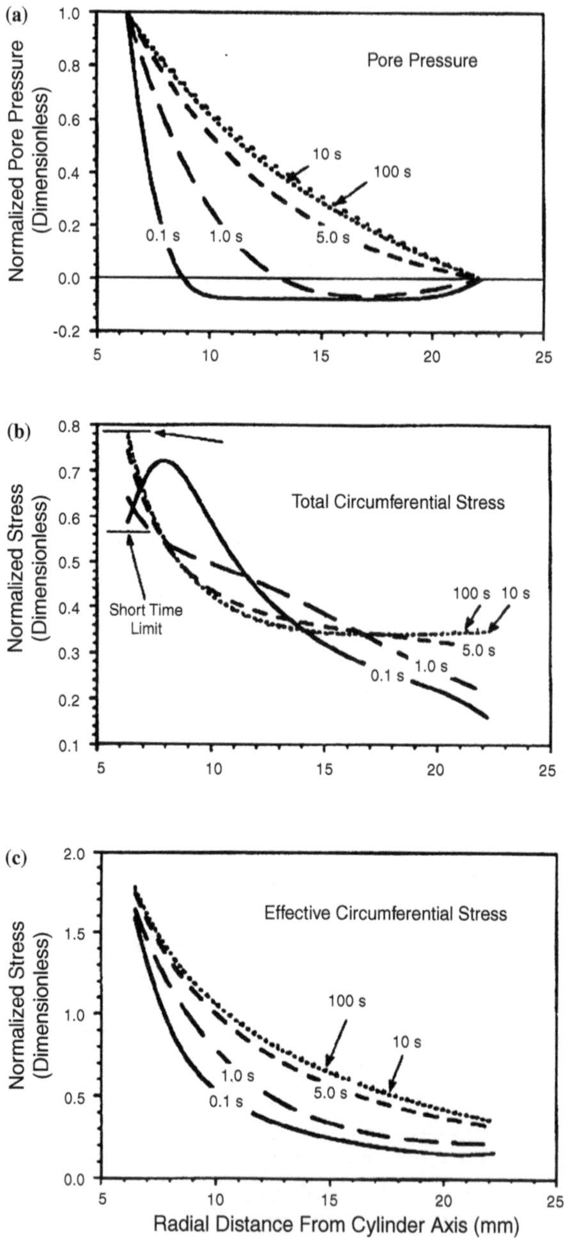

Figure 8.8: Pore pressure (a), circumferential stress (b), and effective circumferential stress (c) in a suddenly pressurized hollow cylinder of inner radius $a = 6.35$ cm and outer radius $b = 22.225$ cm (Schmitt et al., 1993). The material parameters are representative of a kerosene-saturated granite with a permeability of 0.4 microdarcy.

8.7.1 Sudden Line Source

The solution of the diffusion equation for increment of fluid content for sudden injection, V_o', at the origin at time zero (Carslaw and Jaeger, 1959, p. 258) is the two-dimensional equivalent of the solution for the sudden introduction of a point source (Eqn. 5.59),

$$\zeta(r,t) = \frac{V_o'}{4\pi ct} \exp\left(\frac{-r^2}{4ct}\right) \tag{8.100}$$

where $r^2 = x^2 + y^2$.

For irrotational radial flow, the pore pressure is proportional to the increment of fluid content, $\zeta = Sp$ (cf. Section 4.10.4). Hence, substituting this result into Eqn. 8.100 and noting that $c = k/(\mu S)$ yields

$$p(r,t) = \frac{V_o'}{4\pi(k/\mu)t} \exp\left(\frac{-r^2}{4ct}\right) \tag{8.101}$$

Inserting Eqn. 8.100 into Eqn. 8.99 gives

$$u_r = \frac{V_o'}{2\pi}\frac{\gamma}{r}\left\{1 - \exp\left(\frac{-r^2}{4ct}\right)\right\} \tag{8.102}$$

The radial displacement decays inversely with distance for long times. The Cartesian displacement components are obtained from Eqn. 8.102 by multiplying u_r by the direction cosines x_i/r, where x_1 is x and x_2 is y.

8.7.2 Continuous Line Source

The solution of the diffusion equation for ζ for continuous injection Q_o' at the origin beginning at time zero is obtained by superposition of the solution for a sudden impulse source (Eqn. 8.100). The term V_o' is replaced by $Q_o'dt'$, t is replaced by $t - t'$, and the expression is integrated from $t' = 0$ to $t' = t$, with the result that

$$\zeta(r,t) = \frac{Q_o'}{4\pi c} E_1(r^2/4ct) \tag{8.103}$$

where the exponential integral, or "well function" in hydrogeology, is given by

$$E_1(u) = \int_u^\infty \frac{\exp(-\xi)}{\xi} d\xi \tag{8.104}$$

The pressure is obtained from the proportionality $\zeta = Sp$:

$$p(r,t) = \frac{Q'_o}{4\pi(k/\mu)} E_1(r^2/4ct) \tag{8.105}$$

Eqn. 8.105 is the classic Theis solution in well hydraulics for the time-dependent drawdown of a well pumped at a constant volumetric rate $-Q_o$ from a horizontal, confined aquifer of thickness b, where $Q'_o = -Q_o/b$. Thus, the Theis solution is the strictly correct poroelastic solution for assumed plane-strain, radial-flow conditions.

The radial displacement for a continuous source is obtained from the radial displacement for a sudden impulse (Eqn. 8.102) by the same procedure as was used for the increment of fluid content:

$$u_r(r,t) = \frac{Q'_o \gamma}{8\pi c} r \left\{ \frac{4ct}{r^2} \left[1 - \exp(-r^2/4ct)\right] + E_1(r^2/4ct) \right\} \tag{8.106}$$

The Cartesian components of displacement are obtained by multiplying Eqn. 8.106 by the direction cosines x_i/r, where $i = 1, 2$ (cf. Rudnicki, 1986a, Eqn. 50).

The radial displacement is a maximum, and the radial strain is zero at $r_{max} \sim 1.13$ (Helm, 1994, p. 957). For fluid withdrawal from a well, the strain field is compressive for $r < r_{max}$ and is extensional for $r > r_{max}$. Compressional strain occurs when the inward displacement of two nearby points is larger for the farther point, thus leading to a shortening of the line segment connecting them. Extensional strain occurs, conversely, when the inward displacement of the farther point is smaller than for the nearer point, thus leading to a relative lengthening of the line segment connecting them. Horizontal displacements due to a pumping well are discussed further in Chapter 9.

9

Axisymmetry

9.0 CHAPTER OVERVIEW

Axially symmetric problems are independent of the azimuth θ but contain the two spatial coordinates r and z. Although u_r and u_z are the only nonzero displacements for nontorsional problems, there are four nonzero stress and strain components: rr, rz, zz, and $\theta\theta$. Axisymmetric problems discussed in this chapter include (1) vertical deformations due to pumping from a well, (2) reverse well fluctuations, (3) laboratory pulse-transient testing of a cylindrical core, and (4) elastostatic and time-dependent land subsidence due to fluid extraction from a circular reservoir.

9.1 GOVERNING EQUATIONS

For axial symmetry, the four nonzero strain components in cylindrical coordinates are

$$\epsilon_{rr} = \frac{\partial u_r}{\partial r} \tag{9.1}$$

$$\epsilon_{\theta\theta} = \frac{u_r}{r} \tag{9.2}$$

$$\epsilon_{zz} = \frac{\partial u_z}{\partial z} \tag{9.3}$$

$$\epsilon_{rz} = \frac{1}{2}\left(\frac{\partial u_r}{\partial z} + \frac{\partial u_z}{\partial r}\right) \tag{9.4}$$

where u_r is the displacement in the radial direction, and u_z is the displacement in the z-direction. The volumetric strain is therefore given by

$$\begin{aligned}\epsilon &= \epsilon_{rr} + \epsilon_{\theta\theta} + \epsilon_{zz} \\ &= \frac{\partial u_r}{\partial r} + \frac{u_r}{r} + \frac{\partial u_z}{\partial z}\end{aligned} \tag{9.5}$$

The force equilibrium equations (with no body forces) specialized to axisymmetry (cf. Jaeger and Cook, 1976, p. 121) are

$$\frac{\partial \sigma_{rr}}{\partial r} + \frac{\partial \sigma_{rz}}{\partial z} + \frac{\sigma_{rr} - \sigma_{\theta\theta}}{r} = 0 \tag{9.6}$$

$$\frac{\partial \sigma_{rz}}{\partial r} + \frac{\partial \sigma_{zz}}{\partial z} + \frac{\sigma_{rz}}{r} = 0 \tag{9.7}$$

The Navier-type equations for displacement and ζ are obtained as in Chapter 4 (cf. Eqn. 4.21) by substituting the constitutive equations in cylindrical coordinates into Eqns. 9.6 and 9.7 (cf. Rajapakse and Senjuntichai, 1993):

$$\begin{aligned} G\left(\frac{\partial^2 u_r}{\partial r^2} + \frac{1}{r}\frac{\partial u_r}{\partial r} - \frac{u_r}{r^2} + \frac{\partial^2 u_r}{\partial z^2}\right) + \frac{G}{1-2\nu_u}\frac{\partial \epsilon}{\partial r} - BK_u\frac{\partial \zeta}{\partial r} &= 0 \\ G\left(\frac{\partial^2 u_z}{\partial r^2} + \frac{1}{r}\frac{\partial u_z}{\partial r} + \frac{\partial^2 u_z}{\partial z^2}\right) + \frac{G}{1-2\nu_u}\frac{\partial \epsilon}{\partial z} - BK_u\frac{\partial \zeta}{\partial z} &= 0 \end{aligned} \tag{9.8}$$

The third governing equation constraining u_r, u_z, and ζ is the homogeneous diffusion equation (Eqn. 4.67) for ζ.

If p is used in place of ζ, the drained constants replace the undrained constants, α replaces BK_u, and the nonhomogeneous pressure diffusion equation (Eqn. 4.65) replaces the homogeneous diffusion equation for ζ. The axisymmetric displacement equations with pressure as the fluid-coupling variable (Hsieh, 1994) are

$$G\left(\frac{\partial^2 u_r}{\partial r^2} + \frac{1}{r}\frac{\partial u_r}{\partial r} - \frac{u_r}{r^2} + \frac{\partial^2 u_r}{\partial z^2}\right) + \frac{G}{1-2\nu}\frac{\partial \epsilon}{\partial r} - \alpha\frac{\partial p}{\partial r} = 0 \tag{9.9}$$

$$G\left(\frac{\partial^2 u_z}{\partial r^2} + \frac{1}{r}\frac{\partial u_z}{\partial r} + \frac{\partial^2 u_z}{\partial z^2}\right) + \frac{G}{1-2\nu}\frac{\partial \epsilon}{\partial z} - \alpha\frac{\partial p}{\partial z} = 0 \tag{9.10}$$

The axisymmetric form of the fluid continuity equation is

$$\frac{k}{\mu}\left(\frac{\partial^2 p}{\partial r^2} + \frac{1}{r}\frac{\partial p}{\partial r} + \frac{\partial^2 p}{\partial z^2}\right) - \alpha\frac{\partial \epsilon}{\partial t} = S_\epsilon\frac{\partial p}{\partial t} \tag{9.11}$$

9.2 PUMPING FROM A WELL

The radial strain measured near a pumping well shows that horizontal deformations are associated with fluid extraction from an aquifer (Fig. 9.1).

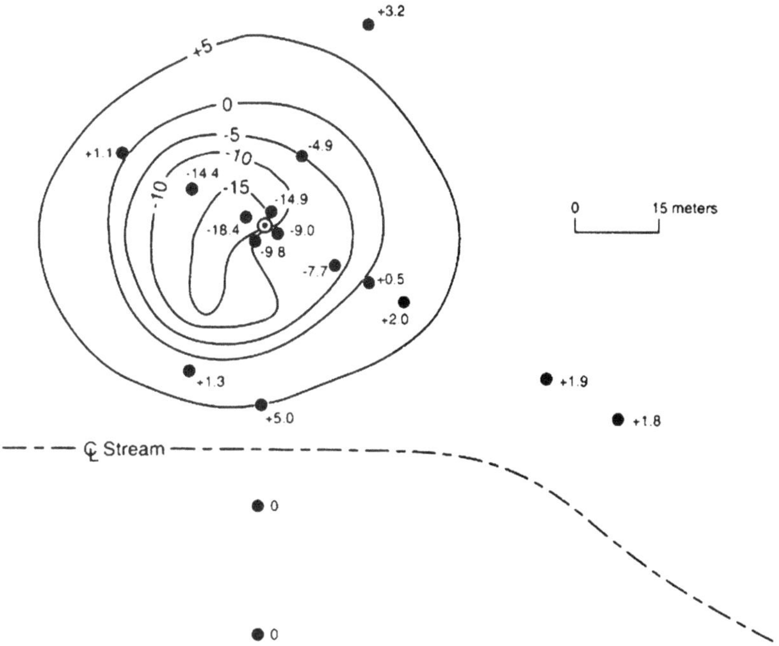

Figure 9.1: Radial strain (in units of 10^{-6}) contours near a pumping well (indicated by a ⊙ near the center of the contours) (Wolff, 1970; redrawn by Helm, 1994). Compressive strains (negative values) occur near the well, and extensional strains (positive values) occur farther away from the well.

Helm (1994) suggested that movement on preexisting fractures may be related to these strains. He calculated horizontal deformations under the assumptions of plane strain and purely radial flow (cf. Section 8.7). Hsieh and Cooley (1995) pointed out that the plane-strain assumption can lead to an overestimate of the radial displacement, because vertical deformations increase the storage capacity of an aquifer. In dimensionless variables, the radial displacement, Eqn. 8.106, becomes

$$u_{rD} = \frac{4t_D}{r_D}\left[1 - \exp\left(-\frac{r_D^2}{4t_D}\right)\right] + r_D E_1\left(\frac{r_D^2}{4t_D}\right) \qquad (9.12)$$

where dimensionless time, radius, and radial displacement are defined by $t_D = ct/b^2$, $r_D = r/b$ and $u_{rD} = 8\pi c u_r/(\gamma Q)$, respectively, and where b is the aquifer thickness and Q is the injection rate (negative for pumping).[1]

[1] Hsieh and Cooley (1995) gave $1/r_D$ rather than r_D as the factor multiplying $E_1(r_D^2/4t_D)$ in Eqn. 9.12. Their figure (Fig. 9.2) is consistent with Eqn. 9.12.

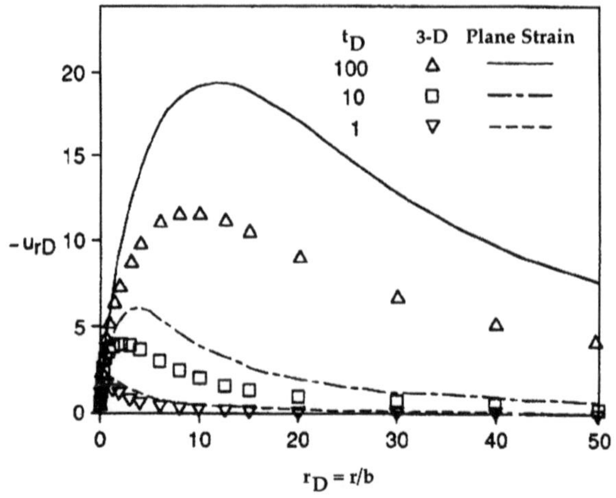

Figure 9.2: Radial profiles of dimensionless horizontal displacement $-u_{rD}$ at dimensionless times $t_D = 1$, 10, and 100 for a pumping well in a vertically deforming aquifer (symbols) versus one in plane strain (solid and dashed lines) (Hsieh and Cooley, 1995, Fig. 4). The negative sign in front of u_{rD} signifies that displacements are toward the well for pumping.

The axisymmetric results for a near-surface aquifer free to deform both vertically and horizontally are compared with the plane-strain results in Figure 9.2.

Gambolati (1977) computed numerically the drawdown versus time for an axisymmetrically deforming shallow aquifer. The cone of depression deepens more slowly than that obtained from the Theis solution (Fig. 9.3) because horizontal deformation contributes to storage capacity. The largest deviations from the Theis solution are for shallow aquifers whose average depth is less than their thicknesses. If the axisymmetric solution is fit to the Theis type curve, the inferred uniaxial storage coefficient overestimates the actual value by approximately 40% for a shallow aquifer because it is being forced to account for horizontal deformation.

9.3 REVERSE WATER-LEVEL FLUCTUATIONS

Reverse water-level fluctuations in wells are rises in water level in an adjacent hydrogeologic unit when a main aquifer is pumped down or, conversely, declines in water level during recovery from pumping. Reverse water-level fluctuations are superficially contrary to intuition. The poroelastic explanation is that drawing down an aquifer produces time-dependent volumetric

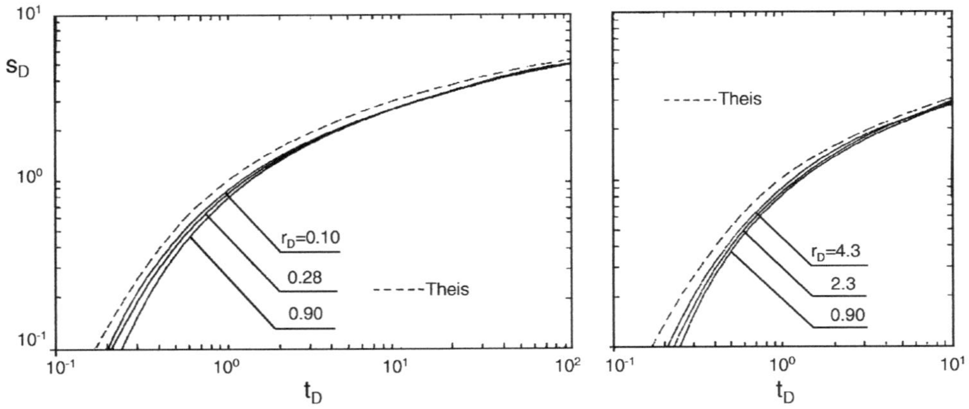

Figure 9.3:. Dimensionless average drawdown ($s_D = 4\pi T\bar{s}/|Q|$) versus dimensionless time ($t_D = ct/r^2$) for several distances ($r_D = r/b$), where $\bar{s}$ is the average drawdown over the aquifer thickness b (Gambolati, 1977, p. 65). The curves apply to aquifers whose average depths are 0.55 times their thicknesses. The deviations from the Theis solution are greatest between $r = 0.9b$ and $r = 2.3b$.

contraction and, hence, induced increases in pore pressure in the aquifer, adjacent confining layers, and adjacent aquifers. The reverse water-level behavior is eventually canceled by pore-pressure diffusion.

9.3.1 Field Observations

The different field situations described in the hydrogeologic literature can be placed in two categories:

1. The reverse water-level response occurs in an aquifer separated from the pumped aquifer by a confining layer (Verruijt, 1969; Yelderman, 1983). Verruijt (1969) called this behavior the *Noordbergum* effect after a village in the Netherlands where a water-level rise of a few centimeters in amplitude persisted for several hundred seconds.
2. The reverse water-level response occurs in a confining layer of the pumped aquifer (Wolff, 1970; Langguth and Treskatis, 1989). In the example described by Wolff, the reverse well fluctuations occurred in a clay bed underlying the pumped aquifer in field tests at Salisbury, Maryland. Langguth and Treskatis used nested pressure transducers to make careful water-level measurements in the marl confining bed overlying the sand aquifer being pumped. At a fixed distance from the pumping well, the reverse water-level fluctuations persisted for longer times at greater vertical distances from the sandy aquifer (Fig. 9.4).

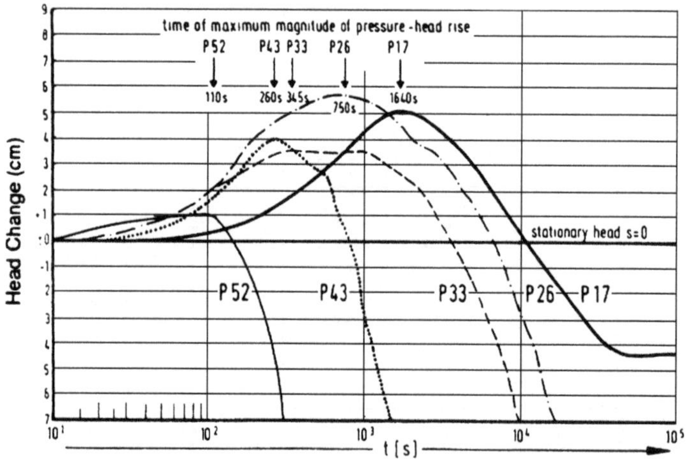

Figure 9.4: Reverse water-level fluctuations observed by Langguth and Treskatis (1989) in a confining bed (Bottrop marl) overlying the pumped aquifer (Haltern sands). The interface is at a depth of 56 meters. The pressure transducers are in the marl at a radial distance of 47 meters. The pressure transducer number is its location in meters below the surface. The arrows indicate the time at which the reverse well fluctuation is greatest.

TABLE 9.1.
Poroelastic Constants in Reverse Water-Level Simulations (Hsieh, 1996)

Layer	Hydraulic Conductivity (m/s)	Specific Storage (m^{-1})	G(GPa)	ν	ϕ
Pumped aquifer	10^{-4}	1.2×10^{-5}	0.3	0.25	0.3
Confining bed	10^{-7}	1.1×10^{-4}	0.03	0.25	0.4

9.3.2 Numerical Simulations

Hsieh (1996) developed a finite-element code to simulate reverse water-level fluctuations for the two situations just described. Aquifer and confining-layer properties (Table 9.1) are typical of relatively unconsolidated sands and clays, respectively, and the water compressibility is taken to be 0.44 GPa^{-1}.

The three-layer model consists of an aquifer confined above and below. The deformation after 10 minutes of pumping is shown highly exaggerated in Figure 9.5. The heads at several depths in the overlying confining layer are shown in Figure 9.6 for a fixed distance from the pumping well. The reverse-well heads are larger and delayed longer for locations vertically farther from the aquifer, except for the shallowest depth due to drainage to the water table (cf. Fig. 9.4). With increasing time, heads decline as pressure diffusion propagates from the aquifer into the confining layer.

9.4 PULSE-DECAY TEST

The measurement of permeability in low-permeability rocks is often made using a transient pulse-decay test (e.g., Brace et al., 1968; Zoback and Byerlee, 1975; Hsieh et al., 1981; Neuzil et al., 1981; Trimmer, 1982; Wang and Hart, 1993). A pressure step, $\Delta p = p_o$, is applied to a reservoir at one end of a cylindrical sample. Measuring the buildup of pressure in a second reservoir at the other end of the sample allows one to determine the permeability and the specific storage (Fig. 9.7). The experiment is normally analyzed as a one-dimensional pore-pressure diffusion problem in the axial direction. However, the passage of a pore-pressure pulse through a cylindrical rock sample induces poroelastic strains, which are in turn coupled to the pore-pressure field. Walder and Nur (1986) recognized that the measured permeability might be affected by poroelastic effects. They measured strains during the pore-pressure decay and looked for sample-size

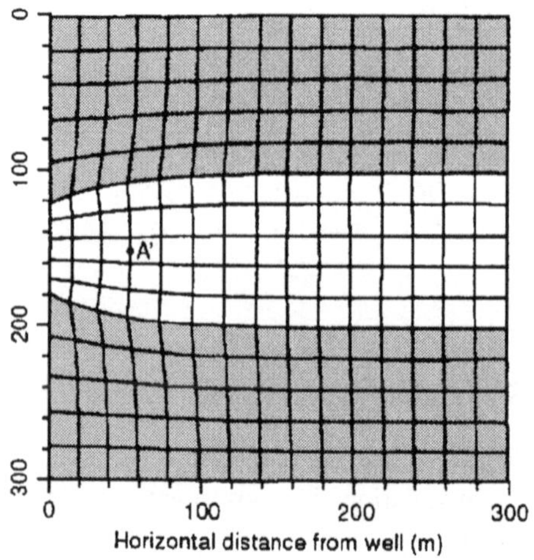

Figure 9.5: Deformation after 10 minutes for a well pumping in a three-layer aquifer-aquitard system. Confining beds are shown slightly shaded. The aquifer is the unshaded region. Before pumping began, each quadrilateral was a 20 × 20 m square along a regular grid. The aquifer has contracted about 1 mm immediately adjacent to the well, and the point A' has moved approximately 0.1 mm toward the well (Hsieh, 1996, Fig. 2*b*). The displacement is exaggerated 40,000 times.

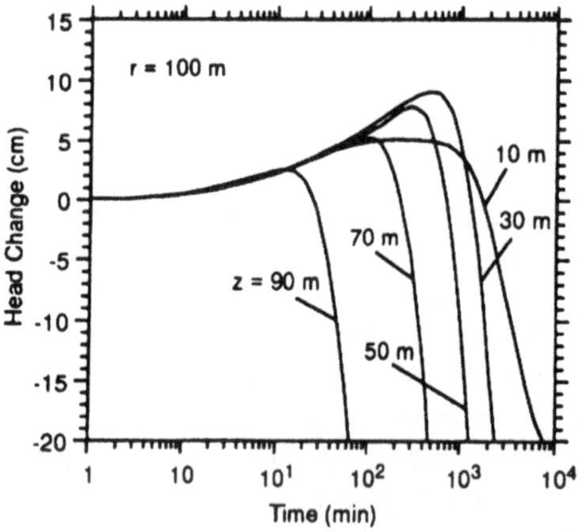

Figure 9.6: Reverse water-level fluctuations at various depths below land surface in the 100-meter-thick confining layer overlying a 100-meter-thick aquifer. The pumping well is located 100 meters away (Hsieh, 1996, Fig. 4).

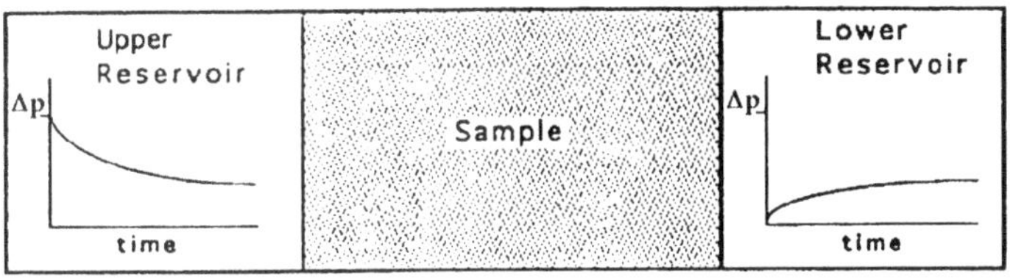

Figure 9.7: Generic pulse transient test to measure the permeability and storage coefficient (Wang and Hart, 1993).

dependence. A consequence of poroelastic coupling is that the pressure field is axisymmetric, not uniaxial; that is, pore pressure varies as functions of both radius and distance along the axis of the cylinder. The relationship between the axisymmetric poroelastic solution and the one-dimensional fluid diffusion solution for a pulse-decay measurement is analogous to that between the poroelastic and Theis solutions for a pumping well.

No analytical solution exists for the fully coupled boundary value problem.[2] The uncoupled thermoelastic problem has been solved analytically by Lee (1966). Hart and Wang (1998) analyzed the coupled problem numerically for a pressure pulse $\Delta p = p_o$ at $z = 0$ and a noflow boundary at $z = L$. The sample is jacketed and surrounded by confining pressure, which is constant throughout the transient pore-pressure buildup, leading to boundary conditions on the sample surfaces,

$$\sigma_{rr} = \sigma_{rz} = 0 \quad \text{at } r = a \tag{9.13}$$

$$\sigma_{zz} = \sigma_{rz} = 0 \quad \text{at } z = 0 \text{ and } z = L \tag{9.14}$$

where a = sample radius and L = sample length. Axial symmetry requires the radial displacement to be zero for $r = 0$:

$$u_r = 0 \text{ at } r = 0 \tag{9.15}$$

Finally, the sample is prevented from translating by fixing the vertical displacement across the bottom of the sample.

Numerical results were obtained using the finite-element codes ABAQUS (Hibbitt, Karlsson and Sorenson, Inc., 1998) or BIOT2 (Hsieh, 1994). At early times, the input pore pressure expands the top of the cylinder, which in turn causes expansion and distortion of a larger volume of the cylinder. A horizontal radial line in the initial cylinder does not remain horizontal, because u_z is nonzero. After a sufficiently long time, the cylinder is again undistorted

[2] Adachi and Detournay (1997) solved the poroelastic problem analytically for an oscillating pore pressure in the approximation of a long cylinder.

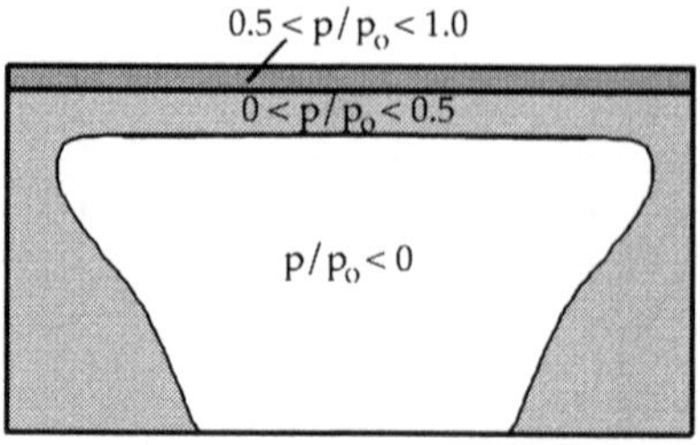

Figure 9.8: Hourglass-shaped region of negative pore pressure in front of the pore-pressure buildup pulse at an early time. Model sample is 2.54 cm in radius and 3.09 cm in length and has the poroelastic properties of Berea sandstone (Hart and Wang, 1998, Fig. 3).

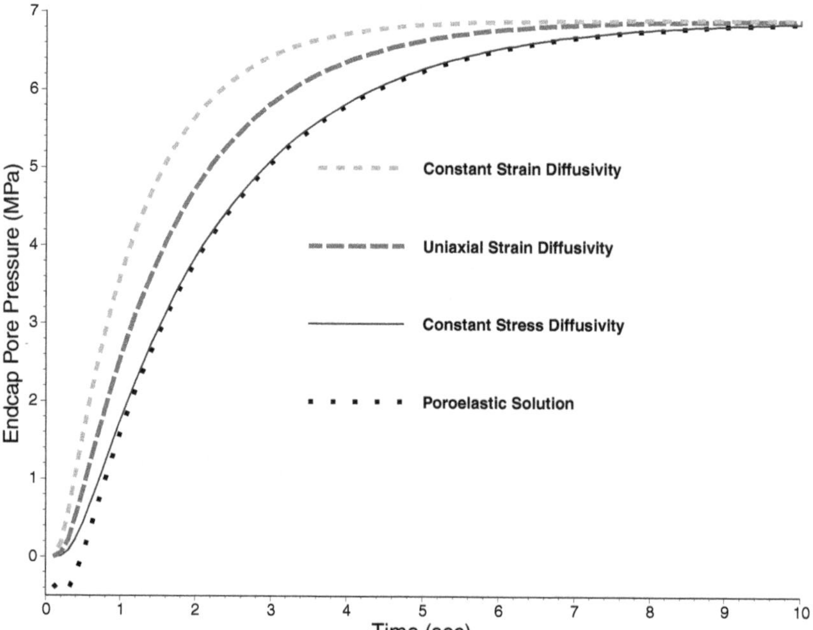

Figure 9.9: Calculated pore pressure versus time in a zero-volume endcap for a step-input pore-pressure pulse of 6.9 MPa. The model is a cylinder 7.62 cm in radius and 5.08 cm long with the poroelastic properties of Berea sandstone (K = 6.6 GPa, K_f = 2.3 GPa, ν = 0.17, α = 0.764, B = 0.676, $k = 10^{-16}$ m^2). The dashed and solid curves represent the solution of the pressure diffusion equation for diffusivities based on constant strain and one-dimensional and three-dimensional storage coefficients, respectively. The individual dots represent the poroelastic solution on the axis of the sample.

but larger by the free strain $(\alpha/K)p_o$. The strain in each element in a horizontal row is slightly different such that the induced pore pressure also varies slightly as a function of radius. A region ahead of the pore-pressure pulse expands and induces a decline in pore pressure (Fig. 9.8). For a relatively short sample, this region can extend to the end of the sample, as shown by negative charges in pore pressure for short times (Fig. 9.9). Also plotted in Figure 9.9 are the solutions of the one-dimensional pressure-diffusion equation for hydraulic diffusivities corresponding to storage coefficients for conditions of constant strain, uniaxial strain, and constant stress. The poroelastic solution is approximated by the curve for constant-stress diffusivity after the initial negative pressures, as would be expected given constant-stress boundary conditions.

The strain induced by the passage of the pore-pressure pulse is shown in Figure 9.10 for the circumferential and axial directions on the cylindrical surface at the midpoint of the sample length. For short times, the main pore-pressure buildup is near the inlet end and has not reached the middle of the sample. The negative axial strain, however, indicates that poroelastic effects have shortened the sample. Such behavior can be predicted only from a

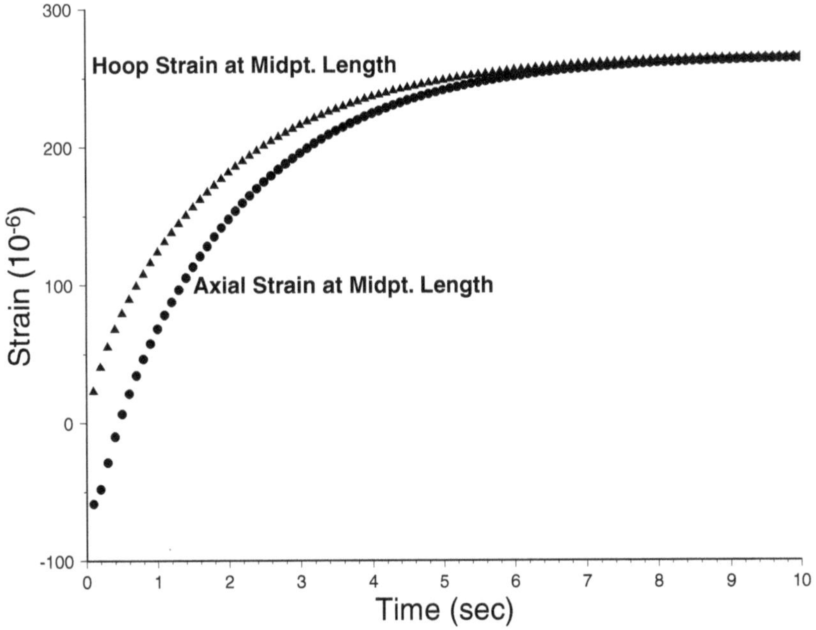

Figure 9.10: Calculated circumferential (hoop) and axial strains on the surface of a cylinder of Berea sandstone at the midpoint of its length. The input pore-pressure step and properties are the same as in Figure 9.9.

fully coupled analysis. At long times, a uniform pore pressure $\Delta p = p_o$ develops throughout the sample, and both the axial and circumferential strains approach the free-strain value, $\alpha p_o/(3K)$.

9.5 ELASTOSTATIC SUBSIDENCE OF A HALF SPACE

Segall (1992) obtained elastostatic Green's functions for displacement due to a pressure source within a flat, circular ring. The surface subsidence for any axially symmetric pressure distribution can then be obtained by integration. The displacements and stresses are those in instantaneous mechanical equilibrium with the pressure distribution. In particular, the solution for a thin, circular disk is presented.

The solution method is first to obtain the infinite-space Green's function for potential and then to obtain the displacement using Eqn. 5.35. Poisson's equation for a circular-ring pressure source is

$$\nabla^2 \Phi = c_m \delta(r - r')\delta(z - z') \tag{9.16}$$

where r' is the radius of the ring and z' is the depth of the ring. Eqns. A.6 and A.23 (see Appendix A) give

$$\nabla^2 \bar{\Phi} = -(k^2 + \xi^2)\bar{\Phi} \tag{9.17}$$

where the overbar denotes the combined Fourier cosine and Hankel transforms of Φ. Taking also the combined Fourier cosine and Hankel transforms of the right-hand side of Eqn. 9.16 and inverting gives

$$\Phi(r, z; r', z') = -\frac{2c_m r'}{\pi} \int_o^\infty \int_o^\infty \frac{k J_o(kr) J_o(kr') \cos \xi z \cos \xi z'}{k^2 + \xi^2} d\xi dk \tag{9.18}$$

Similarly, using the Fourier sine transform gives

$$\Phi(r, z; r', z') = -\frac{2c_m r'}{\pi} \int_o^\infty \int_o^\infty \frac{k J_o(kr) J_o(kr') \sin \xi z \sin \xi z'}{k^2 + \xi^2} d\xi dk \tag{9.19}$$

Adding Eqns. 9.18 and 9.19 gives

$$\begin{aligned} \Phi(r, z; r', z') &= -\frac{c_m r'}{\pi} \int_o^\infty \int_o^\infty \frac{k J_o(kr) J_o(kr') \cos \xi (z - z')}{k^2 + \xi^2} d\xi dk \\ &= -\frac{c_m r'}{2} \int_o^\infty J_o(kr) J_o(kr') \exp(-k|z - z'|) dk \end{aligned} \tag{9.20}$$

where the second equality in Eqn. 9.20 follows from the definite integral (e.g., Peirce and Foster, 1956, No. 505):

$$\int_o^\infty \frac{\cos ax}{1+x^2}dx = \frac{\pi}{2}\exp(-a) \quad a > 0 \tag{9.21}$$

Using Eqn. 9.20 in Eqn. 5.35 and multiplying by the magnitude of the pressure source p_o give the radial and vertical displacements:

$$\begin{aligned} u_r^{\text{ring}}(r,z) = \frac{c_m p_o r'}{2} \int_o^\infty & kJ_1(kr)J_o(kr') \\ & \times\{\exp(-k|z-z'|) \\ & \quad + (3-4\nu-2kz)\exp(-k(z+z'))\}dk \end{aligned} \tag{9.22}$$

$$\begin{aligned} u_z^{\text{ring}}(r,z) = -\frac{c_m p_o r'}{2} \int_o^\infty & kJ_o(kr)J_o(kr') \\ & \times\left\{-\frac{|z-z'|}{z-z'}\exp(-k|z-z'|)\right. \\ & \quad \left. + (3-4\nu+2kz)\exp(-k(z+z'))\right\}dk \end{aligned} \tag{9.23}$$

The displacements at the free surface ($z = 0$) are

$$u_r^{\text{ring}}(r,0) = 2c_m p_o(1-\nu)r' \int_o^\infty kJ_1(kr)J_o(kr')\exp(-kz')dk \tag{9.24}$$

$$u_z^{\text{ring}}(r,0) = -2c_m p_o(1-\nu)r' \int_o^\infty kJ_o(kr)J_o(kr')\exp(-kz')dk \tag{9.25}$$

The integrals in Eqns. 9.24 and 9.25 can be written in the form of elliptic integrals (Geertsma, 1973b; Segall, 1992). Eqns. 9.24 and 9.25 are portrayed as radial profiles in Figure 9.11 for a source radius $r' = 3$ and depths of $z' = 1$, 2 and 4. For $z' = 1$, points on the surface within a radius of about $r = 2.6$ show inward displacement. For $z' = 4$, all the displacements at the surface are outward.

The half-space solution for surface displacements above a circular disk of radius a and thickness h, which undergoes a uniform pressure change p_o, can be obtained by integrating Eqns. 9.24 and 9.25 from 0 to a with respect to r', and from $z' - h/2$ to $z' + h/2$ with respect to z. The z-integration is equivalent to multiplication by h. The r'-integration of $kr'J_o(kr')$ is obtained using the following identity (Carslaw and Jaeger, 1959, p. 489):

$$zJ_1'(z) + J_1(z) = zJ_o(z) \tag{9.26}$$

That is,

$$\frac{d}{dz}\left[zJ_1'(z)\right] = zJ_o(z) \tag{9.27}$$

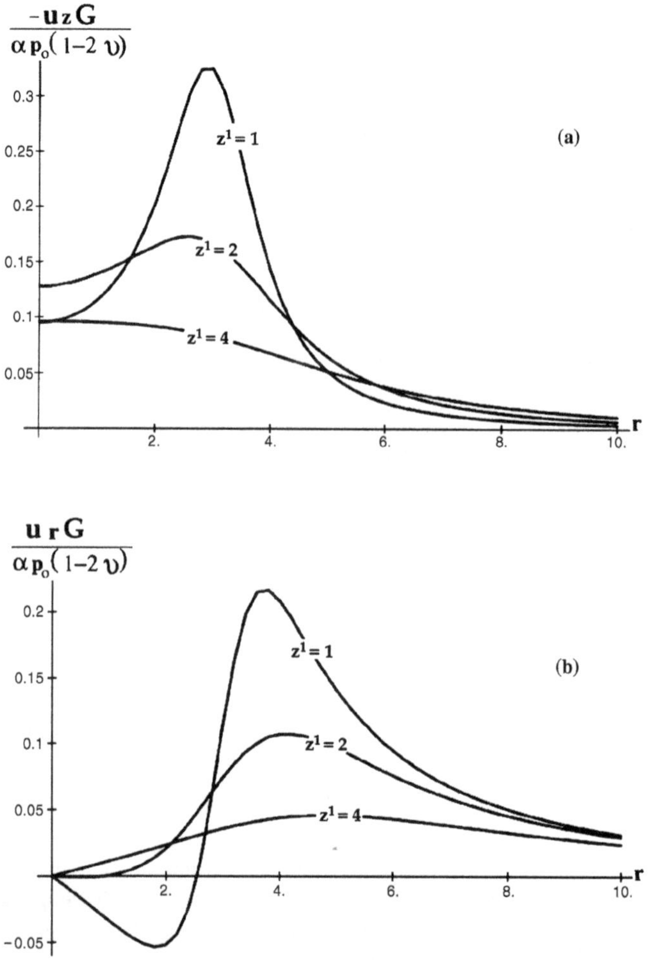

Figure 9.11: Displacements at the surface for a circular ring source of radius $r' = 3$. Profiles are shown for ring depths of $z' = 1$, 2, and 4. (a) Vertical displacement (Segall, 1992, Fig. 2). (b) Radial displacement (Segall, 1992, Fig. 4).

Therefore,

$$\int_0^a kr' J_o(kr')dr' = aJ_1(ka) \tag{9.28}$$

Finally, the displacements for a circular-disk pressure source of radius a and thickness h are

$$u_r^{\text{disk}}(r,0) = 2c_m(1-\nu)p_o ha \int_o^\infty J_1(kr)J_1(ka)\exp(-kz')dk \tag{9.29}$$

$$u_z^{\text{disk}}(r, 0) = -2c_m(1-\nu)p_o ha \int_o^\infty J_o(kr) J_1(ka) \exp(-kz')dk \quad (9.30)$$

Geertsma (1973a) showed that Eqn. 9.30 reduces to Eqn. 5.57 directly over the center of the disk ($r = 0$). Geertsma showed further that the reservoir compaction for a thin disk is uniform across the disk:

$$u_z(r, z' - h/2) - u_z(r, z' + h/2) = -c_m p_o h \quad (9.31)$$

Negative pressure changes ($p_o < 0$) produce positive compaction values. Eqns. 5.57 and 9.31 can be used to obtain the ratio of surface subsidence to compaction for different disk thicknesses and radii.

The Green's functions for the strain components due to a ring pressure source can be obtained by differentiation of the displacements, and the Green's functions for the stress components can be obtained from the constitutive equations. The formulas can be found in Segall (1992).

9.6 TIME-DEPENDENT SUBSIDENCE OF A HALF SPACE

Rajapakse and Senjuntichai (1993) solved the three-dimensional, time-dependent problem of an axisymmetric, ring fluid source, which is constant with time, in a half space with a stress-free and drained surface. The result can be used to obtain the time-dependent solution for sudden withdrawal of a fluid volume from a disk. The solution method is to use the Laplace transform for the time coordinate and the zeroth-order and first-order Hankel transforms for the radial coordinate. The procedural steps are as follows:

1. The Laplace transform followed by the zeroth-order Hankel transform in r is applied to the diffusion equation $c\nabla^2\zeta = \partial\zeta/\partial t$. Assuming axisymmetry and using Eqn. A.23 (see Appendix A) leads to the ordinary differential equation

$$\left[\frac{d^2}{dz^2} - k^2\right]\bar{\zeta} = \frac{s}{c}\bar{\zeta} \quad (9.32)$$

where

$$\bar{\zeta}(k, z, s) \equiv \mathcal{H}_o\{\mathcal{L}\{\zeta(r, z, t)\}\} \quad (9.33)$$

The general solution of Eqn. 9.33 is

$$\bar{\zeta}(k, z, s) = A_o(k, s)e^{lz} + B_o(k, s)e^{-lz} \quad (9.34)$$

where $l = (k^2 + s/c)^{1/2}$, and $A_o(k,s)$ and $B_o(k,s)$ are zeroth-order Fourier coefficients in an expansion with respect to theta. Only the zeroth-order harmonic is nonzero for axisymmetry. The coefficients are determined from the boundary conditions.

2. Eqn. A.23 (see Appendix A) is applied to $\nabla^2\epsilon = \gamma\nabla^2\zeta$ (Eqn. 4.25),

$$\left[\frac{d^2}{dz^2} - k^2\right]\bar{\epsilon} = \gamma\left[\frac{d^2}{dz^2} - k^2\right]\bar{\zeta} \tag{9.35}$$

where

$$\bar{\epsilon}(k,z,s) \equiv \mathcal{H}_o\{\mathcal{L}\{\epsilon(r,z,t)\}\} \tag{9.36}$$

Eqn. 9.34 is then substituted for $\bar{\zeta}$:

$$\left[\frac{d^2}{dz^2} - k^2\right]\bar{\epsilon} = \frac{\gamma s}{c}[A_o(k,s)e^{lz} + B_o(k,s)e^{-lz}] \tag{9.37}$$

The general solution for $\bar{\epsilon}$ is found using the method of undetermined coefficients (Boyce and DiPrima, 1977, p. 115),

$$\begin{aligned}\bar{\epsilon}(k,z,s) = \gamma\left[A_o(k,s)e^{lz} + B_o(k,s)e^{-lz}\right]\\ +C_o(k,s)e^{kz} + D_o(k,s)e^{-kz}\end{aligned} \tag{9.38}$$

where $C_o(k,s)$ and $D_o(k,s)$ are again coefficients determined by boundary conditions.

3. The first-order Hankel transform of the Navier-type equation for $\tilde{u}_r = \mathcal{L}\{u_r\}$ (Eqn. 9.8) is

$$(1-2\nu_u)\left[\frac{d^2\bar{u}_r}{dz^2} - k^2\bar{u}_r\right] = k\bar{\epsilon} - \frac{2B(1+\nu_u)}{3}k\bar{\zeta} \tag{9.39}$$

where $\bar{u}_r \equiv \mathcal{H}_1\{\tilde{u}_r\}$. Eqn. A.25 (see Appendix A) was applied to obtain Eqn. 9.39, such that $\bar{\epsilon}$ and $\bar{\zeta}$ are zeroth-order Hankel transforms, consistent with their definitions in Eqns. 9.33 and 9.36. A particular solution of Eqn. 9.39 is found by the method of undetermined coefficients after substituting Eqns. 9.34 and 9.38 into Eqn. 9.39. Adding the homogeneous solution gives the general solution

$$\begin{aligned}\bar{u}_r(k,z,s) = -\frac{k\gamma}{s/c}\left(A_oe^{lz} + B_oe^{-lz}\right) + a_1z(C_oe^{kz} - D_oe^{-kz})\\ +E_oe^{kz} + F_oe^{-kz}\end{aligned} \tag{9.40}$$

where

$$a_1 = \frac{1}{2(1 - 2\nu_u)} \tag{9.41}$$

and E_o and F_o are additional coefficients to be determined eventually by boundary conditions. Inverting Eqn. 9.40 gives the Laplace transform of u_r:

$$\tilde{u}_r(r, z, s) = \int_0^\infty \left\{ \left[-\frac{k\gamma}{s/c}(A_o e^{lz} + B_o e^{-lz}) + a_1 z (C_o e^{kz} - D_o e^{-kz}) \right] + E_o e^{kz} + F_o e^{-kz} \right\} k J_1(kr) dk \tag{9.42}$$

4. Inserting Eqn. 9.5 into Eqn. 9.36 gives

$$\begin{aligned} \bar{\epsilon}(k, z, s) &= \mathcal{H}_o \left\{ \frac{\partial \tilde{u}_r}{\partial r} + \frac{\tilde{u}_r}{r} + \frac{\partial \tilde{u}_z}{\partial z} \right\} \\ &= k\mathcal{H}_1 \{\tilde{u}_r\} + \frac{d\bar{u}_z}{dz} \end{aligned} \tag{9.43}$$

where $\bar{u}_z \equiv \mathcal{H}_o\{\tilde{u}_z\}$. The second equality in Eqn. 9.43 is obtained using the Hankel transform relation Eqn. A.26 (see Appendix A). Note that a tilde is used in Eqn. 9.43 to represent a Laplace transform. An ordinary first-order linear differential equation can then be obtained for $\bar{u}_z$ by substituting Eqns. 9.38 and 9.40 into Eqn. 9.43. The general solution is

$$\begin{aligned} \bar{u}_z(k, z, s) = \frac{l\gamma}{s/c}(A_o e^{lz} - B_o e^{-lz}) - \left(a_1 z - \frac{a_2}{k}\right) C_o e^{kz} \\ - \left(a_1 z + \frac{a_2}{k}\right) D_o e^{-kz} - E_o e^{kz} + F_o e^{-kz} \end{aligned} \tag{9.44}$$

where

$$a_2 = \frac{3 - 4\nu_u}{2(1 - 2\nu_u)} \tag{9.45}$$

The inverse Hankel transform gives the Laplace transform of u_z:

$$\begin{aligned} \tilde{u}_z(r, z, s) = \int_0^\infty \left\{ \frac{l\gamma}{s/c}(A_o e^{lz} - B_o e^{-lz}) - \left(a_1 z - \frac{a_2}{k}\right) C_o e^{kz} \right. \\ \left. - \left(a_1 z + \frac{a_2}{k}\right) D_o e^{-kz} - E_o e^{kz} + F_o e^{-kz} \right\} \\ \times k J_o(kr) dk \end{aligned} \tag{9.46}$$

5. The Laplace transforms of the stress components are expressed in terms of the coefficients $A_o, \ldots, F_o$ by inserting the Laplace transforms of the displacement components, volumetric strain, and increment of fluid content into the constitutive equations. The Laplace transforms of the stress components σ_{zz} and σ_{zr}, which appear in the boundary conditions, are

$$\frac{\tilde{\sigma}_{zz}(r,z,s)}{2G} = \int_0^\infty \left\{ \frac{k^2\gamma}{s/c}(A_o e^{lz} + B_o e^{-lz}) - (a_1 kz - a_4)C_o e^{kz} + (a_1 kz + a_4)D_o e^{-kz} - kE_o e^{kz} - kF_o e^{-kz} \right\} \times kJ_o(kr)dk \quad (9.47)$$

$$\frac{\tilde{\sigma}_{zr}(r,z,s)}{2G} = \int_0^\infty \left\{ -\frac{kl\gamma}{s/c}(A_o e^{lz} - B_o e^{-lz}) + (2a_1 kz - 1)C_o e^{kz} + (2a_1 kz + 1)D_o e^{-kz} + kE_o e^{kz} - kF_o e^{-kz} \right\} \times kJ_1(kr)dk \quad (9.48)$$

where

$$a_4 = \frac{1 - \nu_u}{1 - 2\nu_u} \quad (9.49)$$

Rajapakse and Senjuntichai (1993) solved the more general problem of an azimuthally varying source. As a result, some subscript numbers for the constants a_i and b_i are missing in the axisymmetric treatment to keep the definitions here the same as in their paper.

6. The zeroth-order Hankel transform of the Laplace transform of the normalized fluid pressure, $\bar{p}/2G$, is obtained from the constitutive equation (cf. Eqn. 2.21):

$$\frac{\bar{p}(k,z,s)}{2G} = -\frac{B(1+\nu_u)}{3(1-2\nu_u)}\bar{\epsilon} + \frac{B^2(1-2\nu)(1+\nu_u)^2}{9(1-2\nu_u)(\nu_u-\nu)}\bar{\zeta} \quad (9.50)$$

where $\bar{p} \equiv \mathcal{H}_o\{\tilde{p}\}$. Substituting Eqns. 9.33 and 9.38 and inverting give

$$\frac{\tilde{p}(r,z,s)}{2G} = \int_0^\infty \{a_5\gamma(A_o e^{lz} + B_o e^{-lz}) - a_4\gamma(C_o e^{kz} + D_o e^{-kz})\} kJ_o(kr)dk \quad (9.51)$$

where

$$a_5 = \frac{B(1-\nu)(1+\nu_u)}{3(\nu_u - \nu)} \quad (9.52)$$

7. Finally, consider a continuous fluid source $\delta(r - r')H(t)$ in the plane $z = z'$, where $H(t)$ is the Heaviside step function. The half space is divided into two problem domains: (1) $0 \le z \le z'$ and (2) $z' \le z < \infty$. The general solutions just given for stress, displacement, and pore pressure apply for each domain, leading to a total of 12 unknown coefficients, $A_o^{(1)}, \ldots, F_o^{(1)}$ and $A_o^{(2)}, \ldots, F_o^{(2)}$, where the superscript indicates the domain number. For the lower domain, the coefficients $A_o^{(2)}$, $C_o^{(2)}$, and $E_o^{(2)}$ must be zero to keep the solution bounded as $z \to \infty$. The remaining nine coefficients are determined from the following boundary and continuity equations:

$$
\begin{aligned}
\tilde{\sigma}_{zz}^{(1)}(r, 0, s) &= 0 \\
\tilde{\sigma}_{zr}^{(1)}(r, 0, s) &= 0 \\
\tilde{p}^{(1)}(r, 0, s) &= 0 \\
\tilde{u}_z^{(1)}(r, z', s) - \tilde{u}_z^{(2)}(r, z', s) &= 0 \\
\tilde{u}_r^{(1)}(r, z', s) - \tilde{u}_r^{(2)}(r, z', s) &= 0 \\
\tilde{p}^{(1)}(r, z', s) - \tilde{p}^{(2)}(r, z', s) &= 0 \\
\tilde{\sigma}_{zz}^{(1)}(r, z', s) - \tilde{\sigma}_{zz}^{(2)}(r, z', s) &= 0 \\
\tilde{\sigma}_{zr}^{(1)}(r, z', s) - \tilde{\sigma}_{zr}^{(2)}(r, z', s) &= 0 \\
\left\{ \frac{k}{\mu} \frac{\partial \tilde{p}^{(1)}(r, z', s)}{\partial z} \right\} - \left\{ \frac{k}{\mu} \frac{\partial \tilde{p}^{(2)}(r, z', s)}{\partial z} \right\} &= \frac{\delta(r - r')}{s}
\end{aligned}
\tag{9.53}
$$

Applying the zeroth-order Hankel transform to $\delta(r - r')$ and inverting give the following integral representation to insert in the fluid-continuity equation in Eqn. 9.53:

$$
\delta(r - r') = \int_o^\infty kr' J_o(kr) J_o(kr') dk \tag{9.54}
$$

The nine nonzero coefficients in the general expressions for stresses, displacements, and pore pressure can then be found by solving the nine algebraic equations represented by Eqn. 9.53. For completeness, all 12 coefficients are presented here:

$$
A_o^{(1)}(k, s) = \frac{e^{-lz'}}{4a_5 l \gamma G(\kappa/\mu) s} r' J_o(kr')
$$

$$
A_o^{(2)}(k, s) = 0
$$

$$
B_o^{(1)}(k, s) = \left(\frac{2b_5 l e^{-kz'} - b_2 e^{-lz'}}{4a_5 b_1 l \gamma G(\kappa/\mu) s} \right) r' J_o(kr')
$$

$$
\begin{aligned}
B_o^{(2)}(k,s) &= B_o^{(1)}(k,s) + A_o^{(1)}(k,s)e^{2lz'} \\
C_o^{(1)}(k,s) &= 0 \\
C_o^{(2)}(k,s) &= 0 \\
D_o^{(1)}(k,s) &= b_5\left(\frac{e^{-kz'} - e^{-lz'}}{2a_4 b_1 \gamma G(\kappa/\mu)s}\right) r' J_o(kr') \\
D_o^{(2)}(k,s) &= D_o^{(1)}(k,s) \\
E_o^{(1)}(k,s) &= \frac{e^{-lz'}}{4a_5 G(\kappa/\mu)s^2/c} r' J_o(kr') \\
E_o^{(2)}(k,s) &= 0 \\
F_o^{(1)}(k,s) &= \left(\frac{b_8 e^{-kz'} - 2b_9 e^{-lz'}}{4a_5 b_1 G(\kappa/\mu)s^2/c}\right) r' J_o(kr') \\
F_o^{(2)}(k,s) &= F_o^{(1)}(k,s) + E_o^{(1)}(k,s)e^{2kz'}
\end{aligned}
\tag{9.55}
$$

where

$$
\begin{aligned}
b_1 &= a_5(s/c) + 2(1-\nu_u)k\gamma(k-l) \\
b_2 &= a_5(s/c) + 2(1-\nu_u)k\gamma(k+l) \\
b_5 &= 2(1-\nu_u)k\gamma \\
b_8 &= 2(1-\nu_u)k\gamma(k+l) + (3-4\nu_u)a_5(s/c) \\
b_9 &= 2(1-\nu_u)(k^2\gamma + a_5(s/c))
\end{aligned}
\tag{9.56}
$$

Substituting the coefficients from Eqn. 9.55 into Eqns. 9.42, 9.46, and 9.51 gives the Laplace transforms of the displacements and fluid pressure. They can be inverted using the Stehfest (1970) algorithm (cf. Appendix A) to obtain time-domain results. Computationally, the semi-infinite integrals involving products of Bessel functions must be evaluated at N discrete values of s for each time t. Rajapakse and Senjuntichai (1993) found that the integrals could be done accurately using an adaptive version of the extended trapezoidal formula with $\Delta k = 0.1$. The time-domain solutions could then be obtained from the Stehfest algorithm with $N \geq 8$.

9.6.1 Instantaneous Ring Source

A circular-ring fluid source, $\zeta = \zeta_o \delta(r - r')\delta(z - z')\delta(t)$, represents the instantaneous injection of a volume of fluid ζ_o per unit aquifer volume into a circle of radius r' lying in the plane $z = z'$. Integrating across the $z = z'$ plane gives $\zeta_o \delta(r - r')\delta(t)$, which is the fluid volume added per unit area in the plane. The time dependence of the instantaneous ring source is represented by $\delta(t)$, whereas that of the continuous ring source is represented by $H(t)$. Because $\mathcal{L}(\delta(t)) = 1$ and $\mathcal{L}(H(t)) = 1/s$, the instantaneous ring-source solution for an increment of fluid content of magnitude ζ_o can be obtained expediently from the continuous ring-source solution by multiplying each coefficient in Eqn. 9.55 by $\zeta_o s$.

The elastostatic solutions for u_r and u_z for a circular ring source are recovered from the time-dependent solutions in the limit $t = 0^+$ by finding $s\mathcal{L}\{u_r\}$ and $s\mathcal{L}\{u_z\}$, respectively, as $s \to \infty$ (cf. Appendix A, Eqn. A.16). The results are identical to Eqns. 9.22 and 9.23, respectively, except that $\gamma\zeta_o$ replaces $c_m p_o$, because a source of increment of fluid content was considered in this section, whereas a pressure source was considered in the previous section.

9.6.2 Continuous Disk Source

The solution for a circular-disk fluid source of radius a and uniform intensity q_o (volume of fluid per unit time per unit area) in the plane $z = z'$ beginning at time zero is readily obtained by integrating the ring solution from 0 to a with respect to r' and multiplying by q_o. The only dependence on r' in Eqn. 9.55 is in the term $r'J_o(kr')$. The identity, Eqn. 9.28, shows that the solution for a circular-disk fluid source can be obtained by replacing $r'J_o(kr')$ in Eqn. 9.55 with $q_o a J_1(ka)/k$. Figure 9.12 shows the dimensionless displacement at the earth's surface directly over the center of a circular reservoir of

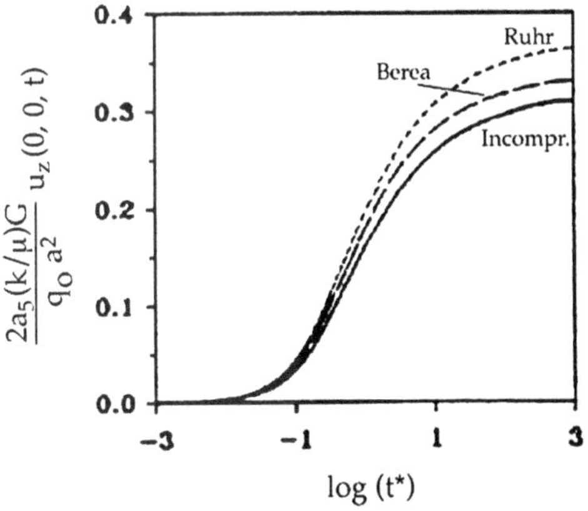

Figure 9.12: Surface displacement over the reservoir axis ($r = z = 0$) versus the logarithm of dimensionless time $t^* = ct/a^2$ for a circular reservoir whose depth is equal to its radius (Rajapakse and Senjuntichai, 1993). Fluid is withdrawn at a uniform rate, $-q_o$.

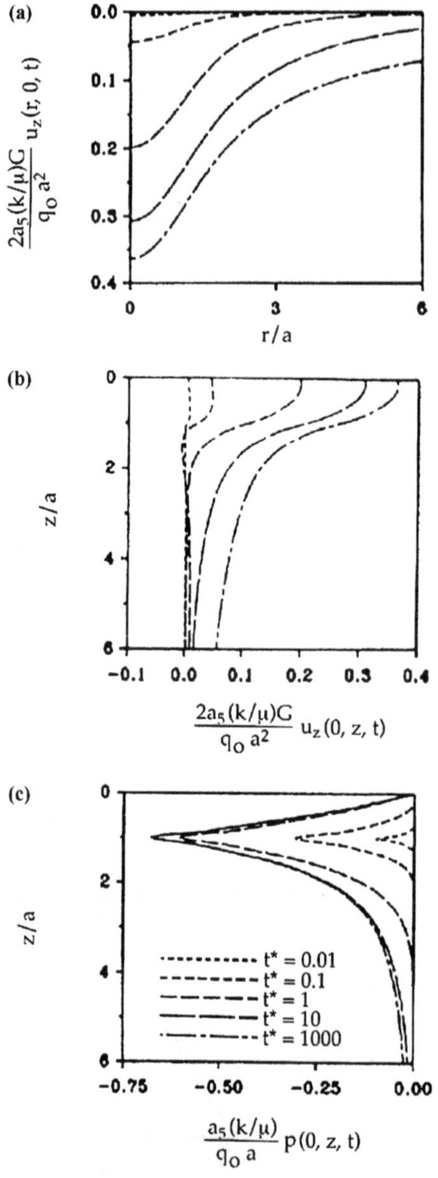

Figure 9.13: Displacement and pore-pressure profiles over a circular reservoir from which fluid is withdrawn at a uniform rate, $-q_o$, at dimensionless times $t^* = ct/a^2$ ranging from $t^* = 0.01$ to $t^* = 1000$ (Rajapakse and Senjuntichai, 1993). Properties are those of Ruhr sandstone. (a) Horizontal profiles of vertical displacement at the free surface. (b) Vertical profiles of vertical displacement. (c) Vertical profiles of fluid pressure.

unit radius at unit depth versus the logarithm of dimensionless time. Results are shown for a material with incompressible constituents, one with the properties of Berea sandstone, and one with the properties of Ruhr sandstone (cf. Appendix C, Tables C.1). The displacement history is characteristic of consolidation behavior with an initial slow increase, an intermediate period of rapid subsidence, and a gradual slowing at later times. Larger vertical displacements are obtained for smaller values of drained Poisson's ratio. Horizontal profiles of vertical displacements at the free surface are shown for several dimensionless times in Figure 9.13*a*. The subsidence bowl deepens and widens, as would be expected. Finally, vertical profiles of vertical displacement and fluid pressure are shown for several dimensionless times in Figure 9.13*b* and *c*, respectively.

9.6.3 Continuous Point Source

The solution for a continuous point-fluid source of intensity Q_o (volume of fluid per unit time) beginning at time zero is obtained from the circular-disk solution by treating it as a continuous disk source of strength $q_o = Q_o/(\pi a^2)$ and taking the limit as $a \to 0$. The circular-disk solution was obtained from Eqn. 9.55 by replacing $r' J_o(kr')$ with $q_o a J_1(ka)/k$. Therefore, the solution for a continuous point source is obtained by replacing $r' J_o(kr')$ with $Q_o/(2\pi)$ because $J_1(ka) \to ka/2$ as $a \to 0$ (Kraut, 1967, p. 270).

9.6.4 Instantaneous Point Source

The solution for an instantaneous point source (slug of volume V_o injected at a point $(0, z')$) is obtained from the solution for a continuous point source by replacing Q_o with sV_o. This same procedure was invoked previously to obtain the solution for an instantaneous ring source from the solution for a continuous ring source. Hence, the result of the previous paragraph shows that the solution for the instantaneous point source is obtained by replacing $r' J_o(kr')$ in Eqn. 9.55 with $sV_o/(2\pi)$.

10

Numerical Methods

10.0 CHAPTER OVERVIEW

Practical poroelastic problems frequently possess complicated geometries and material-property heterogeneities, which may not be idealized readily by an analytical solution. Furthermore, finding analytical solutions can be difficult. Even when they exist, the solutions themselves might require numerical evaluation because they are often in the form of an infinite series or an integral transform. Thus, for a variety of reasons, a general computer code implementing a numerical solution of the governing equations of poroelasticity can be an all-purpose tool for solving a wide range of poroelastic problems. In this chapter the finite-element and boundary-element methods are outlined for the case of plane strain. The discussion assumes familiarity with these methods in problems of standard elasticity (e.g., Burnett, 1987; Smith and Griffiths, 1988; Zienkiewicz and Taylor, 1989; and Fagan, 1992, for the finite-element method; and Banerjee and Butterfield, 1981; Crouch and Starfield, 1983; and Kane, 1994, for the boundary-element method).

10.1 FINITE-ELEMENT METHOD

An approximate solution of the governing partial differential equations and boundary conditions of poroelasticity can be formulated using the Galerkin weighted-residual principle and a finite-element tessellation of the problem domain. The finite-element method allows for material heterogeneity as well as irregular boundaries and distributed mechanical loads and fluid sources (e.g., Fig. 10.1). The procedure for obtaining the matrix equations for nodal displacements and pore pressure is illustrated for problems in plane strain. The force equilibrium equations (Eqns. 7.7 and 7.8) are expressed in terms of

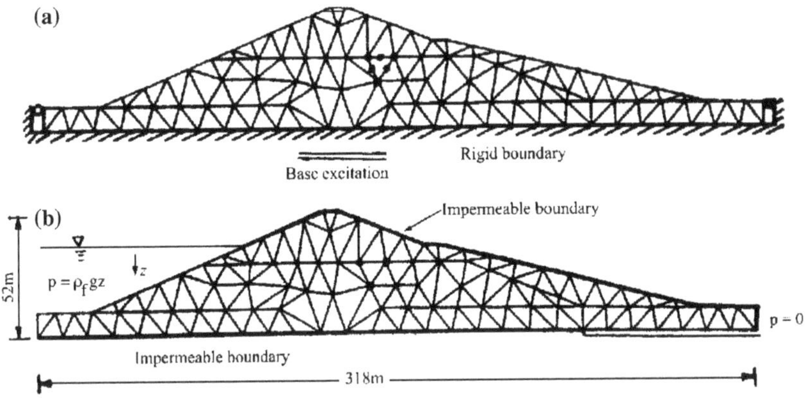

Figure 10.1: Finite-element mesh for analysis of failure of San Fernando dam (after Zienkiewicz et al., 1998, p. 275). (a) Six-node triangular elements for displacement discretization. (b) Three-node triangular elements for pore-pressure discretization.

u, v, and p in a form that maintains their connection to the stress components:

$$\underbrace{\left[\frac{2G(1-\nu)}{1-2\nu}\frac{\partial^2 u}{\partial x^2}+\frac{2G\nu}{1-2\nu}\frac{\partial^2 v}{\partial x\partial y}\right]}_{\frac{\partial(\sigma_{xx}+\alpha p)}{\partial x}}+\underbrace{\left[G\frac{\partial^2 u}{\partial y^2}+G\frac{\partial^2 v}{\partial y\partial x}\right]}_{\frac{\partial \sigma_{yx}}{\partial y}}=\alpha\frac{\partial p}{\partial x} \tag{10.1}$$

$$\underbrace{\left[G\frac{\partial^2 u}{\partial x\partial y}+G\frac{\partial^2 v}{\partial x^2}\right]}_{\frac{\partial \sigma_{xy}}{\partial x}}+\underbrace{\left[\frac{2G\nu}{1-2\nu}\frac{\partial^2 u}{\partial y\partial x}+\frac{2G(1-\nu)}{1-2\nu}\frac{\partial^2 v}{\partial y^2}\right]}_{\frac{\partial(\sigma_{yy}+\alpha p)}{\partial y}}=\alpha\frac{\partial p}{\partial y} \tag{10.2}$$

Galerkin's method is based on the principle of minimization of a weighted residual (Szabo and Lee, 1969). In what follows, their development is combined with the description of the Galerkin method given by Wang and Anderson (1982). The problem domain is tessellated into a mesh of finite elements. A trial solution is defined piecewise over elements by interpolating the nodal values. For the plane-strain problem, the nodal unknowns U_K, V_K, and P_K refer to the x-component of displacement, y-component of displacement, and pore pressure, respectively, of the K-th node. Within an element,

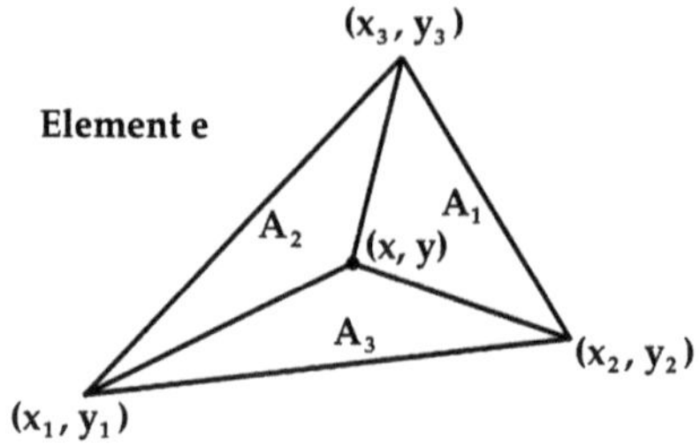

Figure 10.2: Triangular element. The value of the *K*-th basis function for a point (x, y) within the element is A_K/A_e, where A_K is the subarea opposite node *K*, and A_e is the area of the entire triangle.

the interpolations are defined by element basis functions $N_K^e(x, y)$,

$$\hat{u}^e(x, y) = \sum_{K=1}^{m} N_K^e(x, y) U_K$$

$$\hat{v}^e(x, y) = \sum_{K=1}^{m} N_K^e(x, y) V_K$$

$$\hat{p}^e(x, y) = \sum_{K=1}^{m} N_K^e(x, y) P_K \tag{10.3}$$

where an overhead carat denotes the variable as an approximate, trial solution; superscript e designates the element number; and K is the local node number (e.g., 1, 2, 3 for a triangular element; Fig. 10.2). The summations in Eqn. 10.3 are over the m-nodes in element e. The well-known basis functions for a triangular element with nodal coordinates (x_1, y_1), (x_2, y_2), and (x_3, y_3) in counterclockwise order are

$$N_1^e(x, y) = \frac{1}{2A_e}\left[(x_2y_3 - x_3y_2) + (y_2 - y_3)x + (x_3 - x_2)y\right] = \frac{A_1}{A_e}$$

$$N_2^e(x, y) = \frac{1}{2A_e}\left[(x_3y_1 - x_1y_3) + (y_3 - y_1)x + (x_1 - x_3)y\right] = \frac{A_2}{A_e}$$

$$N_3^e(x, y) = \frac{1}{2A_e}\left[(x_1y_2 - x_2y_1) + (y_1 - y_2)x + (x_2 - x_1)y\right] = \frac{A_3}{A_e} \tag{10.4}$$

where

$$2A_e = (x_1y_2 - x_2y_1) + (x_2y_3 - x_3y_2) + (x_3y_1 - x_1y_3) \tag{10.5}$$

A_1, A_2, and A_3 are the subareas opposite the nodes 1, 2, and 3, respectively, and $A_e = A_1 + A_2 + A_3$ is the area of the whole element. The element basis

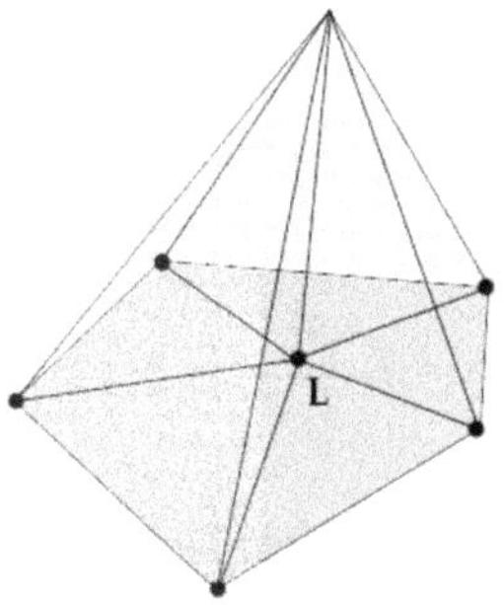

Figure 10.3: Nodal basis function $N_L(x, y)$ (after Wang and Anderson, 1982). The nodal basis function $N_L(x, y)$ is defined as the union of element basis functions for each element in the patch of elements that contains node L. The value of $N_L(x_L, y_L)$ is one and drops linearly to zero at the nodes opposite node L.

functions are zero outside the element. The L-th nodal basis function N_L is defined as the union of element basis functions over each element containing node L (Fig. 10.3). The trial solutions over the entire problem domain are then

$$\hat{u}(x, y) = \sum_{K=1}^{\text{NNODE}} N_K(x, y) U_K$$

$$\hat{v}(x, y) = \sum_{K=1}^{\text{NNODE}} N_K(x, y) V_K$$

$$\hat{p}(x, y) = \sum_{K=1}^{\text{NNODE}} N_K(x, y) P_K \tag{10.6}$$

where NNODE is the total number of nodal points, and K is the unique global number for each node in the problem domain. For higher-order elements, the basis functions in Eqns. 10.3 and 10.6 can be different for each dependent variable. Stability of the finite element solution is improved if the interpolation function for pressure is lower order than that for displacement (Lewis and Schrefler, 1987, p. 79).

In general, a weighted-residual method consists of minimizing the weighted error or residual of the governing equation when the trial solution is inserted. In the Galerkin method the nodal basis functions are also the weighting functions. For each governing equation, one weighted residual equation results for each basis function $N_L(x, y)$. For example, the L-th equation obtained from Eqn. 10.1 is

$$\iint_D \left[\frac{2G(1-\nu)}{1-2\nu} \frac{\partial^2 \hat{u}}{\partial x^2} + \frac{2G\nu}{1-2\nu} \frac{\partial^2 \hat{v}}{\partial x \partial y} + G \frac{\partial^2 \hat{u}}{\partial y^2} + G \frac{\partial^2 \hat{v}}{\partial y \partial x} - \alpha \frac{\partial \hat{p}}{\partial x} \right] N_L(x, y) dx dy = 0 \quad (10.7)$$

where D is the problem domain. Integrating by parts reduces the order of the second derivatives in Eqn. 10.7 by one,

$$\iint_D \frac{\partial^2 \hat{u}}{\partial x^2} N_L dx dy = \int_\Gamma \frac{\partial \hat{u}}{\partial x} N_L n_x d\ell - \iint_D \frac{\partial \hat{u}}{\partial x} \frac{\partial N_L}{\partial x} dx dy \quad (10.8)$$

$$\iint_D \frac{\partial^2 \hat{v}}{\partial x \partial y} N_L dx dy = \int_\Gamma \frac{\partial \hat{v}}{\partial y} N_L n_x d\ell - \iint_D \frac{\partial \hat{v}}{\partial y} \frac{\partial N_L}{\partial x} dx dy \quad (10.9)$$

$$\iint_D \frac{\partial^2 \hat{v}}{\partial y \partial x} N_L dx dy = \int_\Gamma \frac{\partial \hat{v}}{\partial x} N_L n_y d\ell - \iint_D \frac{\partial \hat{v}}{\partial x} \frac{\partial N_L}{\partial y} dx dy \quad (10.10)$$

$$\iint_D \frac{\partial^2 \hat{u}}{\partial y^2} N_L dx dy = \int_\Gamma \frac{\partial \hat{u}}{\partial y} N_L n_y d\ell - \iint_D \frac{\partial \hat{u}}{\partial y} \frac{\partial N_L}{\partial y} dx dy \quad (10.11)$$

where Γ is the boundary of the problem domain, (n_x, n_y) is the unit normal to the boundary, and $d\ell$ is the differential distance along the boundary. Substituting the trial solutions from Eqn. 10.6 into Eqn. 10.7 and using Eqns. 10.8–10.11 give

$$\begin{aligned} &\iint_D \left[\frac{2G(1-\nu)}{1-2\nu} \frac{\partial N_K}{\partial x} \frac{\partial N_L}{\partial x} + G \frac{\partial N_K}{\partial y} \frac{\partial N_L}{\partial y} \right] U_K dx dy \\ &+ \iint_D \left[\frac{2G\nu}{1-2\nu} \frac{\partial N_K}{\partial y} \frac{\partial N_L}{\partial x} + G \frac{\partial N_K}{\partial x} \frac{\partial N_L}{\partial y} \right] V_K dx dy \\ &+ \iint_D \left[\alpha \frac{\partial N_K}{\partial x} N_L \right] P_K dx dy \\ &- \int_\Gamma \underbrace{\left[(\sigma_{xx} + \alpha p) n_x + \sigma_{xy} n_y \right]}_{\text{effective traction, } t'_x} N_L d\ell = 0 \end{aligned} \quad (10.12)$$

where the repeated index K indicates summation from $K = 1$ to $K =$ NNODE. The integrand in the boundary integral term is a weighted average of the x-component of the effective traction on the surface. A second set of NNODE equations is obtained from the force equilibrium equation in the

y-direction:

$$\int\int_D \left[\frac{2G\nu}{1-2\nu}\frac{\partial N_K}{\partial x}\frac{\partial N_L}{\partial y} + G\frac{\partial N_K}{\partial y}\frac{\partial N_L}{\partial x}\right] U_K dxdy$$
$$+ \int\int_D \left[\frac{2G(1-\nu)}{1-2\nu}\frac{\partial N_K}{\partial y}\frac{\partial N_L}{\partial y} + G\frac{\partial N}{\partial x}\frac{\partial N_L}{\partial x}\right] V_K dxdy$$
$$+ \int\int_D \left[\alpha\frac{\partial N_K}{\partial y} N_L\right] P_K dxdy$$
$$- \int_\Gamma \underbrace{\left[\sigma_{xy} n_x + (\sigma_{yy} + \alpha p) n_y\right]}_{\text{effective traction, } t'_y} N_L d\ell = 0 \tag{10.13}$$

Finally, the same procedure is applied to the fluid diffusion equation (Eqn. 7.14):

$$\int\int_D \alpha\left[N_L\frac{\partial N_K}{\partial x}\frac{dU_K}{dt} + N_L\frac{\partial N_K}{\partial y}\frac{dV_K}{dt}\right] dxdy$$
$$+ \int\int_D S_\epsilon N_L N_K \frac{dP_K}{dt} dxdy$$
$$+ \int\int_D \frac{k}{\mu}\left[\frac{\partial N_K}{\partial x}\frac{\partial N_L}{\partial x} + \frac{\partial N_K}{\partial y}\frac{\partial N_L}{\partial y}\right] P_K dxdy$$
$$- \int_\Gamma \underbrace{\left[\frac{\partial p}{\partial x} n_x + \frac{\partial p}{\partial y} n_y\right]}_{-q_n/(k/\mu)} N_L d\ell = 0 \tag{10.14}$$

The integrand in the boundary integral term is a weighted average of the normal flux q_n (positive for outflow) of fluid through the surface.

The integration over each triangular element contributes three nonzero terms to the L-th equation if L is one of its nodes; that is, nonzero coefficients of U_K, V_K, and P_K come from elements in the patch around node L. The usual finite-element procedure is to assemble Eqns. 10.12–10.14 by summing the element-by-element contributions to the coefficients of the global matrix. The 2×NNODE equations represented by Eqns. 10.12 and 10.13 can be written in matrix form as

$$\left[S^{11}\right]\{U\} + \left[S^{12}\right]\{V\} + \left[S^{13}\right]\{P\} = \left\{B^{(1)}\right\} \tag{10.15}$$
$$\left[S^{21}\right]\{U\} + \left[S^{22}\right]\{V\} + \left[S^{23}\right]\{P\} = \left\{B^{(2)}\right\} \tag{10.16}$$

where $\{U\}$, $\{V\}$, and $\{P\}$ are column vectors of the nodal unknowns $U_1, \ldots, U_{\text{NNODE}}$, $V_1, \ldots, V_{\text{NNODE}}$, and $P_1, \ldots, P_{\text{NNODE}}$. The components of the matrices $[S^{ij}]$ are

$$
\begin{aligned}
S_{LK}^{11} &= \int\int_D \left[\frac{2G(1-\nu)}{1-2\nu}\frac{\partial N_K}{\partial x}\frac{\partial N_L}{\partial x} + G\frac{\partial N_K}{\partial y}\frac{\partial N_L}{\partial y}\right] dxdy \\
S_{LK}^{12} &= \int\int_D \left[\frac{2G\nu}{1-2\nu}\frac{\partial N_K}{\partial y}\frac{\partial N_L}{\partial x} + G\frac{\partial N_K}{\partial x}\frac{\partial N_L}{\partial y}\right] dxdy \\
S_{LK}^{13} &= \int\int_D \alpha\frac{\partial N_K}{\partial x} N_L dxdy \\
S_{LK}^{21} &= \int\int_D \left[\frac{2G\nu}{1-2\nu}\frac{\partial N_K}{\partial x}\frac{\partial N_L}{\partial y} + G\frac{\partial N_K}{\partial y}\frac{\partial N_L}{\partial x}\right] dxdy \\
S_{LK}^{22} &= \int\int_D \left[\frac{2G(1-\nu)}{1-2\nu}\frac{\partial N_K}{\partial y}\frac{\partial N_L}{\partial y} + G\frac{\partial N}{\partial x}\frac{\partial N_L}{\partial x}\right] dxdy \\
S_{LK}^{23} &= \int\int_D \alpha\frac{\partial N_K}{\partial y} N_L dxdy \\
B_L^{(1)} &= \int_\Gamma \left[(\sigma_{xx} + \alpha p)n_x + \sigma_{xy}n_y\right] N_L d\ell \\
B_L^{(2)} &= \int_\Gamma \left[\sigma_{xy}n_x + (\sigma_{yy} + \alpha p)n_y\right] N_L d\ell \qquad (10.17)
\end{aligned}
$$

The two unknown nodal displacement components $\{U\}$ and $\{V\}$ in Eqns. 10.15 and 10.16 are written as separate solution vectors, whereas U_K and V_K are often alternated in a single solution vector in many finite-element formulations of elasticity.

The set of equations represented by Eqn. 10.14 can be expressed as a first-order matrix differential equation,

$$
\begin{aligned}
&[A^{(1)}]\left\{\frac{dU}{dt}\right\} + [A^{(2)}]\left\{\frac{dV}{dt}\right\} + [A^{(3)}]\left\{\frac{dP}{dt}\right\} + [A^{(4)}]\{P\} \\
&\quad = \{B^{(3)}\} \qquad (10.18)
\end{aligned}
$$

where the L-th row of $\{dU/dt\}$ is defined to be dU_L/dt, and similarly for $\{dV/dt\}$ and $\{dP/dt\}$. The coefficients of the matrices $[A^{(1)}]$, $[A^{(2)}]$, $[A^{(3)}]$, and $[A^{(4)}]$ are

$$
A_{LK}^{(1)} = \int\int_D \alpha\frac{\partial N_K}{\partial x} N_L dxdy \qquad (10.19)
$$

$$
A_{LK}^{(2)} = \int\int_D \alpha\frac{\partial N_K}{\partial y} N_L dxdy \qquad (10.20)
$$

$$A_{LK}^{(3)} = \int\int_D S_\epsilon N_L N_K dxdy \tag{10.21}$$

$$A_{LK}^{(4)} = \int\int_D \frac{k}{\mu}\left[\frac{\partial N_K}{\partial x}\frac{\partial N_L}{\partial x} + \frac{\partial N_K}{\partial y}\frac{\partial N_L}{\partial y}\right] dxdy \tag{10.22}$$

$$B_L^{(3)} = \int_\Gamma \left[\frac{\partial p}{\partial x} n_x + \frac{\partial p}{\partial y} n_y\right] N_L d\ell \tag{10.23}$$

The simultaneous solution of Eqns. 10.15, 10.16, and 10.18 is made using a finite-difference approximation for the time derivatives in Eqn. 10.18:

$$\left\{\frac{dU}{dt}\right\} = \frac{1}{\Delta t}\left(\{U\}^{t+\Delta t} - \{U\}^t\right)$$

$$\left\{\frac{dV}{dt}\right\} = \frac{1}{\Delta t}\left(\{V\}^{t+\Delta t} - \{V\}^t\right)$$

$$\left\{\frac{dP}{dt}\right\} = \frac{1}{\Delta t}\left(\{P\}^{t+\Delta t} - \{P\}^t\right) \tag{10.24}$$

A weighted average of $\{U\}^{t+\Delta t}$ and $\{U\}^t$ is substituted for $\{U\}$, and similarly for $\{V\}$ and $\{P\}$, in Eqns. 10.15 and 10.16. A weight of one at time $t + \Delta t$ is a fully implicit time-stepping approximation, and a weight of one at time t is a fully explicit time-stepping approximation. The resulting set of $3\times$NNODE linear equations can then be solved simultaneously at each time step. Finite-element computer codes for poroelastic problems have been described by Sandhu and Wilson (1969), Szabo and Lee (1969), Noorishad et al. (1982), Lewis and Schrefler (1987), Smith and Griffiths (1988), and Hsieh (1994). A very general three-dimensional code, ABAQUS, is available commercially for a variety of computers from Hibbitt, Karlsson, and Sorensen, Inc. (1998).

10.2 BOUNDARY-ELEMENT METHOD

The boundary-element method can be broken down into three main steps:

1. The starting point is applying the reciprocity theorem to two independent stress and strain states. The divergence theorem yields a boundary integral equation containing tractions, displacements, pore pressures, and fluid fluxes.
2. An impulsive point force in the x-direction is chosen for one of the states. The free-space Green's functions for the impulsive point force are substituted into the boundary integral to yield one boundary integral equation. The same procedure is repeated for an impulsive point force in the y-direction to yield a second boundary integral equation. The

procedure is repeated a third time for an instantaneous fluid source as one of the states to yield a third boundary integral equation.

3. The boundary integrals are discretized in a finite-element manner. Inserting boundary conditions produces a set of algebraic equations for the nodal unknowns. These steps are outlined later for the time-dependent poroelastic problem in plane strain. The Laplace transform is used to obtain a formulation in the spatial variables x and y and the transform variable s (Cheng and Detournay, 1988, 1998). This formulation is the so-called direct method in contrast with the indirect method based on distributions of sources and dipoles on the boundary.

10.2.1 Reciprocity Theorem

The existence of the strain energy density (Eqn. 1.13) leads to the poroelastic extension of the Betti-Rayleigh reciprocal theorem (cf. Nowacki, 1986, for the case of thermoelasticity) between two independent states, 1 and 2, which exist at two different times, t and $t - \tau$:

$$\sigma_{ij}^{(1)}(\vec{x}, t)\epsilon_{ij}^{(2)}(\vec{x}, t - \tau) + p^{(1)}(\vec{x}, t)\zeta^{(2)}(\vec{x}, t - \tau)$$
$$= \sigma_{ij}^{(2)}(\vec{x}, t - \tau)\epsilon_{ij}^{(1)}(\vec{x}, t) + p^{(2)}(\vec{x}, t - \tau)\zeta^{(1)}(\vec{x}, t) \qquad (10.25)$$

Integrating Eqn. 10.25 with respect to time leads to a convolution integral of each term. Because the Laplace transform of a convolution of two functions is the product of their Laplace transforms (Appendix A, Eqn. A.14), the reciprocity relation becomes

$$\tilde{\sigma}_{ij}^{(1)}(\vec{x}, s)\tilde{\epsilon}_{ij}^{(2)}(\vec{x}, s) + \tilde{p}^{(1)}(\vec{x}, s)\tilde{\zeta}^{(2)}(\vec{x}, s)$$
$$= \tilde{\sigma}_{ij}^{(2)}(\vec{x}, s)\epsilon_{ij}^{(1)}(\vec{x}, s) + \tilde{p}^{(2)}(\vec{x}, s)\tilde{\zeta}^{(1)}(\vec{x}, s) \qquad (10.26)$$

The Laplace transform of the force equilibrium equation (Eqn. 4.10) is

$$\frac{\partial \tilde{\sigma}_{ij}}{\partial x_j} = -\widetilde{F}_i \qquad (10.27)$$

Cheng and Detournay (1998, Eqn. A5) used Eqn. 10.27 to express $\tilde{\sigma}_{ij}^{(1)}\tilde{\epsilon}_{ij}^{(2)}$ as

$$\tilde{\sigma}_{ij}^{(1)}\tilde{\epsilon}_{ij}^{(2)} = \frac{\partial}{\partial x_j}\left(\tilde{\sigma}_{ij}^{(1)}\tilde{u}_i^{(2)}\right) + \tilde{u}_i^{(2)}\widetilde{F}_i^{(1)} \qquad (10.28)$$

The Laplace transform of the continuity equation (Eqn. 4.60) is

$$s\tilde{\zeta}(\vec{x}, s) - s\zeta(\vec{x}, 0) = \widetilde{Q}(\vec{x}, s) - \frac{\partial \tilde{q}_k}{\partial x_k} \qquad (10.29)$$

Taking a volume integral of Eqn. 10.26 over a problem domain D bounded by the surface Γ, using Eqns. 10.28 and 10.29, applying the divergence theorem, using Darcy's law without the fluid body-force term, and assuming that $\zeta^{(1)}(\vec{x},0)=\zeta^{(2)}(\vec{x},0)=0$ give

$$\int_\Gamma \left[\tilde{t}_i^{(1)}(\vec{x},s)\tilde{u}_i^{(2)}(\vec{x},s)-\tilde{t}_i^{(2)}(\vec{x},s)\tilde{u}_i^{(1)}(\vec{x},s)\right]dS$$
$$-\frac{1}{s}\int_\Gamma \left[\tilde{p}^{(1)}(\vec{x},s)\tilde{q}^{(2)}(\vec{x},s)-\tilde{p}^{(2)}(\vec{x},s)\tilde{q}^{(1)}(\vec{x},s)\right]dS$$
$$+\int_D \left[\widetilde{F}_i^{(1)}(\vec{x},s)\tilde{u}_i^{(2)}(\vec{x},s)-\widetilde{F}_i^{(2)}(\vec{x},s)\tilde{u}_i^{(1)}(\vec{x},s)\right]dV$$
$$+\frac{1}{s}\int_D \left[\tilde{p}^{(1)}(\vec{x},s)\widetilde{Q}^{(2)}(\vec{x},s)-\tilde{p}^{(2)}(\vec{x},s)\widetilde{Q}^{(1)}(\vec{x},s)\right]dV=0 \quad (10.30)$$

where $t_i=\sigma_{ik}n_k$ is the boundary traction, and $q=q_kn_k$ is the fluid flux normal to the boundary.

10.2.2 Free-Space Green's Functions

Eqn. 10.30 can be used to express the displacement at an interior point of the problem domain as a boundary integral if state 2 is an impulse point force of unit strength in the J-direction applied at point $\vec{d}$ at time zero—that is, $F_{iJ}^{(2)}=\delta_{iJ}\delta(\vec{x}-\vec{d})\delta(t)$. The subscript J is used in addition to the component index i to indicate the direction of the point force. The Laplace transform is $\widetilde{F}_{iJ}^{(2)}=\delta_{iJ}\delta(\vec{x}-\vec{d})$. Substituting into Eqn. 10.30, designating the free-space Green's function for state 2 with an asterisk and dropping the superscript for state 1, and assuming no body forces or fluid sources for state 1 give

$$\frac{\theta}{2\pi}\tilde{u}_J(\vec{d},s)$$
$$=\int_\Gamma \left[\tilde{u}_{iJ}^*(\vec{x}-\vec{d},s)\tilde{t}_i(\vec{x},s)-\tilde{t}_{iJ}^*(\vec{x}-\vec{d},s)\tilde{u}_i(\vec{x},s)\right]d\ell$$
$$-\frac{1}{s}\int_\Gamma \left[\tilde{q}_J^*(\vec{x}-\vec{d},s)\tilde{p}(\vec{x},s)-\tilde{p}_J^*(\vec{x}-\vec{d},s)\tilde{q}(\vec{x},s)\right]d\ell \quad (10.31)$$

where $\tilde{t}_{iJ}^*$, $\tilde{u}_{iJ}^*$, $\tilde{p}_J^*$, and $\tilde{q}_J^*$ are the Laplace transforms of the free-space Green's functions for the impulse point force, and ℓ is the distance along the boundary Γ. The angle θ is 2π for an interior point, π for a point along a smooth boundary, and the interior angle at a corner along the boundary (Fig. 10.4). Eqn. 10.31 is the poroelastic extension of *Somigliana's identity* (e.g., Kane, 1994, p. 144). The Laplace transforms of the Green's functions for a unit point force applied at $\vec{d}$ in the J-direction have been given by

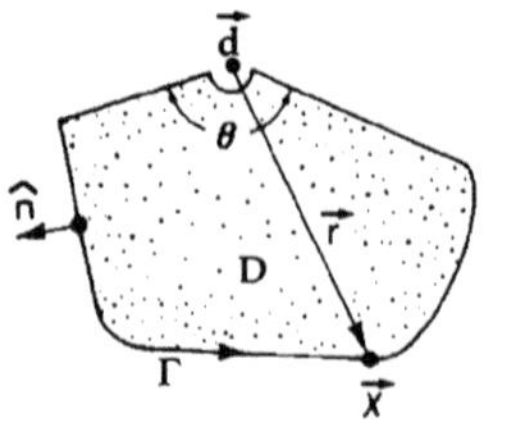

Figure 10.4: Definition of the angle θ for a point on the boundary. The angle measure assumes that it lies within the problem domain (Cheng and Detournay, 1988, Fig. 1*a*).

Cheng and Detournay (1988),

$$\tilde{u}^*_{iJ}(\vec{r},s) = -\frac{3-4\nu_u}{8\pi G(1-\nu_u)}\delta_{iJ}\ln r + \frac{1}{8\pi G(1-\nu_u)}\frac{r_i r_J}{r^2} + \frac{\nu_u - \nu}{4\pi G(1-\nu)(1-\nu_u)} \times\left\{\delta_{iJ}\left[\xi^{-2} - \xi^{-1}K_1(\xi)\right] + \frac{r_i r_J}{r^2}\left[K_2(\xi) - 2\xi^{-2}\right]\right\}$$

$$\tilde{t}^*_{iJ}(\vec{r},s) = \frac{1-2\nu_u}{4\pi(1-\nu_u)}\frac{n_i r_J - n_J r_i - \delta_{iJ}n_k r_k}{r^2} - \frac{1}{2\pi(1-\nu_u)}\frac{r_i r_J r_k n_k}{r^4} + \frac{\nu_u-\nu}{2\pi(1-\nu)(1-\nu_u)} \times\sqrt{\frac{s}{c}}\left\{\delta_{iJ}\frac{r_k n_k}{r}\left[\xi^{-1}K_2(\xi) - 2\xi^{-3}\right] + \frac{r_i r_J r_k n_k}{r^3}\left[8\xi^{-3} - K_3(\xi)\right] + \frac{n_i r_J}{r}\left[K_3(\xi) - 3\xi^{-1}K_2(\xi) - 2\xi^{-3}\right] + \frac{n_J r_i}{r}\left[\xi^{-1} - 2\xi^{-3}\right]\right\}$$

$$\tilde{q}^*_J(\vec{r},s) = \frac{3(\nu_u-\nu)}{4\pi BGs(1-\nu)(1+\nu_u)}\left\{n_J\left[\xi^{-1}K_1(\xi) - \xi^{-2}\right] + \frac{r_J r_k n_k}{r^2}\left[2\xi^{-2} - K_2(\xi)\right]\right\}$$

$$\tilde{p}^*_J(\vec{r},s) = \frac{\gamma}{2\pi}\sqrt{\frac{s}{c}}\frac{r_J}{r}\left[\xi^{-1} - K_1(\xi)\right] \tag{10.32}$$

where $\vec{r} = \vec{x} - \vec{d}$, $r = |\vec{x} - \vec{d}|$, and $\xi = r\sqrt{s/c}$.

An equation similar to Eqn. 10.31 is obtained by choosing a sudden point source, $\widetilde{Q}^{(2)} = \delta(\vec{x} - \vec{d})$,

$$\frac{\theta}{2\pi}\tilde{p}(\vec{d},s) = -s\int_\Gamma\left[\tilde{u}^*_i(\vec{x}-\vec{d},s)\tilde{t}_i(\vec{x},s) - \tilde{t}^*_i(\vec{x}-\vec{d},s)\tilde{u}_i(\vec{x},s)\right]d\ell + \int_\Gamma\left[\tilde{q}^*(\vec{x}-\vec{d},s)\tilde{p}(\vec{x},s) - \tilde{p}^*(\vec{x}-\vec{d},s)\tilde{q}(\vec{x},s)\right]d\ell \tag{10.33}$$

where $\tilde{t}_i^*$, $\tilde{u}_i^*$, $\tilde{p}^*$, and $\tilde{q}^*$ are the Laplace transforms of the free-space Green's functions for the impulse fluid source given by Cheng and Detournay (1988):

$$\tilde{u}_i^*(\vec{r}, s) = \frac{\gamma}{2\pi s}\sqrt{\frac{s}{c}}\frac{r_i}{r}\left[\xi^{-1} - K_1(\xi)\right]$$

$$\tilde{t}_i^*(\vec{r}, s) = \frac{\gamma G}{2\pi c}\Big\{n_i\left[2\xi^{-2} - K_o(\xi) - K_2(\xi)\right] - \frac{r_i r_k n_k}{r^2}\left[4\xi^{-2} - 2K_2(\xi)\right]\Big\}$$

$$\tilde{q}^*(\vec{r}, s) = \frac{1}{2\pi}\sqrt{\frac{s}{c}}\frac{r_k n_k}{r}K_1(\xi)$$

$$\tilde{p}^*(\vec{r}, s) = \frac{1}{2\pi(k/\mu)}K_o(\xi) \tag{10.34}$$

10.2.3 Discretization

The integrands in Eqns. 10.31 and 10.33 are approximated by discretizing each solution variable at nodal points and employing basis or interpolation functions as for finite elements. The nodal basis function $N_\alpha^e(\chi)$ is defined for node α in a canonical element e with χ as the distance coordinate (Fig. 10.5). The basis functions are also used to map the nodal coordinates of the canonical element onto the actual element. The solution variables are thus defined parametrically over an actual element. This type of element is

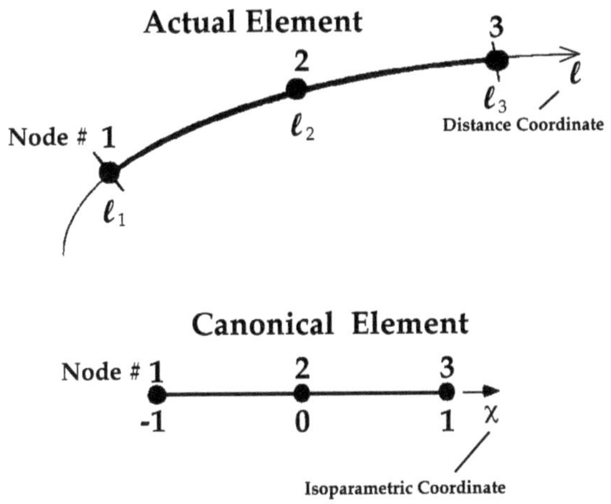

Figure 10.5: Actual and canonical three-node element (after Kane, 1994, Fig. 3.9, p. 51).

called isoparametric, because the same basis functions are used for the coordinate mapping as for the interpolation of the solution variables. The canonical three-node element shown in Figure 10.5 is a straight-line segment of length 2 with end nodes 1 and 3 at $\chi = -1$ and $\chi = +1$, and midside node 2 at $\chi = 0$. The three quadratic interpolation functions are

$$\begin{aligned} N_1^e(\chi) &= -\frac{1}{2}\chi(1-\chi) \\ N_2^e(\chi) &= (1+\chi)(1-\chi) \\ N_3^e(\chi) &= \frac{1}{2}\chi(1+\chi) \end{aligned} \tag{10.35}$$

Note that $N_1^e(-1) = 1$ and $N_2^e(-1) = N_3^e(-1) = 0$; in other words, the weighting function is one at the node corresponding to its interpolation function, and it is zero at the other two nodes. Similarly, $N_2^e(0) = 1$ and $N_3^e(+1) = 1$. The interpolations using local node numbers α within element e are then

$$\begin{aligned} \ell^e(\chi) &= \sum_{\alpha=1}^{3} N_\alpha^e(\chi)\ell_\alpha^e \\ \tilde{u}^e(\chi, s) &= \sum_{\alpha=1}^{3} N_\alpha^e(\chi)\widetilde{U}_\alpha^e(s) \\ \tilde{v}^e(\chi, s) &= \sum_{\alpha=1}^{3} N_\alpha^e(\chi)\widetilde{V}_\alpha^e(s) \\ \tilde{t}_x^e(\chi, s) &= \sum_{\alpha=1}^{3} N_\alpha^e(\chi)\widetilde{X}_\alpha^e(s) \\ \tilde{t}_y^e(\chi, s) &= \sum_{\alpha=1}^{3} N_\alpha^e(\chi)\widetilde{Y}_\alpha^e(s) \\ \tilde{p}^e(\chi, s) &= \sum_{\alpha=1}^{3} N_\alpha^e(\chi)\widetilde{P}_\alpha^e(s) \\ \tilde{q}^e(\chi, s) &= \sum_{\alpha=1}^{3} N_\alpha^e(\chi)\widetilde{Q}_\alpha^e(s) \end{aligned} \tag{10.36}$$

where ℓ_α^e, $\widetilde{U}_\alpha^e$, $\widetilde{V}_\alpha^e$, $\widetilde{X}_\alpha^e$, $\widetilde{Y}_\alpha^e$, $\widetilde{P}_\alpha^e$, and $\widetilde{Q}_\alpha^e$ refer to the nodal values of the boundary distance and the Laplace transforms of the displacement components, traction components, pore pressure, and normal fluid flux, respectively. The nodal values of all the variables except distance are functions of the Laplace transform variable s.

The integration over each element can now be performed explicitly in terms of the variable χ. Because the basis functions N_α^e map the interval $[-1, 1]$ onto the element, the differential arc length $d\ell$ equals $J^e d\chi$, where J^e is the Jacobian for element e (e.g., Kane, 1994, p. 47):

$$J^e = \sqrt{\left(\frac{dx}{d\chi}\right)^2 + \left(\frac{dy}{d\chi}\right)^2} \tag{10.37}$$

The derivatives in the Jacobian are given by

$$\frac{dx}{d\chi} = \sum_{\alpha=1}^{3} \frac{dN_\alpha^e}{d\chi} x_\alpha$$

$$\frac{dy}{d\chi} = \sum_{\alpha=1}^{3} \frac{dN_\alpha^e}{d\chi} y_\alpha \tag{10.38}$$

Choosing the source to be located at node K (i.e., $\vec{d} = \vec{x}_K$) in Eqns. 10.31 and 10.33 and substituting Eqns. 10.35 and 10.37 yield three linear combinations of the nodal values of the solution variables,

$$
\begin{aligned}
&-\sum_{e=1}^{\text{NELEM}} \int_{-1}^{1} \Bigg[\Bigg(\tilde{u}_{1J}^*(\vec{r}_K, s) \sum_{\alpha=1}^{3} N_\alpha^e(\chi) \widetilde{X}_\alpha^e(s) + \tilde{u}_{2J}^*(\vec{r}_K, s) \sum_{\alpha=1}^{3} N_\alpha^e(\chi) \widetilde{Y}_\alpha^e(s) \Bigg) \\
&-\Bigg(\tilde{t}_{1J}^*(\vec{r}_K, s) \sum_{\alpha=1}^{3} N_\alpha^e(\chi) \widetilde{U}_\alpha^e(s) + \tilde{t}_{2J}^*(\vec{r}_K, s) \sum_{\alpha=1}^{3} N_\alpha^e(\chi) \widetilde{V}_\alpha^e(s) \Bigg) \Bigg] J^e(\chi) d\chi \\
&+\frac{1}{s} \sum_{e=1}^{\text{NELEM}} \int_{-1}^{1} \Bigg[\tilde{q}_J^*(\vec{r}_K, s) \sum_{\alpha=1}^{3} N_\alpha^e(\chi) \widetilde{P}_\alpha^e(s) \\
&\qquad\qquad - \tilde{p}_J^*(\vec{r}_K, s) \sum_{\alpha=1}^{3} N_\alpha^e(\chi) \widetilde{Q}_\alpha^e(s) \Bigg] J^e(\chi) d\chi \\
&+\frac{\theta}{2\pi} \left\{ \begin{matrix} \widetilde{U}_K(s) & \text{if } J = 1 \\ \widetilde{V}_K(s) & \text{if } J = 2 \end{matrix} \right\} = 0
\end{aligned} \tag{10.39}
$$

$$
\begin{aligned}
s\sum_{e=1}^{\mathrm{NELEM}}\int_{-1}^{1}\Bigg[&\left(\tilde{u}_1^*(\vec{r}_K,s)\sum_{\alpha=1}^{3}N_\alpha^e(\chi)\widetilde{X}_\alpha^e(s)+\tilde{u}_2^*(\vec{r}_K,s)\sum_{\alpha=1}^{3}N_\alpha^e(\chi)\widetilde{Y}_\alpha^e(s)\right)\\
&-\left(\tilde{t}_1^*(\vec{r}_K,s)\sum_{\alpha=1}^{3}N_\alpha^e(\chi)\widetilde{U}_\alpha^e(s)+\tilde{t}_2^*(\vec{r}_K,s)\sum_{\alpha=1}^{3}N_\alpha^e(\chi)\widetilde{V}_\alpha^e(s)\right)\Bigg]J^e(\chi)d\chi\\
-\sum_{e=1}^{n}\int_{-1}^{1}&\left[\tilde{q}^*(\vec{r}_K,s)\sum_{\alpha=1}^{3}N_\alpha^e(\chi)\widetilde{P}_\alpha^e(s)-\tilde{p}^*(\vec{r}_K,s)\sum_{\alpha=1}^{3}N_\alpha^e(\chi)\widetilde{Q}_\alpha^e(s)\right]J^e(\chi)d\chi\\
+\frac{\theta}{2\pi}\widetilde{P}_K(s)=0 &\qquad (10.40)
\end{aligned}
$$

where $\vec{r}_K \equiv \vec{x}(\ell^e(\chi)) - \vec{x}_K$, and NELEM is the number of elements.

10.2.4 Algebraic Equations

A system of 3×NNODE algebraic equations is developed from Eqns. 10.39 and 10.40 by successively choosing the source point $\vec{x}_K$, to be at each of the NNODE boundary nodal points. This number of equations is obtained because Eqn. 10.39 represents two equations per node, one for each of the two coordinate directions for the point force, and Eqn. 10.40 represents one equation per node for a point fluid source. The structure of Eqns. 10.39 and 10.40 is that they are the K -th triplet of equations associated with node K being the source point. The boundary integrals are performed piecewise over boundary elements. The functions within the integrands are defined by the interpolation functions using nodal values. Common nodes between adjacent elements generate coefficients that are summed (assembled) in the same manner as in the finite method. The 3×NNODE algebraic equations contain the 6×NNODE nodal values—$\widetilde{U}_K$, $\widetilde{V}_K$, $\widetilde{X}_K$, $\widetilde{Y}_K$, $\widetilde{P}_K$, $\widetilde{Q}_K$—of $\tilde{u}$, $\tilde{v}$, $\tilde{t}_x$, $\tilde{t}_y$, $\tilde{p}$, and $\tilde{q}$, respectively. Because either traction or displacement, and either fluid pressure or fluid flux, is specified at each point on the boundary, half of the nodal values are specified as boundary conditions, which brings the number of unknowns to 3×NNODE.

If the six nodal variables are grouped into two 3-component vectors, one consisting of the displacement variables and one consisting of the stress variables, Eqns. 10.39 and 10.40 can be expressed as

$$
\sum_{e=1}^{\mathrm{NELEM}}\sum_{\alpha=1}^{3}[f^{K\alpha}]^e\begin{Bmatrix}\widetilde{U}_\alpha^e\\ \widetilde{V}_\alpha^e\\ \widetilde{Q}_\alpha^e\end{Bmatrix}-\sum_{e=1}^{\mathrm{NELEM}}\sum_{\alpha=1}^{3}[g^{K\alpha}]^e\begin{Bmatrix}\widetilde{X}_\alpha^e\\ \widetilde{Y}_\alpha^e\\ \widetilde{P}_\alpha^e\end{Bmatrix}=0 \qquad (10.41)
$$

where the subscript or superscript α refers to local element node numbers. In addition, $[f^{K\alpha}]^e$ and $[g^{K\alpha}]^e$ are 3×3 element matrices whose entries are

$$
\begin{aligned}
f_{ij}^{K\alpha,e} &= \int_{-1}^{1} \tilde{t}_{ji}^{*}(\vec{r}_K, s) N_\alpha^e(\chi) J^e(\chi) d\chi + \frac{\theta}{2\pi}\delta_{ij} I && i, j = 1, 2 \\
f_{i3}^{K\alpha,e} &= -\frac{1}{s}\int_{-1}^{1} \tilde{p}_i^{*}(\vec{r}_K, s) N_\alpha^e(\chi) J^e(\chi) d\chi && i = 1, 2 \\
f_{3j}^{K\alpha,e} &= s\int_{-1}^{1} \tilde{t}_j^{*}(\vec{r}_K, s) N_\alpha^e(\chi) J^e(\chi) d\chi && j = 1, 2 \\
f_{33}^{K\alpha,e} &= -\int_{-1}^{1} \tilde{p}^{*}(\vec{r}_K, s) N_\alpha^e(\chi) J^e(\chi) d\chi
\end{aligned}
\tag{10.42}
$$

and

$$
\begin{aligned}
g_{ij}^{K\alpha,e} &= \int_{-1}^{1} \tilde{u}_{ji}^{*}(\vec{r}_K, s) N_\alpha^e(\chi) J^e(\chi) d\chi && i, j = 1, 2 \\
g_{i3}^{K\alpha,e} &= -\frac{1}{s}\int_{-1}^{1} \tilde{q}_i^{*}(\vec{r}_K, s) N_\alpha^e(\chi) J^e(\chi) d\chi && i = 1, 2 \\
g_{3j}^{K\alpha,e} &= s\int_{-1}^{1} \tilde{u}_j^{*}(\vec{r}_K, s) N_\alpha^e(\chi) J^e(\chi) d\chi && j = 1, 2 \\
g_{33}^{K\alpha,e} &= -\int_{-1}^{1} \tilde{q}^{*}(\vec{r}_K, s) N_\alpha^e(\chi) J^e(\chi) d\chi + \frac{\theta}{2\pi} I
\end{aligned}
\tag{10.43}
$$

where $I = 1$ if global node K is local node α in element e. The first row of $[f^{K\alpha}]^e$ and $[g^{K\alpha}]^e$ is based on an x-direction point force, the second row on a y-direction point force, and the third row on a point fluid source. The first two columns of $[f^{K\alpha}]^e$ are the coefficients of $\widetilde{U}_\alpha^e$ and $\widetilde{V}_\alpha^e$, respectively, and the third column of $[f^{K\alpha}]^e$ is the coefficient of $\widetilde{Q}_\alpha^e$. The first two columns of $[g^{K\alpha}]^e$ are the coefficients of $\widetilde{X}_\alpha^e$ and $\widetilde{Y}_\alpha^e$, respectively, and the third column of $[g^{K\alpha}]^e$ is the coefficient of $\widetilde{P}_\alpha^e$. The integration over the entire boundary Γ is obtained by assembling the contributions from each element into a global system of algebraic equations in the usual finite-element manner. For three-node quadratic elements, contributions to the common node from two adjacent elements are added. Eqns. 10.39 and 10.40—the K-th row triplet of the global system—can then be expressed symbolically in terms of the triplet of nodal variables at each node L,

$$
\sum_{L=1}^{\text{NNODE}} [F^{KL}]_{3\times 3} \begin{Bmatrix} \widetilde{U}_L \\ \widetilde{V}_L \\ \widetilde{Q}_L \end{Bmatrix} - \sum_{L=1}^{\text{NNODE}} [G^{KL}]_{3\times 3} \begin{Bmatrix} \widetilde{X}_L \\ \widetilde{Y}_L \\ \widetilde{P}_L \end{Bmatrix} = 0 \tag{10.44}
$$

for $K = 1, 2, 3, \ldots,$ NNODE. The entries of the global coefficient matrices $[F]$ and $[G]$ are the 3×3 $[F^{KL}]$ and $[G^{KL}]$ matrices. The structure of the global matrix $[F]$ is illustrated in Figure 10.6.

Singularities occur in the integrands that make up the diagonal matrix entries $[F^{KK}]$ and $[G^{KK}]$ because these terms contain the Green's function response at the node where the source is located. Specifically, the entries F_{11}^{KK}, F_{12}^{KK}, F_{21}^{KK}, F_{22}^{KK}, and G_{33}^{KK} contain strong (nonintegrable) singularities; the other entries contain logarithmic singularities, which can be evaluated by Gaussian quadrature. The strongly singular F_{ij}^{KK} $(i, j = 1,2)$ terms are the coefficients of the displacement components at node K due to point forces at node K. The coefficients are functions only of the element geometry and material properties; they are independent of the boundary conditions of the problem. As a result, they can be evaluated by the "trick" of applying a

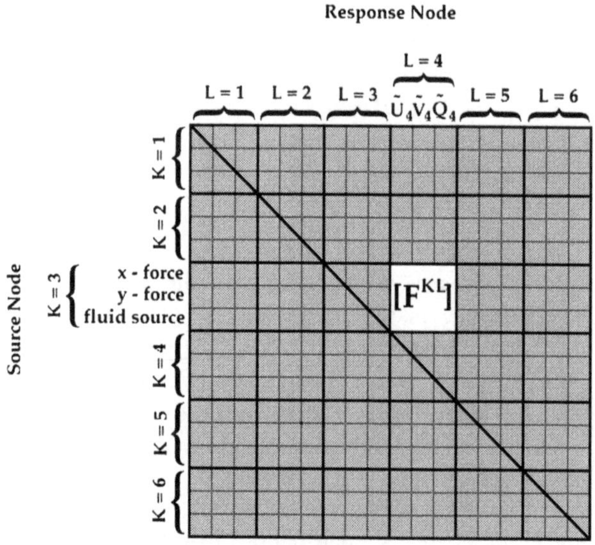

Figure 10.6: Structure of the 18×18 global coefficient matrix $[F]$ for a boundary containing six nodes (after Kane, 1994, p. 195). The building blocks of the global matrix are 3×3 matrices $[F^{KL}]$. The K-th triplet of rows of $[F]$ represents the equations for the source node being at K. The first row of the triplet is associated with an x-component point force, the second row with a y-component point force, and the third row with a fluid source. The three columns of $[F^{KL}]$ are the coefficients of the node L responses $\widetilde{U}_L$, $\widetilde{V}_L$, and $\widetilde{Q}_L$. For sequentially numbered nodes, each of the odd-numbered three-column blocks is the sum of contributions from the two elements on either side of node L.

rigid-body translation. Uniform displacements mean that stresses and surface tractions are zero everywhere, as are all the fluid variables. For example, a rigid body motion $U_1 = U_2 = \cdots = U_{\text{NNODE}} \neq 0$ and $V_1 = V_2 = \cdots = V_{\text{NNODE}} = 0$ inserted into the top row of Eqn. 10.44 gives

$$\sum_{L=1}^{\text{NNODE}} F_{11}^{KL} = 0 \tag{10.45}$$

The permutations of rigid body motions in the different coordinate directions in Eqn. 10.44 lead to expressions for four of the strongly singular entries in $[F^{KK}]$,

$$F_{ij}^{KK} = -\sum_{\substack{L=1 \\ L\neq K}}^{\text{NNODE}} F_{ij}^{KL} \tag{10.46}$$

where i and j each take on the values 1 and 2.

The strongly singular G_{33}^{KK} term is the coefficient of the pore pressure at node K due to a pressure source at node K. In this case the "trick" is to apply a uniform undrained, hydrostatic compression $-P_c$. The fluid flux is zero because of the assumed undrained condition, and the Laplace transform of the pore pressure, traction components, and displacement components are all uniform and calculated easily,

$$\begin{aligned}
\widetilde{U}_L &= -\frac{(1-2\nu_u)x_L}{2G}\frac{P_c}{s} \\
\widetilde{V}_L &= -\frac{(1-2\nu_u)y_L}{2G}\frac{P_c}{s} \\
\widetilde{Q}_L &= 0 \\
\widetilde{P} &= \frac{2B(1+\nu_u)}{3}\frac{P_c}{s} \\
\widetilde{X}_L &= -\frac{P_c}{s}n_x^L \\
\widetilde{Y}_L &= -\frac{P_c}{s}n_y^L
\end{aligned} \tag{10.47}$$

where (n_x^L, n_y^L) is the unit normal to the boundary at node L. Inserting Eqns. 10.47 into Eqn. 10.44 gives

$$G_{33}^{KK} = -\sum_{\substack{L=1 \\ L\neq K}}^{\text{NNODE}} G_{33}^{KL} + \frac{3}{2B(1+\nu_u)} \sum_{\substack{L=1 \\ L\neq K}}^{\text{NNODE}} \left\{n_x^L G_{31}^{KL} + n_y^L G_{32}^{KL}\right\}$$

$$- \frac{3(1-2\nu_u)}{4BG(1+\nu_u)} \sum_{\substack{L=1 \\ L\neq K}}^{\text{NNODE}} \left\{x_L F_{31}^{KL} + y_L F_{32}^{KL}\right\} \qquad (10.48)$$

The algebraic equations must be solved for every discrete value of the Laplace transform variable s. Once the boundary nodal unknowns are obtained, the Laplace transform of the displacement components and pore pressure at an interior point can be recovered by applying Eqns. 10.39 and 10.40. According to Cheng and Detournay (1988), Schapery's (1962) method yields satisfactory results for the inversion back to the time domain for fewer than 10 values of s. The results for several classical poroelastic problems using this boundary element formulation were presented by Cheng and Detournay (1988).

Appendix A

Integral Transforms

A.1 PROPERTIES OF FOURIER TRANSFORMS

1. The Fourier transform of f with respect to x is defined by

$$\bar{f}(\lambda) \equiv \mathcal{F}\{f(x)\} \equiv \int_{-\infty}^{\infty} f(x) \exp(-\imath\lambda x)\, dx \tag{A.1}$$

2. The inverse Fourier transform is given by

$$f(x) = \frac{1}{2\pi} \int_{-\infty}^{\infty} \bar{f}(\lambda) \exp(\imath\lambda x)\, d\lambda \tag{A.2}$$

3. The Fourier transform of the first derivative of f is given by

$$\mathcal{F}\left\{\frac{df}{dx}\right\} = -\imath\lambda\bar{f} \tag{A.3}$$

4. The Fourier transform of the second derivative of f is given by

$$\mathcal{F}\left\{\frac{d^2 f}{dx^2}\right\} = -\lambda^2\bar{f} \tag{A.4}$$

5. The Fourier cosine transform of a function $f(x)$ is defined by

$$\bar{f}(\xi) \equiv \mathcal{F}\{f(x)\} = \sqrt{\frac{2}{\pi}} \int_{o}^{\infty} f(x) \cos(\xi x)\, dx \tag{A.5}$$

   The inverse transform is symmetrical.

6. The Fourier cosine transform of the second derivative of $f(x)$ is

$$\mathcal{F}\left\{\frac{\partial^2 f}{\partial x^2}\right\} = -\xi^2\mathcal{F}\{f\} \tag{A.6}$$

A.2 PROPERTIES OF LAPLACE TRANSFORMS

Several theorems involving the Laplace transform can be found in Carslaw and Jaeger (1959, p. 299), LePage (1961, p. 374), and Oppenheim et al. (1983, p. 603).

1. The Laplace transform of $f(t)$ is defined by

$$\mathcal{L}\{f(t)\} \equiv \tilde{f}(s) \equiv \int_0^\infty e^{-st} f(t)\,dt \tag{A.7}$$

where the tilde indicates the Laplace transform, and s is the transform variable.

2. The Laplace transform of a constant $aH(t)$, where $H(t)$ is the Heaviside function, is a/s:

$$\mathcal{L}\{a\} = \frac{a}{s} \tag{A.8}$$

3. The Laplace transform of the sum of two functions, $f_1(t)$ and $f_2(t)$, is the sum of the Laplace transforms:

$$\begin{aligned}\mathcal{L}\{f_1(t) + f_2(t)\} &= \mathcal{L}\{f_1(t)\} + \mathcal{L}\{f_2(t)\} \\ &= \tilde{f}_1(s) + \tilde{f}_2(s)\end{aligned} \tag{A.9}$$

4. The Laplace transform of the time derivative of $f(t)$ is given by

$$\mathcal{L}\left\{\frac{df}{dt}\right\} = s\mathcal{L}\{f\} - f(0) \tag{A.10}$$

Note that incorporation of the initial condition $f(0)$ is built into this theorem for the time derivative.

5. The Laplace transform of the n-th derivative of $f(y, t)$ with respect to y is the n-th derivative of the Laplace transform:

$$\mathcal{L}\left\{\frac{\partial^n f}{\partial y^n}\right\} = \frac{\partial^n \tilde{f}}{\partial y^n} \tag{A.11}$$

6. The shift theorem gives the inverse Laplace transform of $\tilde{f}(s + a)$:

$$\mathcal{L}^{-1}\{\tilde{f}(s + a)\} = e^{-at} f(t) \tag{A.12}$$

7. The time-scaling theorem gives the inverse Laplace transform of $\tilde{f}(s/c)$, where c is a constant:

$$\mathcal{L}^{-1}\{\tilde{f}(s/c)\} = |c| f(ct) \quad \text{(A.13)}$$

8. The convolution theorem states that the Laplace transform of the convolution integral of two functions is the product of the Laplace transforms:

$$\mathcal{L}\left\{\int_0^t f_1(\tau) f_2(t-\tau)d\tau\right\} = \mathcal{L}\{f_1(t)\}\mathcal{L}\{f_2(t)\} \quad \text{(A.14)}$$

9. The inverse Laplace transform of $s^{-1}\tilde{f}(s)$ can be obtained from Eqns. A.14 and A.8 to be an integral with respect to time of $f(t)$:

$$\mathcal{L}^{-1}\left\{\frac{1}{s}\tilde{f}(s)\right\} = \int_0^t f(\tau)d\tau \quad \text{(A.15)}$$

where $\mathcal{L}^{-1}\{\tilde{f}(s)\} = f(t)$.

10. The initial-value theorem states that the limit as $t \to 0^+$ of $f(t)$ can be obtained as the limit as $s \to \infty$ of $s\tilde{f}(s)$:

$$f(0^+) = \lim_{s\to\infty} s\tilde{f}(s) \quad \text{(A.16)}$$

11. The final-value theorem states that limit as $t \to \infty$ of $f(t)$ can be obtained as the limit as $s \to 0$ of $s\tilde{f}(s)$:

$$f(\infty) = \lim_{s\to 0} s\tilde{f}(s) \quad \text{(A.17)}$$

A.3 STEHFEST ALGORITHM FOR LAPLACE TRANSFORM INVERSION

The Stehfest (1970) algorithm for numerically evaluating the inverse Laplace transform of a function $P(s)$ is based on the equation

$$F = \frac{\ln 2}{t}\sum_{i=1}^{N} V_i P\left(\frac{\ln 2}{t} i\right) \quad \text{(A.18)}$$

F is the inverse transform evaluated at time t, N must be even, and the coefficients V_i depend on N:

$$V_i = (-1)^{((N/2)+i)} \sum_{k=\left[\frac{i+1}{2}\right]}^{\min(i,N/2)} \frac{k^{N/2}(2k)!}{((N/2)-k)!k!(k-1)!(i-k)!(2k-i)!} \tag{A.19}$$

where the lower limit of the summation is the greatest integer on dividing $i+1$ by two. Note that there is a typographical error in the expression for V_i in Stehfest's original paper (Moench and Ogata, 1984, p. 150).

A.4 PROPERTIES OF HANKEL TRANSFORMS

Useful relations among Hankel transforms are given by Sneddon (1972, p. 310 ff).

1. The m-th order Hankel transform of $F(r)$ is defined by

$$\mathcal{H}_m\{F(r)\} = \bar{F}(k) = \int_0^\infty F(r) J_m(kr)\, r dr \tag{A.20}$$

where $J_m(kr)$ is the Bessel function of the first kind of order m. The inverse transform is defined symmetrically as

$$F(r) = \int_0^\infty \overline{F}(k) J_m(kr)\, k dk \tag{A.21}$$

2. A general m-th order relationship turns out to be especially useful for $m = 0$ and $m = 1$:

$$\mathcal{H}_m\left\{\frac{\partial^2 F}{\partial r^2} + \frac{1}{r}\frac{\partial F}{\partial r} - \frac{m^2}{r^2}F\right\} = -k^2\mathcal{H}_m\{F\} \tag{A.22}$$

In particular, for $m = 0$,

$$\mathcal{H}_o\left\{\frac{\partial^2 F}{\partial r^2} + \frac{1}{r}\frac{\partial F}{\partial r}\right\} = -k^2\mathcal{H}_o\{F\} \tag{A.23}$$

The quantity in braces on the left-hand side of Eqn. A.23 is the same as the radial derivative terms of the Laplacian in cylindrical coordinates.

For $m = 1$,

$$\mathcal{H}_1\left\{\frac{\partial^2 F}{\partial r^2} + \frac{1}{r}\frac{\partial F}{\partial r} - \frac{F}{r^2}\right\} = -k^2\mathcal{H}_1\{F\} \tag{A.24}$$

3. The first-order Hankel transform of $\partial F/\partial r$ is related to the zeroth-order Hankel transform of F:

$$\mathcal{H}_1\left\{\frac{\partial F}{\partial r}\right\} = -k\mathcal{H}_o\{F\} \tag{A.25}$$

4. Finally, the zeroth-order Hankel transform of $\partial F/\partial r + F/r$ is related to the first-order Hankel transform of F:

$$\mathcal{H}_o\left\{\frac{\partial F}{\partial r} + \frac{F}{r}\right\} = k\mathcal{H}_1\{F\} \tag{A.26}$$

The quantity in braces on the left-hand side of Eqn. A.26 is the sum of radial and circumferential strain for $F = u_r$.

Appendix B

Relations Among Poroelastic Constants

TABLE B.1.
Elastic Constant Relationships for an Isotropic Material (Birch, 1966)

K	E	λ	ν	K_ν	G
$\lambda + \frac{2G}{3}$	$G\frac{3\lambda + 2G}{\lambda + G}$	—	$\frac{\lambda}{2(\lambda + G)}$	$\lambda + 2G$	—
—	$9K\frac{K - \lambda}{3K - \lambda}$	—	$\frac{\lambda}{3K - \lambda}$	$3K - 2\lambda$	$\frac{3}{2}(K - \lambda)$
—	$\frac{9KG}{3K + G}$	$K - \frac{2G}{3}$	$\frac{3K - 2G}{2(3K + G)}$	$K + \frac{4G}{3}$	—
$\frac{EG}{3(3G - E)}$	—	$G\frac{E - 2G}{3G - E}$	$\frac{E}{2G} - 1$	$G\frac{4G - E}{3G - E}$	—
—	—	$3K\frac{3K - E}{9K - E}$	$\frac{3K - E}{6K}$	$3K\frac{3K + E}{9K - E}$	$\frac{3KE}{9K - E}$
$\lambda\frac{1 + \nu}{3\nu}$	$\lambda\frac{(1 + \nu)(1 - 2\nu)}{\nu}$	—	—	$\lambda\frac{1 - \nu}{\nu}$	$\lambda\frac{1 - 2\nu}{2\nu}$
$G\frac{2(1 + \nu)}{3(1 - 2\nu)}$	$2G(1 + \nu)$	$G\frac{2\nu}{1 - 2\nu}$	—	$G\frac{2 - 2\nu}{1 - 2\nu}$	—
—	$3K(1 - 2\nu)$	$3K\frac{\nu}{1 + \nu}$	—	$3K\frac{1 - \nu}{1 + \nu}$	$3K\frac{1 - 2\nu}{2 + 2\nu}$
$\frac{E}{3(1 - 2\nu)}$	—	$\frac{E\nu}{(1 + \nu)(1 - 2\nu)}$	—	$\frac{E(1 - \nu)}{(1 + \nu)(1 - 2\nu)}$	$\frac{E}{2 + 2\nu}$

TABLE B.2.
Comparison of Notation Used in Poroelastic Literature (after Green and Wang, 1986)

Biot (1941a)	Biot (1955)	Biot and Willis (1957)	Rice and Cleary (1976)	This Book
$-\frac{1}{3}(\sigma_x + \sigma_y + \sigma_z)$	$-\frac{1}{3}(\sigma_{xx} + \sigma_{yy} + \sigma_{zz}) - \sigma$	$-\frac{1}{3}(\tau_{xx} + \tau_{yy} + \tau_{zz}) = p'$	$-\frac{1}{3}\sigma_{kk}$	$-\sigma$
σ	$p = \frac{-\sigma}{f}$	p	p	p
θ	$f(e - \epsilon)$	$\zeta = f(e - \epsilon)$	$(m - m_o)/\rho_o$	$\zeta = dm_f/\rho_{f_o}$
H	$\frac{\frac{2}{3}N + A - Q^2/R}{f(1 + Q/R)}$	$(\kappa - \delta)^{-1}$	$\left(\frac{1}{K} - \frac{1}{K'_s}\right)^{-1}$	H
R	—	$(\gamma - \delta + \kappa)^{-1}$	$\left[\left(\frac{1}{K} - \frac{1}{K'_s}\right) + \left(\frac{v_o}{K_f} - \frac{v_o}{K''_s}\right)\right]^{-1}$	R
$\frac{1}{R} - \frac{1}{H}$	—	γ	$\left(\frac{v_o}{K_f} - \frac{v_o}{K''_s}\right)$	$\frac{1}{R} - \frac{1}{H}$
—	$\frac{2}{3}N + A - Q^2/R^{-1}$	κ	$\frac{1}{K}$	$\frac{1}{K}$
—	$\frac{1 - (1 + Q/R)f}{\frac{2}{3}N + A - Q^2/R}$	δ	$\frac{1}{K'_s}$	$\frac{1}{K'_s}$
—	f	f	v_o	ϕ
—	—	c	$\frac{1}{K_f}$	$\frac{1}{K_f}$
—	—	$c - \frac{\gamma}{f}$	$\frac{1}{K''_s}$	$\frac{1}{K_\phi}$
—	—	—	$\frac{1}{K_s}$	$\frac{1}{K_s}$

TABLE B.3.
Useful Poroelastic Constant Relationships

$$\eta = \frac{\alpha(1-2\nu)}{2(1-\nu)}$$

$$\gamma = \frac{B(1+\nu_u)}{3(1-\nu_u)} = \frac{\eta}{GS}$$

$$B = \frac{3(\nu_u-\nu)}{\alpha(1-2\nu)(1+\nu_u)} = \frac{3(\nu_u-\nu)}{2\eta(1-\nu)(1+\nu_u)}$$

$$M = \frac{2G(\nu_u-\nu)}{\alpha^2(1-2\nu_u)(1-2\nu)} = \frac{BK_u}{\alpha}$$

$$S = \frac{(1-\nu_u)(1-2\nu)}{M(1-\nu)(1-2\nu_u)}$$

$$S_\sigma = \frac{\alpha}{KB}$$

$$S_\epsilon = \frac{\alpha}{K_u B}$$

Appendix C

Representative Poroelastic Constants

TABLE C.1.
Example Poroelastic Moduli for Different Rock Types*

Rock	G(GPa)	ν	K(GPa)	K_s'(GPa)	α	ν_u	K_u(GPa)	B	c(m^2/s)	ϕ	k(mD)
Berea sandstone	6	0.20	8.0	36	0.79	0.33	16.0	0.62	1.6×10^0	0.19	1.9×10^2
Boise sandstone	4.2	0.15	4.6	42	0.85	0.31	8.3	0.50	4.0×10^{-1}	0.26	8.0×10^2
Ohio sandstone	6.8	0.18	8.4	31	0.74	0.28	13.0	0.50	3.9×10^{-2}	0.19	5.6×10^0
Pecos sandstone	5.9	0.16	6.7	39	0.83	0.31	14.0	0.61	5.4×10^{-3}	0.20	8.0×10^{-1}
Ruhr sandstone	13	0.12	13	36	0.65	0.31	30.0	0.88	5.3×10^{-3}	0.02	2.0×10^{-1}
Weber sandstone	12	0.15	13	36	0.64	0.29	25.0	0.73	2.1×10^{-2}	0.06	1.0×10^0
Tennessee marble	24	0.25	40	50	0.19	0.27	44.0	0.51	1.3×10^{-5}	0.02	1.0×10^{-4}
Charcoal granite	19	0.27	35	45	0.27	0.30	41.0	0.55	7.0×10^{-6}	0.02	1.0×10^{-4}
Westerly granite	15	0.25	25	45	0.47	0.34	42.0	0.85	2.2×10^{-5}	0.01	4.0×10^{-4}
Clay	—	—	0.062	∞	1	—	6.2	0.99	—	—	—
Mudstone	—	—	2.13	42	0.95	—	10.1	0.83	—	—	—
Kayenta sandstone	—	—	9.1	37.9	0.76	—	18.5	0.67	—	—	—
Limestone	—	—	33.3	107	0.69	—	40.2	0.25	—	—	—
Hanford basalt	—	—	46.7	59	0.23	—	45.4	0.12	—	—	—
Berea sandstone	5.6	0.17	6.6	28.9	0.77	0.34	15.8	0.75	1.5×10^0	0.19	1.9×10^2
Indiana limestone	12.1	0.26	21.2	72.6	0.71	0.33	31.2	0.46	—	0.13	—

* The top section of the table is a combination of measured and calculated values compiled by Detournay and Cheng (1993), who incorporated values previously presented by Rice and Cleary (1976). Typically, drained moduli are measured values, and the solid-grain modulus is taken from handbooks of mineral properties. The undrained moduli are computed assuming that the solid-grain modulus K_s is equal to both the unjacketed bulk modulus K_s' and the unjacketed pore incompressibility K_ϕ. The fluid-bulk modulus was assumed to be $K_f = 3.3$ GPa. The second section of the table compiled by Palciauskas and Domenico (1989) and reprinted in Domenico and Schwartz (1998, p. 171) also combined measurement and calculation. The bottom set of values for Berea sandstone and Indiana limestone is a best-fit set of constants based on eight different measurements by Hart and Wang (1995) in which water ($K_f = 2.3$ GPa) was the pore fluid.

Symbols

c consolidation coefficient (Eqn. 1.1), hydraulic diffusivity (Eqn. 3.65)
c_m Geertsma's uniaxial poroelastic expansion coefficient (Eqn. 3.71)
g acceleration of gravity
h Darcy's head (Eqn. 4.49)
k permeability (Eqn. 4.52)
m_f fluid mass content (Eqn. 1.5)
p fluid (pore) pressure
q_i i-th component of fluid flux (Eqn. 4.43)
r, θ, z cylindrical coordinates
t time
t_i i-th component of traction
u, v, w rectangular components of displacement
u_x, u_y, u_z rectangular components of displacement
x, y, z rectangular coordinates

B Skempton's coefficient (Eqn. 1.17)
E Young's modulus (Eqn. 2.11)
G Shear modulus (Eqn. 2.33)
H reciprocal of poroelastic expansion coefficient (Eqn. 1.8)
$H(x)$ Heaviside step function (Eqn. 7.136)
K bulk modulus (Eqn. 1.8)
K_f fluid bulk modulus
K_p drained pore bulk modulus (Eqn. 3.9)
K_s' unjacketed bulk modulus (Eqn. 3.7)
K_u undrained bulk modulus (Eqn. 2.22)
K_v uniaxial drained bulk modulus (Eqn. 3.55)
$K_v^{(u)}$ uniaxial undrained bulk modulus (Eqn. 3.56)
K_ϕ unjacketed pore bulk modulus (Eqn. 3.14)
M Biot modulus, reciprocal of constant strain storage coefficient (Eqn. 1.19)
P_c confining pressure
P_d differential pressure (Eqn. 3.1)
R Biot modulus, reciprocal of constant stress storage coefficient (Eqn. 1.8)
S uniaxial specific storage (Eqn. 3.45)
S_s hydrogeological uniaxial specific storage (Eqn. 3.45)

S_γ unjacketed specific storage, specific storage at constant differential pressure (Eqn. 3.60)
S_σ unconstrained specific storage, three-dimensional specific storage, specific storage at constant stress (Eqn. 1.9)
S_ϵ constrained specific storage, specific storage at constant strain (Eqn. 1.19)
V bulk volume
V_f fluid volume
V_p pore volume
V_s solid volume

α Biot-Willis coefficient (Eqn. 1.24)
γ loading efficiency (Eqn. 3.76)
δ_{ij} Kronecker delta (Eqn. 2.37)
$\delta(x)$ Dirac delta function (Eqn. 5.24)
ϵ volumetric strain (Eqn. 1.11)
ϵ_{ij} strain tensor component (Eqn. 2.7)
η poroelastic stress coefficient (Eqn. 3.47)
λ drained Lamé's modulus (Eqn. 2.42)
λ_u undrained Lamé's modulus (Eqn. 2.54)
μ viscosity (Eqn. 4.52)
ν drained Poisson's ratio (Eqn. 3.30)
ν_u undrained Poisson's ratio (Eqn. 3.31)
ϕ porosity
ρ bulk density
ρ_f fluid density
σ mean stress
σ_{ij} stress tensor component
ζ increment of fluid content (Eqn. 2.9)
Φ Hubbert's force potential (Chap. 4, Eqn. 4.46), displacement potential (Chap. 5, Eqn. 5.1)

References

Abousleiman, Y., A. H.-D. Cheng, L. Cui, E. Detournay, and J.-C. Roegiers (1996). Mandel's problem revisited. *Géotechnique 46*, 187–195.

Abramowitz, M., and I. A. Stegun (eds.) (1964). *Handbook of Mathematical Functions. Appl. Math. Ser. 55*. National Bureau of Standards, Washington, D.C.

Adachi, J. I., and E. Detournay (1997). A poroelastic solution of the oscillating pore pressure method to measure permeabilities of "tight" rocks. *Int. J. Rock Mech. Min. Sci. Geomech. Abstr. 34:3–4*, Paper No. 062.

Banerjee, P. K., and R. Butterfield (1981). *Boundary Element Methods in Engineering Science*. McGraw-Hill, London.

Bear, J. (1972). *Dynamics of Fluids in Porous Media*. Elsevier, New York.

Beaumont, C., and J. Berger (1975). An analysis of tidal strain observations for the United States of America. 1, The laterally homogeneous tide. *Bull. Seismol. Soc. Am. 65*, 1615–1629.

Berge, P. A. (1998). Pore compressibility in rocks. In *Poromechanics*, ed. J.-F. Thimus, Y. Abousleiman, A.H.-D. Cheng, O. Coussy, and E. Detournay, pp. 351–356. Balkema, Rotterdam.

Berge, P. A., and J. G. Berryman (1995). Realizability of negative pore compressibility in poroelastic composites. *ASME J. Appl. Mech. 62*, 1053–1062.

Berge, P. A., H. F. Wang, and B. P. Bonner (1993). Pore pressure buildup coefficient in synthetic and natural sandstones. *Int. J. Rock Mech. Min. Sci. Geomech. Abstr. 30*, 1135–1141.

Berger, J., and C. Beaumont (1976). An analysis of tidal strain observations for the United States of America. 2, The inhomogeneous tide. *Bull. Seismol. Soc. Am. 66*, 1821–1846.

Berryman, J. G. (1992). Effective stress for transport properties of inhomogeneous porous rock. *J. Geophys. Res. 97*, 17409–17424.

Berryman, J. G. (1993). Effective-stress rules for pore-fluid transport in rocks containing two minerals. *Int. J. Rock Mech. Min. Sci. Geomech. Abstr. 30*, 1165–1168.

Berryman, J. G., and G. W. Milton (1992). Exact results in linear thermomechanics of fluid-saturated porous media. *Appl. Phys. Letters 61*, 2030–2032.

Biot, M. A. (1941a). General theory of three-dimensional consolidation. *J. Appl. Phys. 12*, 155–164.

Biot, M. A. (1941b). Consolidation settlement under a rectangular load distribution. *J. Appl. Phys. 12*, 426–430.

Biot, M. A. (1955). Theory of elasticity and consolidation for a porous anisotropic solid. *J. Appl. Phys. 26*, 182–185.

Biot, M. A. (1956a). Theory of propagation of elastic waves in a fluid-saturated porous solid. I. Low-frequency range. *J. Acoustical Soc. America 28*, 168–178.

Biot, M. A. (1956b). Theory of propagation of elastic waves in a fluid-saturated porous solid. II. Higher frequency range. *J. Acoustical Soc. America 28*, 179–191.

Biot, M. A. (1956c). Thermoelasticity and irreversible thermodynamics. *J. Appl. Phys. 27*, 240–253.

Biot, M. A. (1956d). General solutions of the equations of elasticity and consolidation for a porous material. *J. Appl. Mech. 78*, 91–96.

Biot, M. A. (1962). Mechanics of deformation and acoustic propagation in porous media. *J. Appl. Phys. 33*, 1482–1498.

Biot, M. A. (1973). Nonlinear and semilinear rheology of porous solids. *J. Geophys. Res. 78*, 4924–4937.

Biot, M. A., and D. G. Willis (1957). The elastic coefficients of the theory of consolidation. *J. Appl. Mech. 24*, 594–601.

Birch, F. (1966). Compressibility: Elastic Constants. In *Handbook of Physical Constants—Revised Edition*, ed. S.P. Clark, Jr., pp. 97–173, *Geological Society of America Memoir 97*. Geological Society of America, New York.

Boley, B. A., and J. H. Weiner (1985). *Theory of Thermal Stresses*, reprint ed., Krieger, Malabar, Florida.

Booker, J. R. (1974). Time dependent strain following faulting of a porous medium. *J. Geophys. Res. 79*, 2037–2044.

Boyce, W. E., and R. C. DiPrima (1977). *Elementary Differential Equations and Boundary Value Problems*, 3rd ed. John Wiley, New York.

Brace, W. F., J. B. Walsh, and W. T. Frangos (1968). Permeability of granite under high pressure. *J. Geophys. Res. 73*, 2225–2236.

Bredehoeft, J. D., and B. B. Hanshaw (1968). On the maintenance of anomalous fluid pressures: I. Thick sedimentary sequences. *Geol. Soc. Am. Bull. 79*, 1097–1106.

Brown, R.J.S., and J. Korringa (1975). On the dependence of the elastic properties of a porous rock on the compressibility of the pore fluid. *Geophysics 40*, 608–616.

Burnett, D. S. (1987). *Finite Element Analysis: From Concepts to Applications*. Addison-Wesley, Reading, Massachusetts.

Carrier, G. F., M. Krook, and C. E. Pearson (1966). *Functions of a Complex Variable: Theory and Technique*. McGraw-Hill, New York.

Carroll, M. M. (1979). An effective stress law for anisotropic elastic deformation. *J. Geophys. Res. 84*, 7510–7512.

Carroll, M. M. (1980). Mechanical response of fluid-saturated porous materials. In *Theoretical and Applied Mechanics*, ed. F.P.J. Rimrott and B. Taborrok, pp. 251–262. North-Holland, Amsterdam.

Carslaw, H. S., and J. C. Jaeger (1959). *Conduction of Heat in Solids*, 2nd ed. Oxford Univ. Press, UK.

Cheng, A. H.-D. (1997). Material coefficients of anisotropic poroelasticity. *Int. J. Rock Mech. Min. Sci. 34*, 199–205.

Cheng, A. H.-D., Y. Abousleiman, and J.-C. Roegiers (1993). Review of some poroelastic effects in rock mechanics. *Int. J. Rock Mech. Min. Sci. Geomech. Abstr. 30*, 1119–1126.

Cheng, A. H.-D., and E. Detournay (1988). A direct boundary element method for plane strain poroelasticity. *Int. J. Num. Analyt. Methods Geomechanics 12*, 551–572.

Cheng, A. H.-D., and E. Detournay (1998). On singular integral equations and fundamental solutions of poroelasticity. *Int. J. Solids Structures 35*, 4521–4555.

Cheng, A. H.-D., E. Detournay, and Y. Abousleiman (1998). A tribute to Maurice A. Biot. *Int. J. Solids Structures 35*, 4515–4517.

Cleary, M. P. (1977). Fundamental solutions for a fluid-saturated porous solid. *Int. J. Solids Structures 13*, 785–806.

Craig, R. F. (1997). *Soil Mechanics*, 6th ed. E & FN Spon, London.

Crouch, S. L., and A. M. Starfield (1983). *Boundary Element Methods in Solid Mechanics: With Applications in Rock Mechanics and Geological Engineering*. Allen and Unwin, London.

Cryer, C. W. (1963). A comparison of the three-dimensional consolidation theories of Biot and Terzaghi. *Quart. J. Mech. Appl. Math. 16*, 401–412.

Darcy, H. (1856). *Les Fontaines de la Ville de Dijon*. Victor Dalmont, Paris.

de Boer, R. (1996). Highlights in the historical development of the porous media theory. *Appl. Mech. Rev. 49*, 201–262.

de Boer, R., and W. Ehlers (1990). The development of the concept of effective stresses. *Acta Mechanica 83*, 77–92.

de Boer, R., R. L. Schiffman, and R. E. Gibson (1996). The origins of the theory of consolidation: The Terzaghi-Fillunger dispute. *Géotechnique 46*, 175–186.

Detournay, E., and J. L. Carvalho (1989). Application of the pressurized hollow poroelastic cylinder solution to the interpretation of laboratory rock burst experiments. In *Rock Mechanics As a Guide for Efficient Utilization of Natural Resources*, ed. A. W. Khair, pp. 377–383. A. A. Balkema, Rotterdam, Netherlands.

Detournay, E., and Cheng A. H.-D. (1988). Poroelastic response of a borehole in a non-hydrostatic stress field. *Int. J. Rock Mech. Min. Sci. Geomech. Abstr. 25*, 171–182.

Detournay, E., and Cheng A. H.-D. (1993). Fundamentals of poroelasticity. In *Comprehensive Rock Engineering: Principles, Practice and Projects*, Vol. 2, ed. J. A. Hudson, pp. 113–171. Pergamon Press, Oxford, UK.

Domenico, P. A. (1983). Determination of bulk rock properties from groundwater level fluctuations. *Bull. Assoc. Eng. Geologists 83*, 283–287.

Domenico, P. A., and M. D. Mifflin (1965). Water from low permeability sediments and land subsidence. *Water Resources Res. 4*, 563–576.

Domenico, P. A., and F. W. Schwartz (1998). *Physical and Chemical Hydrogeology*, 2nd ed. John Wiley, New York.

Engelder, T., and A. Lacazette (1990). Natural hydraulic fracturing. In *Proceedings of the International Symposium on Rock Joints*, ed. N. Barton and O. Stephansson, pp. 35–43. A. A. Balkema, Rotterdam, Netherlands.

Erdélyi, A., W. Magnus, F. Oberhettinger, and F. Tricomi (1954). *Tables of Integral Transforms*, Vol. 4, McGraw-Hill, New York.

Fagan, M. J. (1992). *Finite Element Analysis: Theory and Practice*. Longman Scientific and Technical, Essex, England.

Fang, W. W., M. G. Langseth, and P. J. Schultheiss (1993). Analysis and application of in situ pore pressure measurements in marine sediments. *J. Geophys. Res. 98*, 7921–7938.

Freeze, R. A. (1985). Historical correspondence between C. V. Theis and C. I. Lubin. *Eos, Trans. AGU 66*, 442.

Freeze, R. A. (1994). Henry Darcy and the fountains of Dijon. *Ground Water 32*, 23–30.

Furbish, D. J. (1991). The response of water level in a well to a time series of atmospheric loading under confined conditions. *Water Resources Res. 27*, 557–568.

Gambolati, G. (1974). Second-order theory of flow in three-dimensional deforming media. *Water Resources Res. 10*, 1217–1228.

Gambolati, G. (1977). Deviations from the Theis solution in aquifers undergoing three-dimensional consolidation. *Water Resources Res. 13*, 62–68.

Gambolati, G., and R. A. Freeze (1973). Mathematical simulation of the subsidence of Venice. I. Theory. *Water Resources Res. 9*, 721–733.

Gambolati, G., P. Gatto, and R. A. Freeze (1974). Mathematical simulation of the subsidence of Venice. II. Results. *Water Resources Res. 10*, 563–577.

Gassmann, F. (1951). Über die Elastizität poröser Medien. *Vierteljahrschrift Naturforsch. Ges. Zürich 96*, 1–23.

Geertsma, J. (1966). Problems of rock mechanics in petroleum production engineering. In *Proc. 1st Cong. Int. Soc. Rock Mech.*, Vol. 1, pp. 585–594, Lisbon.

Geertsma, J. (1973a). A basic theory of subsidence due to reservoir compaction: The homogeneous case. *Verhandelingen Kon. Ned. Geol. Mijnbouwk. Gen. 28*, 43–62.

Geertsma, J. (1973b). Land subsidence above compacting oil and gas reservoirs. *J. Pet. Tech. 25*, 734–744.

Gibson, R. E. (1958). The progress of consolidation in a clay layer increasing in thickness with time. *Géotechnique 8*, 171–182.

Gibson, R. E., and J. McNamee (1963). A three-dimensional problem of the consolidation of a semi-infinite clay stratum. *Quart. J. Mech. Appl. Math. 16*, 115–127.

Gibson, R. E., R. L. Schiffman, and S. L. Pu (1970). Plane strain and axially symmetric consolidation of a clay layer on a smooth impervious base. *Quart. J. Mech. Appl. Math. 23*, 505–520.

Gould, P. M. (1983). *Introduction to Linear Elasticity*. Springer-Verlag, New York.

Green, D. H., and H. F. Wang (1986). Fluid pressure response to undrained compression in saturated sedimentary rock. *Geophysics 51*, 948–956.

Green, D. H., and H. F. Wang (1990). Specific storage as a poroelastic coefficient. *Water Resources Res. 26*, 1631–1637.

Haimson, B. C., and C. Fairhurst (1969). Hydraulic fracturing in porous-permeable materials. *J. Pet. Technol. 21*, 811–817.

Hantush, M. S. (1960). Modification of the theory of leaky aquifers. *J. Geophys. Res. 65*, 3713–3725.

Hart, D. J., and H. F. Wang (1995). Laboratory measurements of a complete set of poroelastic moduli for Berea sandstone and Indiana limestone. *J. Geophys. Res. 100*, 17741–17751.

Hart, D. J., and H. F. Wang (1998). Poroelastic effects during a laboratory transient pore pressure test. In *Poromechanics*, ed. J.-F. Thimus, Y. Abousleiman, A. H.-D. Cheng, O. Coussy, and E. Detournay, pp. 579–582. Balkema, Rotterdam.

Helm, D. C. (1994). Horizontal aquifer movement in a Theis-Thiem confined system. *Water Resources Res. 30*, 953–964.

Hibbitt, Karlsson, and Sorensen, Inc. (1998). *ABAQUS/Standard User's Manual*, Vols. I–III, Version 5.8. Pawtucket, Rhode Island.

Hsieh, P. A. (1994). *Guide to BIOT2: A Finite Element Model to Simulate Axisymmetric/Plane-Strain Solid Deformation and Fluid Flow in a Linearly Elastic Porous Medium*. U.S. Geological Survey, Menlo Park, California.

Hsieh, P. A. (1996). Deformation-induced changes in hydraulic head during groundwater withdrawal. *Ground Water 34*, 1082–1089.

Hsieh, P. A., and R. L. Cooley (1995). Horizontal aquifer movement in a Theis-Theim [*sic*] confined system; discussion and reply. *Water Resources Res. 31*, 3107–3118.

Hsieh, P. A., J. V. Tracy, C. E. Neuzil, J. D. Bredehoeft, and S. E. Silliman (1981). A transient laboratory method for determining the hydraulic properties of 'tight' rocks—I. Theory. *Int. J. Rock Mech. Min. Sci. Geomech. Abstr. 18*, 245–252.

Hubbert, M. K. (1940). The theory of groundwater motion. *J. Geol. 48*, 785–944.

Hubbert, M. K. (1956). Darcy's law and the field equations of the flow of underground fluids. *Pet. Trans. AIME 207*, 222–239.

Hubbert, M. K., and D. G. Willis (1957). Mechanics of hydraulic fracturing. *J. Pet. Tech. 9*, 153–168.

Hughes, D. S., and C. E. Cooke Jr. (1953). The effect of pressure on the reduction of pore volume of consolidated sandstones. *Geophysics 18*, 298–309.

Iverson, R. M., and J. J. Major (1986). Groundwater seepage vectors and the potential for hillslope failure and debris flow mobilization. *Water Resources Res. 22*, 1543–1548.

Jackson, J. D. (1999). *Classical Electrodynamics*, 3rd ed. John Wiley, New York.

Jacob, C. E. (1940). On the flow of water in an elastic artesian aquifer. *Trans. Am. Geophys. Union 22*, 783–787.

Jacob, C. E. (1950). Flow of groundwater. In *Engineering Hydraulics*, pp. 321–386, John Wiley, New York.

Jaeger, J. C., and N.G.W. Cook (1976). *Fundamentals of Rock Mechanics*, 2nd ed. John Wiley, New York.

Johnson, A. G., R. L. Kovach, and A. Nur (1973). Pore pressure changes during creep events on the San Andreas fault. *J. Geophys. Res. 78*, 851–857.

Kane, J. H. (1994). *Boundary Element Analysis in Engineering Continuum Mechanics*. Prentice-Hall, Englewood Cliffs, New Jersey.

King, F. H. (1892). *Fluctuations in the Level and Rate of Movement of Ground-Water on the Wisconsin Agricultural Experiment Station Farm and at Whitewater, Wisconsin*. U.S. Department of Agriculture, Bull. No. 5. U.S. Department of Agriculture, Washington, D.C.

Kraut, E. A. (1967). *Fundamentals of Mathematical Physics*. McGraw-Hill, New York.

Kümpel, H.-J. (1991). Poroelasticity: Parameters reviewed. *Geophys. J. Int. 105*, 783–799.

Lambe, T. W., and R. V. Whitman (1979). *Soil Mechanics, SI Version*. John Wiley, New York.

Langguth, H. R., and C. Treskatis (1989). Reverse water level fluctuations in semiconfined aquifer systems—"Rhade effect." *J. Hydrology 109*, 79–93.

Laurent, J., M. J. Boutéca, J.-P. Sarda, and D. Bary (1993). Pore-pressure influence in the poroelastic behavior of rocks: Experimental studies and results. *SPE Formation Eval. 8*, 117–122.

Lee, C. W. (1966). Thermoelastic stresses in thick-walled cylinders under axial temperature gradient. *J. Appl. Mech. 33*, 467–469.

LePage, W. R. (1961). *Complex Variables and the Laplace Transform for Engineers*. McGraw-Hill, New York.

Lewallen, K. T., and H. F. Wang (1998). Consolidation of a double-porosity medium. *Int. J. Solids and Structures 35*, 4845–4867.

Lewis, R. W., and B. A. Schrefler (1987). *The Finite Element Method in the Deformation and Consolidation of Porous Media.* John Wiley, London.

Love, A.E.H. (1944). *A Treatise on the Mathematical Theory of Elasticity*, 4th ed. Dover, New York.

MacMillan, W. D. (1930). *The Theory of the Potential.* McGraw-Hill, New York.

Mandel, J. (1953). Consolidation des sols (étude mathématique). *Géotechnique 3*, 287–299.

Mason, D. P., A. Solomon, and L. O. Nicolaysen (1991). Evolution of stress and strain during the consolidation of a fluid-saturated porous elastic sphere. *J. Appl. Phys. 70*, 4724–4740.

McNamee, J., and R. E. Gibson (1960a). Displacement functions and linear transforms applied to diffusion through porous elastic media. *Quart. J. Mech. Appl. Math. 13*, 98–111.

McNamee, J., and R. E. Gibson (1960b). Plane strain and axially symmetric problems of the consolidation of a semi-infinite clay stratum. *Quart. J. Mech. Appl. Math. 13*, 210–227.

Means, W. D. (1976). *Stress and Strain: Basic Concepts of Continuum Mechanics for Geologists.* Springer-Verlag, New York.

Mei, C., and M. Foda (1982). Boundary layer theory of waves in a poroelastic sea bed. In *Soil Mechanics—Transient and Cyclic Loads*, ed. G. N. Pande and O. C. Zienkiewicz, pp. 17–35. John Wiley, New York.

Meinzer, O. E. (1928). Compressibility and elasticity of artesian aquifers. *Econ. Geol. 23*, 263–291.

Melan, E. (1932). Der Spannungszustand der durch eine Einzelkraft im Innern beanspruchten Halbscheibe. *Z. Angew. Math. Mech. 12*, 343–346. [Correction: (1940). *Z. Angew. Math. Mech. 20*, 368.]

Melchior, P. J. (1983). *The Tides of the Planet Earth*, 2nd ed. Pergamon, New York.

Mindlin, R. D. (1936). Force at a point in the interior of a semi-infinite solid. *Physics 7*, 195–202.

Mindlin, R. D., and D. H. Cheng (1950a). Nuclei of strain in the semi-infinite solid. *J. Appl. Phys. 21*, 926–930.

Mindlin, R. D., and D. H. Cheng (1950b). Thermoelastic stress in the semi-infinite solid. *J. Appl. Phys. 21*, 931–933.

Moench, A., and A. Ogata (1984). Analysis of constant discharge wells by numerical inversion of Laplace transform solutions. In *Groundwater Hydraulics*, ed. J. Rosenshein and G. D. Bennett, Water Resources Monograph Ser., vol. 9, pp. 146–170. American Geophysical Union, Washington, D.C.

Muskat, M. (1937). *The Flow of Homogeneous Fluids through Porous Media.* McGraw-Hill, New York.

Neuzil, C. E., C. Cooley, S. E. Silliman, J. D. Bredehoeft, and P. A. Hsieh (1981). A transient laboratory method for determining the hydraulic properties of 'tight' rocks—II. Application. *Int. J. Rock Mech. Min. Sci. Geomech. Abstr. 18*, 253–258.

Neuzil, C. E., and D. W. Pollock (1983). Erosional unloading and fluid pressures in hydraulically "tight" rocks. *J. Geol. 91*, 179–193.

Noorishad, J., M. S. Ayatollahi, and P. A. Witherspoon (1982). A finite-element method for coupled stress and fluid flow analysis in fractured rock masses. *Int. J.*

Rock Mech. Min. Sci. Geomech. Abstr. 19, 185–193.

Norris, A. N. (1992). On the correspondence between poroelasticity and thermoelasticity. *J. Appl. Phys. 71*, 1138–1141.

Nowacki, W. (1986). *Thermoelasticity*, 2nd ed. Pergamon, Oxford, UK.

Nur, A., and J. D. Byerlee (1971). An exact effective stress law for elastic deformation of rock with fluids. *J. Geophys. Res. 76*, 6414–6419.

Oppenheim, A. V., A. S. Willsky, and I. T. Young (1983). *Signals and Systems*. Prentice-Hall, Englewood Cliffs, New Jersey.

Orange, D. L., and N. A. Breen (1992). The effects of fluid escape on accretionary wedges. 2. Seepage force, slope failure, headless submarine canyons, and vents. *J. Geophys. Res. 97*, 9277–9295.

Palciauskas, V. V., and P. A. Domenico (1989). Fluid pressures in deforming porous rocks. *Water Resources Res. 25*, 203–213.

Peirce, B. O., and R. M. Foster (1956). *A Short Table of Integrals*, 4th ed. Ginn, New York.

Pratt, W. E., and D. W. Johnson (1926). Local subsidence of the Goose Creek oil field. *J. Geology 34*, 577–590.

Rajapakse, R.K.N.D., and T. Senjuntichai (1993). Fundamental solutions for a poroelastic half-space with compressible constituents. *J. Appl. Mech. 60*, 847–856.

Rendulic, L. (1936). Prenziffer und Porenwasserdruck in Tonen. *Der Bauingenieur 17*, 559–564.

Rice, J. R., and M. P. Cleary (1976). Some basic stress diffusion solutions for fluid saturated elastic porous media with compressible constituents. *Rev. Geophys. Space Phys. 14*, 227–241.

Roeloffs, E. A. (1988). Fault stability changes induced beneath a reservoir with cyclic variations in water level. *J. Geophys. Res. 93*, 2107–2124.

Roeloffs, E. A. (1995). Poroelastic techniques in the study of earthquake-related hydrologic phenomena. *Adv. Geophys. 37*, 135–195.

Roeloffs, E., and J. W. Rudnicki (1984/85). Coupled deformation-diffusion effects on water-level changes due to propagating creep events. *Pure Appl. Geophys. 122*, 560–582.

Rojstaczer, S., and D. C. Agnew (1989). The influence of formation material properties on the response of water levels in wells to earth tides and atmospheric loading. *J. Geophys. Res. 94*, 12403–12411.

Rudnicki, J. W. (1986a). Fluid mass sources and point forces in linear elastic diffusive solids. *Mech. Mater. 5*, 383–393.

Rudnicki, J. W. (1986b). Slip on an impermeable fault in a fluid-saturated rock mass. In *Earthquake Source Mechanics*, ed. S. Das, J. Boatwright, and C. H. Scholz, Geophys. Monograph Ser., Vol. 37, pp. 81–89. American Geophysical Union, Washington, D.C.

Rudnicki, J. W. (1987). Plane strain dislocations in linear elastic diffusive solids. *J. Appl. Mech. 54*, 545–552.

Rudnicki, J. W., and T.-C. Hsu (1988). Pore pressure changes induced by slip on permeable and impermeable faults. *J. Geophys. Res. 93*, 3275–3285.

Rudnicki, J. W., and E. A. Roeloffs (1990). Plane-strain shear dislocations moving steadily in linear elastic diffusive solids. *J. Appl. Mech., 57*, 32–39.

Sandhu, R. S., and E. L. Wilson (1969). Finite-element analysis of seepage in elastic media. *J. Engineering Mech. Div.*, ASCE Proc., *95* (No. EM3), 641–652.

Schapery, R. A. (1962). Approximate methods of transform inversion for viscoelastic stress analysis. *Proc. 4th U.S. National Congress on Appl. Mech. 2*, 1075–1085.

Schmitt, D. R., R. J. Tait, and H. Spann (1993). Solution for pore pressure and stress in a porous hollow cylinder: Application to a laboratory experiment. *Int. J. Rock Mech. Min. Sci. Geomech. Abstr. 30*, 1057–1060.

Schmitt, D. R., and M. D. Zoback (1992). Diminished pore pressure in low-porosity crystalline rock under tensional failure: Apparent strengthening by dilatancy. *J. Geophys. Res. 97*, 273–288.

Schultheiss, P. J. (1990). In-situ pore-pressure measurements for a detailed geotechnical assessment of marine sediments: State of the art. In *Geotechnical Engineering of Ocean Waste Disposal, ASTM STP 1087*, ed. K. R. Demars and R. C. Chaney, pp. 190–205. American Society for Testing and Materials, Philadelphia.

Segall, P. (1985). Stress and subsidence resulting from subsurface fluid withdrawal in the epicentral region of the 1983 Coalinga earthquake. *J. Geophys. Res. 90*, 6801–6816.

Segall, P. (1989). Earthquakes triggered by fluid extraction. *Geology 17*, 942–946.

Segall, P. (1992). Induced stresses due to fluid extraction from axisymmetric reservoirs. *Pure Appl. Geophys. 139*, 535–560.

Segall, P., J.-R. Grasso, and A. Mossop (1994). Poroelastic stressing and induced seismicity near the Lacq gas field, southwestern France. *J. Geophys. Res. 99*, 15423–15438.

Sen, B. (1951). Note on the stresses produced by nuclei of thermo-elastic strain in a semi-infinite elastic solid. *Quart. Appl. Math. 8*, 365–369.

Skempton, A. W. (1960). Terzaghi's discovery of effective stress. In *From Theory to Practice in Soil Mechanics*, ed. L. Bjerrum, A. Casagrande, R. B. Peck, and A. W. Skempton, pp. 42–53. John Wiley, New York.

Smith, I. M., and D. V. Griffiths (1988). *Programming the Finite Element Method*, 2nd ed. John Wiley, New York.

Sneddon, I. N. (1951). *Fourier Transforms*. McGraw-Hill, New York.

Sneddon, I. N. (1972). *The Use of Integral Transforms*. McGraw-Hill, New York.

Stehfest, H. (1970). Numerical inversion of Laplace transforms. *Commun. ACM 13*, 47–49.

Streltsova, T. D. (1988). *Well Testing in Heterogeneous Formations*. John Wiley, New York.

Szabo, B. A., and G. C. Lee (1969). Derivation of stiffness matrices for problems in plane elasticity by Galerkin's method. *Int. J. Num. Methods Engineering 1*, 301–310.

Telford, W. M., L. P. Geldart, and R. E. Sheriff (1990). *Applied Geophysics*, 2nd ed. Cambridge Univ. Press, London.

Terzaghi, K. (1925). *Erdbaumechanik auf bodenphysikalischer Grundlage*. Deuticke, Leipzig.

Terzaghi, K. (1943). *Theoretical Soil Mechanics*. John Wiley, New York.

Theis, C. V. (1935). The relation between the lowering of the piezometric surface and rate and duration of discharge of a well using groundwater storage. *Trans. Am. Geophys. Union 2*, 519–524.

Theis, C. V. (1938). The significance and nature of the cone of depression in groundwater bodies. *Economic Geology 33*, 889–902.

Thimus, J.-F., Y. Abousleiman, A. H.-D. Cheng, O. Coussy, E. Detournay (eds.) (1998). *Collected Papers of M. A. Biot*, CD-ROM. Université Catholique de Louvain-La-Neuve, Louvain-La-Neuve, Belgium.

Thompson, M., and J. R. Willis (1991). A reformulation of the equations of anisotropic poroelasticity. *J. Appl. Mech. ASME 58*, 612–616.

Tolstoy, I. (1986). Maurice A. Biot. *Physics Today 39*, 104–106.

Tolstoy, I. (ed.) (1992). *Acoustics, Elasticity, and Thermodynamics of Porous Media: Twenty-One Papers by M. A. Biot.* Acoustical Society of America, Woodbury, New York.

Trimmer, D. (1982). Laboratory measurements of ultralow permeability of geologic materials. *Rev. Sci. Instrum. 53*, 1246–1254.

Turcotte, D. L., and G. Schubert (1982). *Geodynamics: Applications of Continuum Physics to Geological Problems.* John Wiley, New York.

Van der Kamp, G, and J. E. Gale (1983). Theory of earth tide and barometric effects in porous formations with compressible grains. *Water Resources Res. 19*, 538–544.

Verruijt, A. (1969). Elastic storage of aquifers. In *Flow Through Porous Media*, ed. R.J.M. DeWiest, pp. 331–376. Academic Press, New York.

Verruijt, A. (1971). Displacement functions in the theory of consolidation or in thermoelasticity. *J. Appl. Math. Phys. 22*, 891–898.

Verruijt, A. (1982). Approximations of cyclic pore pressures caused by sea waves in a poro-elastic half-plane. In *Soil Mechanics—Transient and Cyclic Loads*, ed. G. N. Pande and O. C. Zienkiewicz, pp. 37–51. John Wiley, New York.

von Kármán, and M. Biot (1940). *Mathematical Methods in Engineering.* McGraw-Hill, New York.

Walder, J., and A. Nur (1986). Permeability measurement by the pulse-decay method: Effects of poroelastic phenomena and non-linear pore pressure diffusion. *Int. J. Rock Mech. Min. Sci. Geomech. Abstr. 23*, 225–232.

Wang, H. F. (1993). Quasi-static poroelastic parameters in rock and their geophysical applications. *Pure Appl. Geophys. 141*, 269–286.

Wang, H. F. (1997). Effects of deviatoric stress on undrained pore pressure response to fault slip. *J. Geophys. Res. 102*, 17943–17750.

Wang, H. F., and M. P. Anderson (1982). *Introduction to Groundwater Modeling: Finite Difference and Finite Element Methods.* Academic Press, New York.

Wang, H. F., and D. J. Hart (1993). Experimental error for permeability and specific storage from pulse decay measurements. *Int. J. Rock Mech. Min. Sci. Geomech. Abstr. 30*, 1173–1176.

Wang, K., and E. E. Davis (1996). Theory for the propagation of tidally induced pore pressure variations in layered subseafloor formations. *J. Geophys. Res. 101*, 11483–11495.

White, R. R., and A. Clebsch (1994). C. V. Theis, the man and his contributions to hydrogeology. In *Selected Contributions to Ground-Water Hydrology by C. V. Theis, and a Review of His Life and Work*, ed. A. Clebsch, pp. 45–56. U.S. Geological Survey Water-Supply Paper No. 2415.

Williams, A. O., Jr. (1983). Citation to Maurice Anthony Biot. *J. Acoustical Soc. America 74*, insert at p. S32.

Wolff, R. G. (1970). Relationship between horizontal strain near a well and reverse water level fluctuation. *Water Resources Res. 6*, 1721–1728.

Yelderman, J. C., Jr. (1983). *The Relationship of Geologic Conditions to Hydrologic Responses in Fluvial Aquifers with Emphasis on Leaky Artesian Conditions and Storage*. Ph.D. thesis, University of Wisconsin-Madison, 234 pp.

Zienkiewicz, O. C., A.H.C. Chan, M. Pastor, B. A. Schrefler, and T. Shiomi (1998). *Computational Geomechanics with Special Reference to Earthquake Engineering*. John Wiley, Chichester, England.

Zienkiewicz, O. C., and R. L. Taylor (1989). *The Finite Element Method, 4th ed., Vol. 1. Basic Formulation and Linear Problems*. McGraw-Hill, London.

Zimmerman, R. W., W. H. Somerton, and M. S. King (1986). Compressibility of porous rocks. *J. Geophys. Res. 91*, 12765–12777.

Zoback, M. D., and J. D. Byerlee (1975). Permeability and effective stress. *AAPG Bull. 59*, 154–158.

Uncited References

Aoki, T., C. P. Tan, and W. E. Bamford (1993). Effects of deformation and strength anisotropy on borehole failures in saturated shales. *Int. J. Rock Mech. Min. Sci. Geomech. Abstr. 30*, 1031–1034.

Barber, J. R. (1992). *Elasticity, Solid Mechanics and Its Applications*, Vol. 12. Kluwer, Dordrecht, the Netherlands.

Berryman, J. G., and H. F. Wang (1995). The elastic coefficients of double-porosity models for fluid transport in jointed rock. *J. Geophys. Res. 100*, 24611–24627.

Bodvarsson, G. (1970). Confined fluids as strain meters. *J. Geophys. Res. 75*, 2711–2718.

Borchers, J. W. (ed.) (1998). *Land Subsidence Case Studies and Current Research: Proceedings of the Dr. Joseph F. Poland Symposium on Land Subsidence*. Association of Engineering Geologists Special Publ. No. 8. Star Publishing, Belmont, California.

Bowen, R. M. (1981). Green's functions for consolidation problems. *Lett. Appl. Engng. Sci. 19*, 455–466.

Bredehoeft, J. D. (1967). Response of well aquifer systems to earth tides. *J. Geophys. Res. 72*, 3075–3087.

Carroll, M. M., and N. Katsube (1983). The role of Terzaghi effective stress in linearly elastic deformation. *J. Energy Resources Tech. 105*, 509–511.

Cheng, A., B. Gurevich, A. Norris, J. Berryman, K. Tuncay, S. Lopatnikov, and A. M. Maksimov (1999). *Poroelasticity Bibliographical Database*, www.ce.udel.edu/faculty/cheng/poronet/porobib.html.

Chou, P. C., and N. J. Pagano (1967). *Elasticity: Tensor, Dyadic, and Engineering Approaches*. Van Nostrand, Princeton, New Jersey.

Cleary, M. P. (1977). Fundamental solutions for a fluid-saturated porous solid. *Int. J. Solids Structures 13*, 785–806.

Cooley, R. L. (1975). *A Review and Synthesis of the Biot and Jacob-Cooper Theories of Ground-water Motion*. Technical Report Series H-W, Hydrology and Water Resources Publ. No. 25. Desert Research Institute, Reno, Nevada.

Cooper, H. H., Jr. (1966). The equation of groundwater flow in fixed and deforming coordinates. *J. Geophys. Res. 71*, 4785–4790.

Davis, E. E., K. Becker, T. Pettigrew, B. Carson, and R. MacDonald (1992). CORK: A hydrologic seal and downhole observatory for deep-ocean boreholes. *Proc. Ocean Drilling Prog. (Initial Reports) 139*, 43–53.

Engelder, T. (1993). *Stress Regimes in the Lithosphere*. Princeton Univ. Press, Princeton, New Jersey.

Farlow, S. J. (1982). *Partial Differential Equations for Scientists and Engineers*. John Wiley, New York.

Fetter, C. W. (1994). *Applied Hydrogeology*, 3rd ed. Macmillan, New York.

Freeze, R. A., and J. A. Cherry (1979). *Groundwater*. Prentice-Hall, Englewood Cliffs, New Jersey.

Ge, S., and G. Garven (1992). Hydromechanical modeling of tectonically driven groundwater flow with application to the Arkoma foreland basin. *J. Geophys. Res. 97*, 9119–9144.

Geertsma, J., and G. Van Opstal (1973). A numerical technique for predicting subsidence above compacting reservoirs, based on the nucleus of strain concept. *Verhandelingen Kon. Ned. Geol. Mijnbouwk. Gen. 28*, 63–78.

Guéguen, Y., and V. Palciauskas (1994). *Introduction to the Physics of Rocks.* Princeton Univ. Press, Princeton, New Jersey.

Gurevich, A. E., and G. V. Chilingarian (1993). Subsidence over producing oil and gas fields, and gas leakage to the surface. *J. Petr. Science Eng. 9*, 239–250.

Hart, D. J., and H. F. Wang (1999). Pore pressure and confining stress dependence of poroelastic linear compressibilities and Skempton's B coefficient for Berea Sandstone. In *Rock Mechanics for Industry*, ed. B. Amadei, R. L. Kranz, G. A. Scott, and P. H. Smeallie, pp. 365–371. A. A. Balkema, Rotterdam, Netherlands.

Holzer, T. L. (ed.) (1984). *Man-Induced Land Subsidence, Reviews in Engineering Geology*, Vol. 6. Geological Society of America, Boulder, Colorado.

Hsieh, P. A., J. D. Bredehoeft, and S. A. Rojstaczer (1988). Response of well aquifer systems to earth tides: Problem revisited. *Water Resources Res. 24*, 468–472.

Jones, M. (1994). Mechanical principles of sediment deformation. In *The Geological Deformation of Sediments*, ed. A. Maltman, pp. 37–71. Chapman and Hall, London.

Karasudhi, P. (1991). *Foundations of Solid Mechanics, Solid Mechanics and Its Applications*, Vol. 3. Kluwer, Dordrecht, the Netherlands.

Kim, J. M., and R. R. Parizek (1997). Numerical modeling of the Noordbergum effect resulting from groundwater pumping in a layered aquifer system. *J. Hydrology 202*, 231–243.

Kingsbury, H. B. (1984). Determination of material parameters of poroelastic media. In *Fundamentals of Transport Phenomena in Porous Media*, ed. J. Bear and M. Y. Corapcioglu, pp. 579–615. Martinus Nijhoff, Dordrecht, the Netherlands..

Kümpel, H.-J. (1997). Tides in water saturated rock. In *Tidal Phenomena, Lecture Notes in Earth Sciences*, No. 66, ed. H. Wilhelm, W. Zurn, and H.-G. Wenzel, pp. 277–291, Springer, Berlin.

McTigue, D. F. (1986). Thermoelastic response of fluid-saturated porous rock. *J. Geophys. Res. 91*, 9533–9542.

McTigue, D. F. (1990). Flow to a heated borehole in porous, thermoelastic rock: Analysis. *Water Resources Res. 26*, 1763–1774.

Nur, A., and J. R. Booker (1973). Aftershocks caused by pore fluid flow? *Science 175*, 885–887.

Parkus, H. (1976). *Thermoelasticity*, 2nd ed. Springer-Verlag, New York.

Perloff, W. H., and W. Baron (1976). *Soil Mechanics: Principles and Applications.* Ronald Press, New York.

Raghavan, R., and F. G. Miller (1975). Mathematical analysis of sand compaction. In *Compaction of Coarse-Grained Sediments, I*, ed. G. V. Chilingarian and K. H. Wolf, pp. 403–524. Elsevier, Amsterdam.

Rodrigues, J. D. (1983). The Noordbergum effect and characterization of aquitards at the Rio Major Mining Project. *Ground Water 21*, 200–207.

Roeloffs, E. A. (1998). Persistent water level changes in a well near Parkfield, California, due to local and distant earthquakes. *J. Geophys. Res. 103*, 869–889.

Sharp, J. M., Jr., and D. W. Hill (1995). Land subsidence along the northeastern Texas Gulf coast: Effects of deep hydrocarbon production. *Environmental Geology 25*, 181–191.

Skempton, A. W. (1954). The pore pressure coefficients *A* and *B*. *Géotechnique 4*, 143–147.

Verruijt, A. (1977). Generation and dissipation of pore-water pressures. In *Finite Elements in Geomechanics*, ed. G. Gudehus, pp. 293–317. John Wiley, New York.

Wang, H. F., and J. G. Berryman (1996). On constitutive equations and effective stress principles for deformable, double-porosity media. *Water Resources Res. 12*, 3621–3622.

Warpinski, N. R., and L. W. Teufel (1992). Determination of the effective stress law for permeability and deformation in low-permeability rocks. *Society of Petroleum Engineers Formation Evaluation 7*, 123–131.

Weiler, F. C. (1992). Fully coupled thermo-poro-elasto governing equations. In *Computational Mechanics of Porous Materials and Their Thermal Decomposition*, ed. N. J. Salamon and R. M. Sullivan, pp. 1–28. *Applied Mechanics Division* Vol. 136. American Society of Mechanical Engineers, New York.

Wesson, R. L. (1981). Interpretation of changes in water level accompanying fault creep and implications for earthquake prediction. *J. Geophys. Res. 86*, 9259–9267.

Yue, Z. Q., A.P.S. Selvadurai, and K. T. Law (1994). Excess pore pressure in a poroelastic seabed saturated with a compressible fluid. *Can. Geotech. J. 31*, 989–1003.

Author Index

Subject Index

Ingram Content Group UK Ltd.
Milton Keynes UK
UKHW020230110323
418395UK00003B/11/J

9 780691 037462